REED'S
HEAT AND HEAT ENGINES
FOR
ENGINEERS

REED'S
HEAT AND HEAT ENGINES
FOR
ENGINEERS

By
WILLIAM EMBLETON O.B.E.
EXTRA FIRST CLASS ENGINEERS CERTIFICATE
C.Eng.,   M.I.Mech.E.,   F.I.Mar.E.

PUBLISHED BY THOMAS REED PUBLICATIONS LIMITED
SUNDERLAND AND LONDON

First Edition - 1963
Reprinted 1966
Reprinted 1969

Second Edition - 1971 *(in SI units)*
Reprinted 1974
Reprinted 1976
Reprinted 1979

ISBN 0 901281 25 5

© Thomas Reed Publications Limited

Reproduced send printed by photolithography and bound in
Great Britain at The Pitman Press, Bath

PREFACE

This third volume of Reed's Marine Engineering Series covers the Heat and Heat Engines syllabuses for Part A of the examinations for the Second and First Class Certificates of Competency, and is closely related to the syllabuses of the O.N.C. and O.N.D. in engineering. Emphasis is placed on basic principles, and numerous explanatory diagrammatic sketches are included. Parts of the text and certain test examples are marked with the prefix "f" to indicate that they are normally considered to be more suited to the standard required for the first class examination.

Fully worked solutions are given to all examples and test problems, and detailed mathematical steps towards the solution are given where such are considered to be of help to engineers studying on their own without a tutor to turn to for assistance.

Since the first edition of this volume, the work has been completely rewritten in SI units, much new material has been added and redundant matter deleted in an endeavour to keep the book fully up-to-date with modern practice. The study of steam reciprocating engines is now drastically reduced, all reference to their design, determination of dimensions and valve diagrams is omitted, leaving only that part relating to the understanding and efficient running of these engines for the benefit of engineers in charge of the few still remaining in service.

W. EMBLETON
South Shields

CONTENTS

CHAPTER 1

UNITS AND COMMON TERMS

MASS, FORCE, ENERGY, POWER

MASS is the quantity of matter possessed by a body and is proportional to the volume and the density of the body. It is a constant quantity, that is, the mass of a body can only be changed by adding more matter to it or taking matter away from it.

The abbreviation for mass is m and the unit is the kilogramme [kg]. For very large or small quantities, multiples or submultiples of the gramme [g] are used. Large masses are common in engineering and these are measured in megagrammes [Mg]. One megagramme is equal to 10^3 kilogrammes and called a tonne [t].

Mass is proportionally accelerated or retarded by an applied force. To maintain a coherent system of units, a unit of force is chosen which will give unit acceleration to unit mass. This unit of force is called the *newton* [N]. Hence, one newton of force acting on one kilogramme of mass will give it an acceleration of one metre per second per second, therefore:

Accelerating force [N] = mass [kg] $\times$ acceleration [m/s^2]. In symbols:

$$F = m\,a$$

FORCE OF GRAVITY. All bodies are attracted towards each other, the force of attraction depending upon the masses of the bodies and their distances apart. Newton's law of gravitation states that this force of attraction is proportional to the product of the masses of the bodies and inversely proportional to the square of the distance apart.

An important example of this is the mass of the earth which attracts all comparatively smaller bodies towards it, the attractive force by which a body tends to be drawn towards the centre of the earth is the force of gravity and is called the *weight* of the body.

If a body is allowed to fall freely, it will fall with an acceleration of 9·81 m/s^2, this is termed gravitational acceleration and is represented by g. Since one newton is the force which will give one kilogramme of mass an acceleration of one m/s^2, then the force in newtons to give m kg of mass an acceleration of 9·81 m/s^2 is $m \times 9·81$. Hence, at the earth's surface, the gravitational force on a mass of m kg is mg newtons, or in other words:

$$\text{weight [N]} = \text{mass [kg]} \times g \text{ [m/s}^2\text{]}.$$

The further the distance between the centre of gravity of the mass and the centre of gravity of the earth, the less is the attractive force between them. Thus, the weight of a mass measured by a spring balance (not a pair of scales which is merely a means of comparing the weight of one mass with another) will vary slightly at different parts of the earth's surface due to the earth not being a perfect sphere.

If a body is projected in a space-rocket, the attractive force of the earth on the body becomes less as its distance from the earth increases until, in complete outer-space, it becomes nil, that is, it is then weightless. The mass of the body of course remains unchanged.

WORK is done when a force applied on a body causes it to move and is measured by the product of the force and the distance through which the force moves.

The unit of work is the *joule* [J] which is defined as the work done when the point of application of a force of one newton moves through a distance of one metre in the direction in which the force is applied. Hence, one joule is equal to one newton-metre. In symbols, $J = Nm$.

$$\text{Work done [J]} = \text{force [N]} \times \text{distance moved [m]}.$$

The joule is a small unit. Moderate quantities of work may be expressed in kilojoules [1 kJ $= 10^3$ J] and larger quantities in megajoules [1 MJ $= 10^6$ J].

POWER is the rate of doing work, that is, the quantity of work done in a given time. The unit of power is the *watt* [W] which is equal to the rate of one joule of work being done every second. In symbols, $W = J/s = Nm/s$.

$$\text{Power [W]} = \frac{\text{work done [J]}}{\text{time [s]}}$$

The watt is a small unit and only suitable for small powers. For normal powers in engineering, the kilowatt [1 kW $= 10^3$ W] and megawatt [1 MW $= 10^6$ W] are usually more convenient units.

ENERGY is the capacity for doing work and it is measured by the amount of work done. Energy is therefore expressed in the same units as work, that is, joules, kilojoules and megajoules.

Another useful unit of energy is the *kilowatt-hour* [kWh]. This, as its name implies, represents the energy used or the work done when one kilowatt of power is exerted continually for one hour.

$$\begin{aligned}
\text{Energy} &= \text{power} \times \text{time} \\
1\ \text{kWh} &= 1000\ \text{watts} \times 3600\ \text{seconds} \\
&= 1000\ [\text{J/s}] \times 3600\ [\text{s}] \\
&= 3{\cdot}6 \times 10^6\ \text{J} \\
&= 3{\cdot}6\ \text{MJ}.
\end{aligned}$$

The MECHANICAL EFFICIENCY is the ratio of the work got out of a machine to the work put into it, and, as this is done in the same time, it is also the ratio of the output power to the input power. Since no machine is perfect, the output is always less than the input, due to frictional and other losses, therefore the efficiency is always less than unity.

The symbol for efficiency is η, the Greek letter eta, and it may be expressed as a fraction or as a percentage.

$$\eta = \frac{\text{output power}}{\text{input power}}$$

Example. A mass of 1600 kg is lifted by a winch through a height of 25 m in 30 seconds. Calculate (i) the work done, and, if the efficiency of the winch is 60%, find (ii) the input power in kW and (iii) the energy consumed in kWh.

Force [N] to lift mass against gravity

$$\begin{aligned}
&= \text{weight of mass} = mg \\
&= 1600 \times 9{\cdot}81\ \text{newtons}
\end{aligned}$$

$$\begin{aligned}
\text{Work done } [\text{J} = \text{Nm}] &= \text{force } [\text{N}] \times \text{distance } [\text{m}] \\
&= 1600 \times 9{\cdot}81 \times 25 \\
&= 392\ 400\ \text{J} = 392{\cdot}4\ \text{kJ Ans. (i)}
\end{aligned}$$

$$\text{Output power } [\text{kW} = \text{kJ/s}] = \frac{\text{work done } [\text{kJ}]}{\text{time } [\text{s}]}$$

$$= \frac{392{\cdot}4}{30} = 13{\cdot}08\ \text{kW}$$

$$\text{Efficiency} = \frac{\text{output power}}{\text{input power}}$$

$$\text{Input power} = \frac{13 \cdot 08}{0 \cdot 6} = 21 \cdot 8 \text{ kW Ans. (ii)}$$

$$\text{Energy consumed} = \frac{392 \cdot 4}{0 \cdot 6} = 654 \text{ kJ}$$

$$1 \text{ kWh} = 3 \cdot 6 \text{ MJ}$$

$$\therefore 654 \text{ kJ} = \frac{0 \cdot 654}{3 \cdot 6} = 0 \cdot 1817 \text{ kWh Ans. (iii)}$$

Alternatively,

$$\text{Energy [kWh]} = \text{power [kW]} \times \text{time [h]}$$

$$= 21 \cdot 8 \times \frac{30}{3600}$$

$$= 0 \cdot 1817 \text{ kWh.}$$

PRESSURE

Pressure is expressed as the intensity of force, that is, the force acting on unit area. The unit of force is the newton [N] and the unit of area is the square metre [m²], therefore the fundamental unit of pressure is newton per square metre [N/m²]. The symbol representing pressure is usually p.

Pressures of liquids and gases reach high values which are expressed in multiples of the basic unit of force. For example, the steam pressure in small auxiliary boilers is often in the region of 8×10^5 N/m² and in high pressure water-tube boilers it could be 6×10^6 N/m², the former can be conveniently written 800 kN/m² and the latter 6000 kN/m² or 6 MN/m². Another very convenient unit of pressure commonly used is the *bar*, this has the advantage of being easy to "think" in these units since one bar is approximately equal to one atmosphere of pressure (1 atm = 1·013 bar). One bar is 10^5 N/m², which is 100 kN/m², hence the working pressure of the auxiliary boilers given as an example above would be stated as 8 bar, and the working pressure of the water-tube boilers as 60 bar.

Pressures in internal combustion engines vary from a little above or below one bar during the air charging period to about 40 bar ($= 40 \times 10^5$ N/m² $= 4000$ kN/m²) during combustion.

PRESSURE GAUGE. The most common instrument used for measuring high static pressures is the Bourdon type of pressure gauge, illustrated in Fig. 1. This consists of a bronze or steel curved tube of elliptical section, one end of the tube is openly connected to the source of pressure, the other end is sealed. The effect of the fluid pressure inside the tube is to tend to straighten it, the higher the pressure the greater the straightening effect. The small movement of the sealed end of the tube is magnified by linking it to a quadrant meshing with a pinion which carries a pointer on its shaft, the pointer moves over a circular scale on the dial of the gauge which is graduated in bars or kN/m^2.

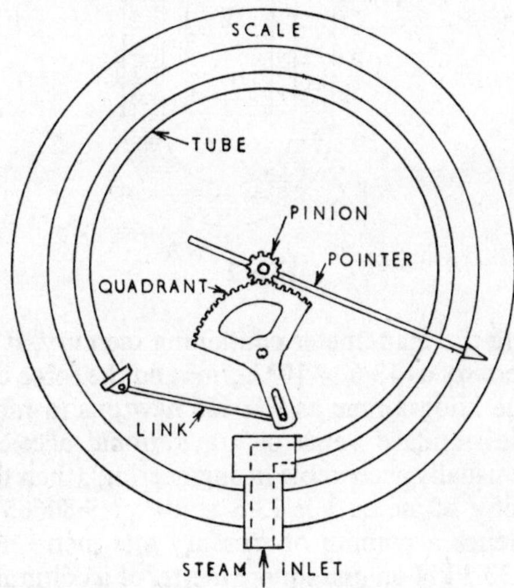

BOURDON PRESSURE GAUGE

Fig. 1

MANOMETER. LOW pressures and vacua are usually measured in millimetres of mercury [mm Hg], very small pressures such as furnace draught are measured in millimetres of water [mm water]. The instrument used is the manometer which is shown in its simplest form in Fig. 2. This is a glass U-tube partially filled with mercury or water, one end is connected to the source of pressure and the other end is open to the atmosphere. The difference between the levels of the liquid in the two legs indicates the difference in pressure between the source and the atmosphere.

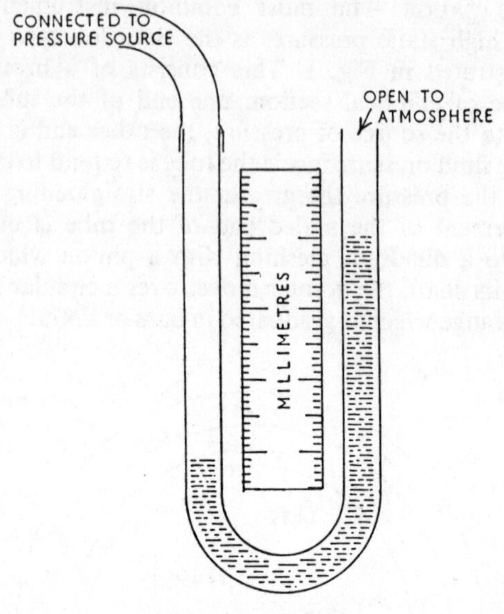

Fig. 2

Considering the manometer containing mercury, if we take the density of mercury as $13 \cdot 6 \times 10^3$ kg/m³ and the force of gravity on a mass of one kilogramme as 9·80665 newtons (a more accurate figure for the standard value of gravitational acceleration than 9·81 which is usually acceptable in engineering), then the weight of one cubic metre of mercury is $13 \cdot 6 \times 10^3 \times 9 \cdot 80665$ newtons = 133·3 kN. Hence a column of mercury one metre high exerts a pressure of 133·3 kN on one square metre, or a column of mercury one millimetre high is equivalent to a pressure of 133·3 N/m².

Similarly, each millimetre of water pressure is equal to 9·80665 N/m² which is usually taken as 9·81 N/m².

Small pressures may also be expressed in millibars [mbar]: One mbar = 1 bar $\times 10^{-3} = 10^5 \times 10^{-3}$ N/m² = 100 N/m².

BAROMETER. The mercurial barometer works on the principle of the atmospheric pressure supporting a column of mercury. Fig. 3 shows the principle involved. A glass tube about one metre long, closed at one end and open at the other, is completely filled with mercury and, with the open end plugged, the tube is inverted and the open end submerged into a vessel containing mercury. When the plug is removed, the level of the mercury falls, leaving a perfect vacuum between the mercury level and the sealed end. The vertical column of mercury left standing up the tube is supported by the outside atmospheric pressure and is therefore a measure of the pressure of the atmosphere. As the atmospheric pressure rises and falls, the level of the supported column of mercury rises and falls accordingly.

For example, if the column of mercury supported by the atmospheric pressure is 760 mm, then the atmospheric pressure will be:

$$760 \times 133\cdot3 = 1\cdot013 \times 10^5 \text{ N/m}^2 = 1\cdot013 \text{ bar or } 101\cdot3 \text{ kN/m}^2.$$

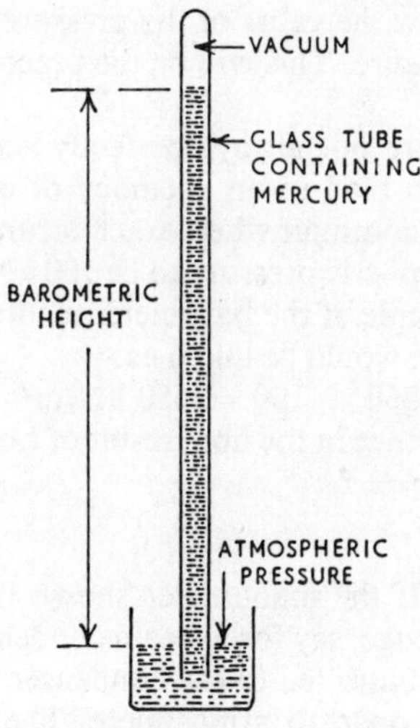

Fig. 3

GAUGE PRESSURE AND ABSOLUTE PRESSURE. Most pressure recording instruments, including the ordinary pressure gauge and the open-ended manometer, measure the pressure from the level of atmospheric pressure. The pressure so recorded is termed the *gauge pressure* and the word "gauge" should follow the units of pressure. Thus, if a steam pressure gauge reads 2000 kN/m² the pressure should be stated as 2000 kN/m² gauge, meaning that this is the pressure of the steam over and above the atmospheric pressure.

The true pressure is measured above a perfect vacuum and called the *absolute pressure* and this is the value which is used in thermodynamic calculations. The absolute pressure is therefore obtained by adding the atmospheric pressure to the gauge pressure, the gauge pressure being read from the pressure gauge and the atmospheric pressure obtained from the barometric reading.

As an example, if the pressure of a fluid is 550 kN/m² gauge, and the barometer stands at 758 mm Hg then,

$$\text{Atmospheric pressure} = 758 \times 133{\cdot}3$$
$$= 1{\cdot}01 \times 10^5 \text{ N/m}^2 = 101 \text{ kN/m}^2$$
$$\text{Absolute pressure} = \text{gauge pressure} + \text{atmospheric pressure}$$
$$= 550 + 101$$
$$= 651 \text{ kN/m}^2$$

This could be written 651 kN/m² absolute. However it is usual to omit the word absolute and take it for granted that if the word gauge does not follow the value of the pressure then it means that it is an absolute pressure. This will be the practice throughout this book.

Pressure gauges are not always perfectly accurate and, in any case, it is difficult to read to an accuracy of one or two kN/m². It is therefore quite common when exact accuracy is not essential to assume the atmospheric pressure to be 100 kN/m².

In the above example, if the barometer reading was not known, the absolute pressure would be taken as:

$$550 + 100 = 650 \text{ kN/m}^2$$

with very little difference in the final result of a calculation.

VACUUM GAUGE. If the manometer shown in Fig. 2 was to be used as a vacuum gauge, say for a steam condenser, the level of the mercury in the leg connected to the condenser will be higher than the level in the leg open to atmosphere. The difference in level indicates the pressure *below* atmospheric and written "mm Hg vacuum".

For example, if the gauge reads 600 mm Hg of vacuum and the barometer stands at 758 mm Hg then:

Pressure below atmospheric
$$= 600 \times 133{\cdot}3 = 8 \times 10^4 \text{ N/m}^2 = 80 \text{ kN/m}^2$$

Atmospheric pressure
$$= 758 \times 133{\cdot}3 = 1{\cdot}01 \times 10^5 \text{ N/m}^2 = 101 \text{ kN/m}^2$$

Therefore absolute pressure in condenser is 80 kN/m² *below* 101 kN/m² which is 21 kN/m², or more simply calculated:

$$\text{Abs. press.} = (\text{barometer mm Hg} - \text{vacuum gauge mm Hg}) \times 133{\cdot}3$$
$$= (758 - 600) \times 133{\cdot}3$$
$$= 158 \times 133{\cdot}3$$
$$= 2{\cdot}1 \times 10^4 \text{ N/m}^2 = 21 \text{ kN/m}^2$$

VOLUME

The basic unit of volume is the cubic metre [m³] and the symbol for volume is V. A common submultiple is the litre (l) this is equal in volume to one cubic decimetre and is used only for fluid measure.

$$1 \text{ m}^3 = 10^3 \text{ dm}^3 \text{ therefore } 10^3 \text{ litres} = 1 \text{ m}^3$$

The millilitre [ml] is 1×10^{-3} litre and therefore equal in volume to one cubic centimetre. Whereas the basic unit of density is kilogramme per cubic metre [kg/m³], densities of liquids are often expressed in grammes per millilitre [g/ml] and densities of solids in grammes per cubic centimetre [g/cm³].

SPECIFIC VOLUME is the volume occupied by unit mass, the symbol is v and the basic unit is cubic metre per kilogramme [m³/kg] thus, the specific volume is the reciprocal of density. In certain cases, specific volume may be expressed in cubic metres per tonne [m³/tonne] and litres per kilogramme [l/kg].

TEMPERATURE

Temperature is an indication of hotness or coldness and therefore is a measure of the intensity of heat.

The most common temperature measuring instrument is the mercurial thermometer. This consists of a glass tube of very fine bore with a bulb at its lower end, the bulb and tube are exhausted of air, partially filled with mercury and hermetically sealed at the top end. When the thermometer is placed in a substance whose temperature is to be measured, the mercury takes up the same temperature and expands (if heated) or contracts (if cooled) and the level, which rises or falls in consequence indicates on the thermometer scale the degree of heat intensity.

The divisions on the thermometer are called degrees and are related to the freezing point and boiling point of pure water at atmospheric pressure.

The Celsius scale (formerly known as Centigrade) is the S I system of measuring and specifying temperatures. Any instruments still in present use which are graduated in the Fahrenheit scale will probably be replaced within two or three years, in the meantime a comparison of the two scales and their relation to each other should be familiar.

Referring to Fig. 4, the point at which pure water freezes into ice is marked zero on the Celsius scale and 32 on the Fahrenheit. The point at which pure water boils into steam at atmospheric pressure is assigned the number 100 on the Celsius scale and 212 on the Fahrenheit. The former is sometimes referred to as the *lower fixed point* or *ice point*, the latter as the *upper fixed point* or *steam point*. The number of degrees into which the thermometer scale is divided between these two fixed points is therefore 100 on the Celsius and $212 - 32 = 180$ on the Fahrenheit, and the sizes of these degrees are continued uniformly above and below the two fixed points for measuring higher and lower temperatures.

The unit representing a temperature reading on the Celsius scale is °C and on the Fahrenheit it is °F, and the symbol for temperature is θ, the Greek letter theta.

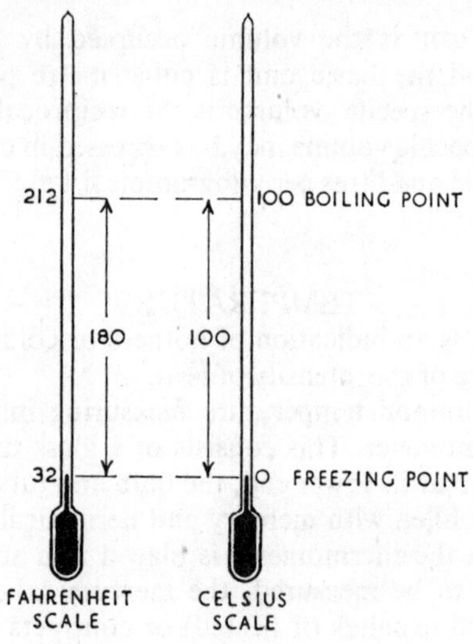

Fig. 4

The advantages of mercury are that it does not wet the bore of the glass and therefore none sticks to the glass as the temperature falls. It can be used over a wide useful range of temperature as its freezing point is low, about −38°C, and boiling point high, about 358°C.

Alcohol is sometimes used but requires to be coloured. Although its freezing point is in the region of −115°C and can be useful for measuring very low temperatures, its boiling point is also low, about 78°C, and therefore has a limited use.

The difference between two temperature readings from the same scale, and the change in temperature through which a body is heated or cooled, is referred to as a *temperature interval*. This may be represented by deg C to distinguish it from a temperature reading, but it is recommended that, after the early stages of study and the subject is more familiar, different methods of expressing the units should not be necessary, temperature intervals and specific points on the temperature scale should both be expressed by °C, the descriptive sentence in which the quantity is contained being sufficient to distinguish between the two.

TO CONVERT A TEMPERATURE INTERVAL FROM FAHRENHEIT TO CELSIUS. Consider the temperature interval between freezing point and natural boiling point of pure water.

Interval on the Fahrenheit scale $= 212 - 32 = 180$ degrees
„ „ „ Celsius „ $= 100 - 0 = 100$ „

In the same proportion,

$$9 \text{ Fahrenheit degrees} = 5 \text{ Celsius degrees}$$

Hence to convert a temperature interval from Fahrenheit to Celsius:

Interval on Celsius scale $= \frac{5}{9} \times$ interval on Fahrenheit scale. For example, if a body is heated through 153°F, this is

$$153 \times \tfrac{5}{9} = 85°C$$

TO CONVERT A TEMPERATURE READING FROM FAHRENHEIT TO CELSIUS. Zero on the Celsius scale corresponds to 32 on the Fahrenheit, thus the two thermometer scales are not graduated from the same numerical value. Zero mark on the Fahrenheit thermometer being 32 degrees below freezing point must therefore be taken into account when converting a specific point on the temperature scale, by first subtracting the 32 from the Fahrenheit reading to bring the two scales to a common base, and then multiplying the result by the ratio $\frac{5}{9}$ to obtain the Celsius reading, thus,

$$\text{Reading on the Celsius scale} = (F - 32) \times \tfrac{5}{9}$$

For example, a temperature of 77°F is equivalent to

$$(77 - 32) \times \tfrac{5}{9} = 25°C$$

ABSOLUTE TEMPERATURE. All gases expand at practically the same rate when heated through the same range of temperature, and contract at the same rate when cooled.

The rate of expansion or contraction of a perfect gas is (very nearly) $\frac{1}{273}$ of its volume at 0°C when heated or cooled at constant pressure through one degree Celsius. Hence, if a gas initially at 0°C could be cooled at constant pressure until its temperature is 273 Celsius degrees below 0°C, the volume would contract until there was nothing left and no further reduction of temperature would be possible, that is, the gas would then have reached its *absolute zero of temperature* (see Fig. 5). In practice of course, it is not possible to cool a gas down to the absolute zero and cause it to disappear. As the absolute zero of temperature is approached the gas will change into a liquid and the laws of gases are then no longer applicable.

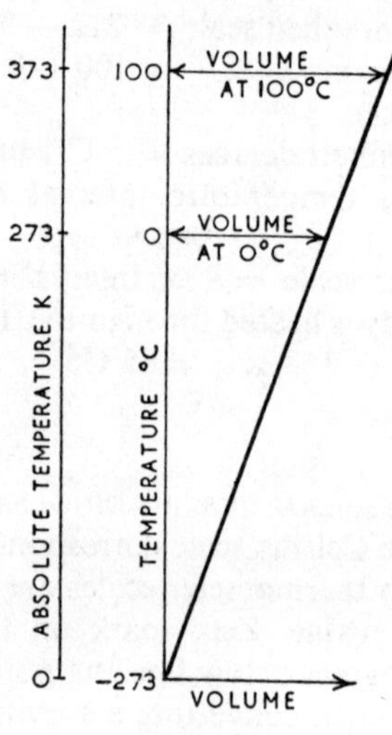

Fig. 5

We see from the above that temperatures can be expressed as *absolute* quantities, that is, stating the degrees of temperature above the level of Absolute Zero, by adding 273 to the ordinary Celsius thermometer reading.

Absolute temperature is often referred to as *thermodynamic temperature*, the symbol for this is T and the unit is the kelvin which is represented by K, thus,

Thermodynamic temperature $=$ Celsius temperature $+$ 273.

In symbols,

$$T \, [\text{K}] = \theta \, [°\text{C}] + 273$$

COMMON TERMS AND DEFINITIONS

VOLUME FLOW is the volume of a fluid flowing past a given point in unit time. The basic unit is cubic metres per second [m³/s], other convenient units are cubic metres per hour [m³/h], cubic metres per minute [m³/min] and litres per hour [l/h].

For example, if a fluid is flowing at a velocity of v metres per second full bore through a pipe of internal diameter d metres, the quantity flowing in cubic metres per second is:

$$\text{Volume flow } [\text{m}^3/\text{s}] = \text{area } [\text{m}^2] \times \text{velocity } [\text{m/s}]$$
$$= 0\cdot7854 \, d^2 \times v$$

MASS FLOW is the mass of fluid flowing past a given point in unit time, the basic unit being kilogrammes per second [kg/s]. Since density is the mass per unit volume, then:

Mass flow [kg/s] = volume flow [m³/s] × density [kg/m³].

Mass flow may also be expressed in other convenient units such as tonnes per hour [tonne/h] and kilogrammes per hour [kg/h].

SWEPT VOLUME or STROKE VOLUME is the volume swept through by a piston in its cylinder of a reciprocating engine, pump, compressor, etc., it is the product of the piston area [m²] and the stroke of the piston [m].

The space left between the piston at its inner dead centre (top of its stroke) and the cylinder head is termed the *clearance volume*. This may be expressed as the actual volume of the clearance space, or as a fraction of the stroke volume.

SYSTEM AND BOUNDARY. A *system* is the term given to the collection of matter under consideration enclosed within a *boundary*, the region outside the boundary is termed the *surroundings*. The boundary may be an imaginary enclosure, or it may be real such as the cylinder wall, cylinder head and piston of an internal combustion engine which encloses the mixture of gases within. Energy is transferred across a boundary from one system to another.

If there is no transfer of matter across the boundary, that is, if no substance can enter the system or leave it during investigation, it is a *closed system*, energy only being transferred across the boundary, and whatever changes take place to the substance is termed a *non-flow* process because no matter flows into or out of it.

If there is a flow of matter through the boundary, it is an *open system.* If the mass flow entering the system is equal to the mass flow leaving so that at any time the quantity of matter within the system is constant, the series of changes to the matter is referred to as a *steady-flow* process.

An example of a closed system within which a non-flow process takes place is the compression of the air in the cylinder of an air compressor.

A turbine is an example of an open system in which a steady-flow process takes place.

A CYCLE is a recurrent period of a complete set of a series of connected processes which a system undergoes, the final state of the system at the end of the cycle being exactly as it was at the beginning of the cycle.

Many of these terms and definitions will be repeated and enlarged upon as the occasion arises throughout the book.

TEST EXAMPLES 1

1. A pump discharges 50 tonne of water per hour to a height of 8 m the overall efficiency of the pumping system being 69%. Calculate the output power and the input power. Calculate also the energy consumed by the pump in 2 hours, expressed in kWh and in MJ.

2. (a) Express a pressure of 20 mm water in N/m^2 and mbars.
   (b) Express a pressure of 750 mm Hg in kN/m^2 and bars.

3. A condenser vacuum gauge reads 715 mm Hg when the barometer stands at 757 mm Hg. State the absolute pressure in the condenser in kN/m^2 and bars.

4. (i) Water is raised in temperature through 162°F, express this increase in temperature in degrees Celsius.

   (ii) The temperature inside a refrigerated cold chamber is —3°F when the outside temperature is 60°F. Express the temperature difference between inside and outside in degrees Celsius.

5. Convert the following temperature readings from °F to °C:
    140°F   5°F   —31°F   —40°F

6. Oil flows full bore at a velocity of 2 m/s through a nest of 16 tubes in a single pass cooler. The internal diameter of the tubes is 30 mm and the density of the oil is 0·85 g/ml. Find the volume flow in litres per second and the mass flow in kilogrammes per minute.

CHAPTER 2

HEAT

Heat is a form of energy associated with the movement of the molecules which constitute the heated body. Heat is transferred from one substance to another by temperature difference between the two substances, it is interchangeable with other forms of energy and can be made available for doing work and producing mechanical and electrical power.

The basic unit of all energy, including heat, is the *joule* [J]. Thus, units of heat are expressed in joules or multiples of the joule, the commonest being kilojoules [kJ] and megajoules [MJ]. The symbol representing quantity of heat is Q.

The SPECIFIC HEAT of a substance is the quantity of heat required to raise the temperature of unit mass of the substance by one degree. The units of specific heat are therefore heat units per unit mass per unit temperature. The symbol for specific heat is c. In most cases involving specific heat the kilojoule is the most convenient size of heat unit for unit mass of one kilogramme, hence the specific heat is usually expressed in kilojoules per kilogramme per kelvin, in symbols this is kJ/kg K. This is the same value as joules per gramme per kelvin [J/g K]. Note also that one Celsius degree of temperature interval on the thermometer scale is the same as one kelvin degree of temperature interval on the absolute scale, the above units can therefore be written kJ/kg°C and J/g°C respectively.

It follows from the above definition of specific heat that the quantity of heat energy transferred to a substance to raise its temperature is the product of the mass of the substance, its specific heat, and its rise in temperature.

In symbols:

$$Q \text{ [kJ]} = m \text{ [kg]} \times c \text{ [kJ/kg K]} \times (T_2 - T_1) \text{ [K]}$$

Different substances have different specific heat values. Also, the specific heat of any one particular substance is not always a constant value over a large range of temperature, the variation in specific heat however, is small, and an average value over the temperature range under consideration may be taken for most practical purposes. For example, the specific heat of water decreases from 4·21 kJ/kg K at 0°C to 4·178 kJ/kg K at 35°C and increases thereafter with rise in temperature, being 4·219 kJ/kg K at 100°C. Between 0°C and 100°C a mean value is usually taken as 4·2 for temperatures within this range.

Example. Calculate the quantity of heat to be transferred to 2·25 kg of brass to raise its temperature from 20°C to 240°C, taking the specific heat of the brass as 0·394 kJ/kg K.

$$\text{Temperature increase} = 240 - 20 = 220°C = 220 \text{ K}$$
$$Q \text{ [kJ]} = m \text{ [kg]} \times c \text{ [kJ/kg K]} \times (T_2 - T_1) \text{ [K]}$$
$$= 2·25 \times 0·394 \times 220$$
$$= 195 \text{ kJ Ans.}$$

The characteristics of gases vary considerably at different temperatures and pressures, and heat may be transferred under an infinite number of different conditions, consequently the specific heat can have an infinite number of different values. Two important conditions are transferring heat to or from a gas which is at constant pressure, and transferring heat while its volume is constant. The specific heat of a gas at constant pressure is represented by c_P and at constant volume by c_V. This is dealt with in detail later.

The MECHANICAL EQUIVALENT OF HEAT is the relationship between mechanical energy and heat energy. This was determined by Dr. Joule, one of the first scientists to demonstrate that heat was a form of energy, using apparatus which generated heat by the expenditure of mechanical work.

In Imperial units the accepted figure was 778 ft lbf of work = 1 Btu of heat. In the SI system of units, the joule is the unit of all forms of energy. As previously stated, one joule is the work done when a force of one newton moves through a distance of one metre in the direction in which the force is applied, thus the work done is one newton-metre and this is equal to one joule [1 N m = 1 J]. Hence the mechanical equivalent of heat is 1 N m/J which is unity.

When work is done in overcoming friction, the mechanical energy expended is converted into heat energy. The force required to overcome sliding friction between two bodies is the product of the coefficient of friction (μ) and the normal force between the surfaces of the bodies (Chap. 6 Vol. 2).

Example. A shaft runs at a rotational speed of 50 rev/s in oil-cooled bearings 178 mm diameter. The force between the surfaces of the shaft journals and bearings is 2·67 kN and the coefficient of friction is 0·04. Find (i) the friction force at the surface of the journals, (ii) the mechanical energy expended in friction per revolution, (iii) the power loss due to friction, (iv) the temperature rise of the oil if the volume flow through the bearings is 18 litre/min, the specific heat of the oil being 2 kJ/kg K and its density 0·9 g/ml.

$$\text{Friction force} = \mu \times \text{normal force between surfaces}$$
$$= 0{\cdot}04 \times 2{\cdot}67 \times 10^3$$
$$= 106{\cdot}8 \text{ N Ans. (i)}$$

Work done to overcome friction per revolution [J = N m]
$$= \text{friction force [N]} \times \text{circumference of journal [m]}$$
$$= 106{\cdot}8 \times \pi \times 0{\cdot}178$$
$$= 59{\cdot}7 \text{ J Ans. (ii)}$$

Power expended [W = J/s]
$$= \text{energy per revolution [J]} \times \text{rev/s}$$
$$= 59{\cdot}7 \times 50$$
$$= 2985 \text{ W} = 2{\cdot}985 \text{ kW Ans. (iii)}$$

Density of oil = 0·9 g/ml = 0·9 kg/litre

Mass flow of oil [kg/s] = volume flow [l/s] × density [kg/l]
$$= \frac{18}{60} \times 0{\cdot}9 = 0{\cdot}27 \text{ kg/s}$$

$$Q \text{ [kJ/s]} = m \text{ [kg/s]} \times c \text{ [kJ/kg K]} \times (T_2 - T_1) \text{ [K]}$$
$$2{\cdot}985 = 0{\cdot}27 \times 2 \times \text{temp. rise}$$

Temp. rise = 5·527 K or 5·527°C Ans. (iv)

The WATER EQUIVALENT of a mass of a substance is the mass of water that would require the same heat transfer as the mass of that substance to cause the same change of temperature.

For example, taking an aluminium vessel of mass 2 kg, and the specific heat of aluminium as 0·912 kJ/kg K:

$$Q = \text{mass} \times \text{specific heat} \times \text{temperature change}$$
$$\text{For water, } Q_W = m_W \times c_W \times (T_2 - T_1)_W$$
$$\text{For aluminium, } Q_A = m_A \times c_A \times (T_2 - T_1)_A$$

Since Q is to be the same quantity of heat,
$$Q_W = Q_A$$
$$m_W \times c_W \times (T_2 - T_1)_W = m_A \times c_A \times (T_2 - T_1)_A$$

and the temperature change is to be the same,
$$m_W \times c_W = m_A \times c_A$$

$$\therefore\ m_W = m_A \times \frac{c_A}{c_W}$$

Taking the specific heat of water as 4·2 kJ/kgK, the water equivalent of this mass of aluminium is

$$m_W = 2 \times \frac{0\cdot912}{4\cdot2}$$

$$= 0\cdot4343\ \text{kg}$$

That is to say, 0·4343 kg of water would require the same amount of heat transferred to it as the 2 kg of aluminium to raise it through the same range of temperature.

It is useful to know the water equivalent of laboratory calorimeters. When water is contained in a vessel, the temperature of the vessel is the same as that of the water inside, and when the temperature of the water is changed, the temperature of the vessel changes with it. The vessel can therefore be considered as an extra mass of water equal to the water equivalent of the vessel

Example. The mass of a copper calorimeter is 0·28 kg and it contains 0·4 kg of water at 15°C. Taking the specific heat of copper as 0·39 kJ/kg K, calculate the heat required to raise the temperature to 20°C.

Water equivalent of calorimeter

$$= 0\cdot28 \times \frac{0\cdot39}{4\cdot2} = 0\cdot026\ \text{kg}$$

Heat received by water and calorimeter,
$$Q\ [\text{kJ}] = m\ [\text{kg}] \times c\ [\text{kJ/kg K}] \times (T_2 - T_1)\ [\text{K}]$$
$$= (0\cdot4 + 0\cdot026) \times 4\cdot2 \times (20 - 15)$$
$$= 0\cdot426 \times 4\cdot2 \times 5$$
$$= 8\cdot946\ \text{kJ}\quad \text{Ans.}$$

When two substances at different temperatures are mixed together, heat will transfer from the hotter substance to the colder until both become the same temperature. Unless otherwise stated, it is assumed that no heat is transferred to or from an outside source during the mixing process and therefore the quantity of heat absorbed by the colder substance is all at the expense of the loss of heat by the hotter substance.

Example. In an experiment to find the specific heat of lead, 0·5 kg of lead shot at a temperature of 51°C is poured into an insulated calorimeter containing 0·25 kg of water at 13·5°C and the resultant temperature of the mixture is 15·5°C. If the water equivalent of the calorimeter is 0·02 kg, find the specific heat of the lead.

Heat received by water and calorimeter when their temperature is raised from 13·5°C to 15·5°C:

$$Q = m \times c \times (T_2 - T_1)$$
$$= (0.25 + 0.02) \times 4.2 \times (15.5 - 13.5)$$
$$= 0.27 \times 4.2 \times 2 \text{ kJ}$$

Heat lost by lead in cooling from 51°C to 15·5°C:

$$Q = m \times c \times (T_3 - T_2)$$
$$= 0.5 \times c \times (51 - 15.5)$$
$$= 0.5 \times c \times 35.5 \text{ kJ}$$

Heat transferred from the the lead is equal to the heat received by the water and calorimeter:

$$0.5 \times c \times 35.5 = 0.27 \times 4.2 \times 2$$
$$c = \frac{0.27 \times 4.2 \times 2}{0.5 \times 35.5}$$
$$= 0.1278 \text{ kJ/kg K} \quad \text{Ans.}$$

LATENT HEAT is the heat which supplies the energy necessary to overcome some of the binding forces of attraction between the molecules of a substance and is responsible for it changing its physical state from a solid into a liquid, or from a liquid into a vapour, the change taking place without any change of temperature.

The process of changing the physical state from a solid into a liquid is called *melting* or *fusion*, and the quantity of heat required to change unit mass of the substance from solid to liquid at the same temperature is the *latent heat of fusion*.

For example, the latent heat of fusion for ice is 335 kJ/kg at 0°C. This means that one kilogramme of ice at 0°C would require 335 kilojoules of heat transferred to it to completely melt it into one kilogramme of water at 0°C. Also, one kilogramme of water at 0°C would require to lose 335 kilojoules of heat to completely freeze it into ice at 0°C.

The process of changing the physical state of a substance from a liquid into a vapour is called *boiling* or *evaporation* and the quantity of heat to bring about this change at constant temperature to unit mass is the *latent heat of evaporation*.

The latent heat of evaporation of water at atmospheric pressure is 2256·7 kJ/kg. This means that one kilogramme of water at 100°C would require 2256·7 kilojoules of heat to completely boil it into one kilogramme of steam at 100°C. Also, one kilogramme of steam at 100°C would require to lose 2256·7 kilojoules of heat to completely condense it into one kilogramme of water at 100°C.

The temperature at which a liquid boils and the latent heat of evaporation depend strictly upon the pressure, the higher the pressure, the higher the boiling point and the smaller the amount of latent heat required to evaporate it. For example, at atmospheric pressure, the temperature at which water boils is 100°C and the latent heat of evaporation is 2256·7 kJ/kg, at a pressure of 15 bar (1500 kN/m^2) the boiling point is 198·3°C and the latent heat 1947 kJ/kg, at 30 bar (3000 kN/m^2) the boiling point is 233·8°C and the latent heat 1795 kJ/kg. These values are obtained from steam tables which are described later.

When heat is transferred to or from a substance which changes only its temperature, and there is no physical change of state, it is sometimes referred to as *sensible heat*. This distinguishes it from latent heat which changes the physical state of the substance without change of temperature.

We shall see later that, for a constant pressure process, the heat energy transferred to a substance is termed *enthalpy*, then, latent heat of fusion is termed *enthalpy of fusion*, and latent heat of evaporation is termed *enthalpy of evaporation*, and so on.

Example. Calculate the heat required to be given to 2 kg of ice at —15°C to change it into steam at atmospheric pressure, taking the values:

$$\text{Specific heat of ice} = 2\text{·}04\ \text{kJ/kg K}$$
$$\text{Latent heat of fusion} = 335\ \text{kJ/kg}$$
$$\text{Specific heat of water} = 4\text{·}2\ \text{kJ/kg K}$$
$$\text{Latent heat of evaporation} = 2256\text{·}7\ \text{kJ/kg}$$

Heat to raise the temperature of the ice from —15°C to its melting point of 0°C, *i.e.* a temperature rise of 15°C:
$$\text{Sensible heat} = m \times c \times \text{temp. rise}$$
$$= 2 \times 2\text{·}04 \times 15$$
$$= 61\text{·}2\ \text{kJ}$$

Heat to change the ice at 0°C into water at 0°C:
$$\text{Latent heat} = 2 \times 335$$
$$= 670 \text{ kJ}$$

Heat to raise the temperature of the water from 0°C to its boiling point of 100°C, *i.e.* a temperature rise of 100°C:
$$\text{Sensible heat} = 2 \times 4 \cdot 2 \times 100$$
$$= 840 \text{ kJ}$$

Heat to evaporate the water at 100°C into steam at 100°C:
$$\text{Latent heat} = 2 \times 2256 \cdot 7$$
$$= 4513 \cdot 4 \text{ kJ}$$
$$\text{Total heat} = 61 \cdot 2 + 670 + 840 + 4513 \cdot 4$$
$$= 6084 \cdot 6 \text{ kJ} \quad \text{Ans.}$$

The above is set out in detail for the student to follow each step, however, once understood, working is simplified and written down more briefly by finding the total heat transfer required per unit mass and then finally multiplying by the total mass, thus,
$$Q = 2(2 \cdot 04 \times 15 + 335 + 4 \cdot 2 \times 100 + 2256 \cdot 7)$$
$$= 2(30 \cdot 6 + 335 + 420 + 2256 \cdot 7)$$
$$= 2 \times 3042 \cdot 3$$
$$= 6084 \cdot 6 \text{ kJ}$$

TEST EXAMPLES 2

1. A casting of mass 5 kg is pulled a distance of 10 m along a horizontal floor, the coefficient of friction between the surfaces of the casting and floor being 0·4. Calculate the heat generated at the surfaces.

2. A water brake coupled to an engine on test absorbs 70 kW of power. Find the heat generated at the brake per minute and the mass flow of fresh water through the brake, in kg/min if the temperature increase of the water is 10°C. Assume all the heat generated is carried away by the cooling water.

3. The journals of a shaft are 380 mm diameter, it runs at 105 rev/min and the coefficient of friction between journals and bearings is 0·02. If the average load on the bearings is 200 kN, find (i) the power lost due to friction at the bearings, (ii) the heat generated per minute at the bearings.

4. The effective radius of the pads in a single collar thrust block in 230 mm and the total load on the thrust block is 240 kN when the shaft is running at 93 rev/min. Taking the coefficient of friction between thrust collar and pads as 0·025, find (i) the power lost due to friction, (ii) the heat generated per hour, (iii) the mass flow of oil in kilogrammes per hour through the block assuming all the heat is carried away by the oil, allowing an oil temperature rise of 20°C and taking the specific heat of the oil as 2 kJ/kg K.

5. To ascertain the temperature of flue gases, 1·8 kg of copper of specific heat 0·395 kJ/kg K was suspended in the flue until it attained the temperature of the gases, and then dropped into 2·27 kg of water at 20°C. If the resultant temperature of the copper and water was 37·2°C, find the temperature of the flue gases.

6. 2·5 kg of brass of specific heat 0·39 kJ/kg K at a temperature of 176°C is dropped into 1·2 litre of water at 14°C. Find the resultant temperature of the mixture.

7. In an experiment to find the specific heat of iron, 2·15 kg of iron cuttings at 100°C are dropped into a vessel containing 2·3 litre of water at 17°C and the resultant temperature of the mixture is 24·4°C. If the water equivalent of the vessel is 0·18 kg, determine the specific heat of the iron.

8. With three different quantities, A, B and C, of the same kind of liquid, of temperatures 9, 21 and 38°C respectively, it is found that when A and B are mixed together the resultant temperature is 17°C, and when B and C are mixed together the resultant temperature is 28°C. Find the resultant temperature (i) if A and C were mixed, (ii) if all three were mixed together.

9. 0·9 kg of ice was put into 10 kg of water at 22°C contained in a calorimeter of water equivalent 0·2 kg, and the resultant temperature of the mixture was 13°C. Calculate the initial temperature of the ice, taking its specific heat as 2·04 kJ/kg K and the latent heat of fusion 335 kJ/kg.

10. 0·5 kg of ice at —5°C is put into a vessel containing 1·8 kg of water at 17°C and mixed together, the result being a mixture of ice and water at 0°C. Calculate the final masses of ice and water, taking the water equivalent of the vessel as 0·148 kg, specific heat of ice 2·04 kJ/kg K, latent heat of fusion 335 kJ/kg.

CHAPTER 3

THERMAL EXPANSION

EXPANSION OF METALS

The effect of increasing the temperature of metals is generally to cause their dimensions to increase. Most metals expand when they are heated and contract when they are cooled, the amount of expansion per degree rise of temperature differs with different metals. Some alloys are manufactured to have a minimum amount of expansion over a considerable working temperature range, these are usually for special purposes such as measuring instruments and gauges.

Although the expansion is in all directions so that there is an increase in all dimensions, it is sometimes only relevant to consider the expansion in one direction.

LINEAR EXPANSION. When a linear dimension is under consideration, the amount that a metal will expand lengthwise is expressed by its *coefficient of linear expansion*. This is the increase in length per unit length per degree increase in temperature. For instance, if the coefficient of linear expansion of copper is given as $1 \cdot 7 \times 10^{-5}/°C$ (which is $0 \cdot 000017$ per degree Celsius) it means that each metre of length will increase in length by $1 \cdot 7 \times 10^{-5}$ metre when heated through one degree Celsius. This coefficient may be represented by α, the Greek letter alpha. Hence, representing the original length by l, and temperature rise by $(\theta_2 - \theta_1)$,

Increase in length $= \alpha \times l \times (\theta_2 - \theta_1)$.

The new length of the metal will then be:

$$\text{new length} = \text{original length} + \text{increase in length}$$
$$= l + \alpha l(\theta_2 - \theta_1)$$
$$= l\{1 + \alpha(\theta_2 - \theta_1)\}$$

Example. A main steam pipe is $6 \cdot 5$ m long when fitted at a temperature of $15°C$. Calculate how much allowance should be made for its increase in length if it is subjected to a steam temperature of $300°C$, taking the coefficient of linear expansion of the material as $1 \cdot 2 \times 10^{-5}/°C$.

$$\text{Increase in length} = \alpha \times l \times (\theta_2 - \theta_1)$$
$$= 1 \cdot 2 \times 10^{-5} \times 6 \cdot 5 \times (300 - 15)$$
$$= 0 \cdot 02223 \text{ m}$$
$$= 22 \cdot 23 \text{ mm} \quad \text{Ans.}$$

Example. A brass liner is 270 mm diameter when the temperature is 17°C. Take the coefficient of linear expansion of the brass as $1 \cdot 9 \times 10^{-5}$/°C and find the temperature to which the liner should be heated in order to increase the diameter by 2 mm.

Diameter is a linear dimension and therefore the same rule can be applied as for length. Note that the diameter and increase in diameter must be expressed in the same units.

$$\text{Increase in diameter} = \alpha \times d \times (\theta_2 - \theta_1)$$
$$2 = 1 \cdot 9 \times 10^{-5} \times 270 \times (\theta_2 - \theta_1)$$
$$\text{Increase in temperature} = \frac{2 \times 10^5}{1 \cdot 9 \times 270} = 389 \cdot 7°C$$

$$\therefore \text{Required temperature} = 17 + 389 \cdot 7$$
$$= 406 \cdot 7°C \quad \text{Ans.}$$

SUPERFICIAL EXPANSION refers to increase in area. The coefficient of superficial expansion is the increase in area per unit area per degree increase in temperature. Therefore, if A represents original area, and $(\theta_2 - \theta_1)$ the increase in temperature, then,

Increase in area = coeff. of superficial expansion $\times A \times$
$(\theta_2 - \theta_1)$.

Consider an area of metal of unit length and unit breadth (Fig. 6) and let this be heated through one degree.

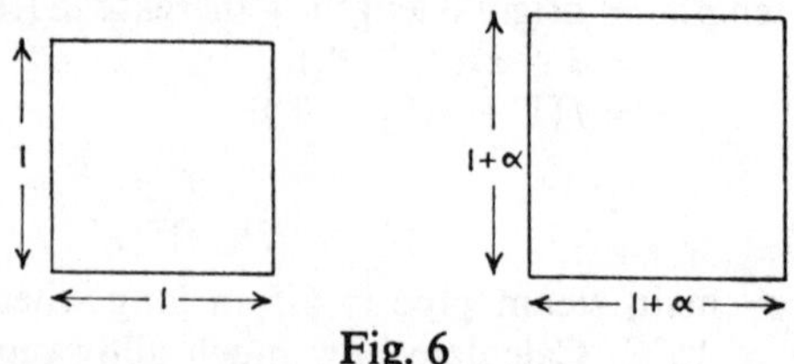

Fig. 6

The length and breadth will each increase by an amount equal to α, the coefficient of linear expansion.

$$\text{original area} = 1 \times 1 = 1$$
$$\text{new length and new breadth} = 1 + \alpha$$
$$\text{new area} = (1 + \alpha)^2 = 1 + 2\alpha + \alpha^2$$
$$\text{Increase in area} = \text{new area} - \text{original area}$$
$$= 1 + 2\alpha + \alpha^2 - 1$$
$$= 2\alpha + \alpha^2$$

α is a very small quantity for any metal (such as about $1 \cdot 2 \times 10^{-5}$ for steel), therefore α^2 being the second order of smallness is a very small quantity indeed and is completely negligible as a quantity to be added for all practical purposes. We can therefore take the increase to be 2α. As this is the increase in area per unit area for one degree increase in temperature, it is the value of the coefficient of superficial expansion, hence,

Coeff. of superficial expansion $= 2 \times$ coeff. of linear expansion therefore,

$$\text{Increase in area} = 2\alpha \times A \times (\theta_2 - \theta_1).$$

CUBICAL EXPANSION refers to the increase in volume. The *coefficient of cubical* (or *volumetric*) *expansion* is the increase in volume per unit volume per degree increase in temperature. Therefore, if V represents the original volume, and $(\theta_2 - \theta_1)$ increase in temperature, then:

Increase in volume $=$ coeff. of cubical expansion $\times V \times$
$$(\theta_2 - \theta_1).$$

Consider a block of metal of unit length, unit breadth, and unit thickness (Fig. 7) and let this be heated through one degree.

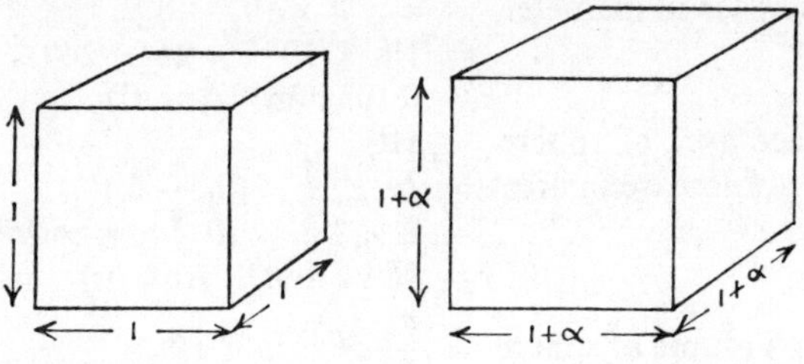

Fig. 7

The length, breadth and thickness will each increase by an amount equal to α, the coefficient of linear expansion.

$$\text{Original volume} = 1 \times 1 \times 1 = 1$$
$$\text{New volume} = (1 + \alpha)^3$$
$$= 1 + 3\alpha + 3\alpha^2 + \alpha^3$$
$$\text{Increase in volume} = \text{new volume} - \text{original volume}$$
$$= 1 + 3\alpha + 3\alpha^2 + \alpha^3 - 1$$
$$= 3\alpha + 3\alpha^2 + \alpha^3$$

α^2 and α^3 being the second and third order of smallness respectively, are negligible quantities for addition, hence the increase may be taken as 3α. As this is the increase in volume per unit volume per degree increase in temperature, it is the value of the coefficient of cubical expansion.

Coeff. of cubical expansion $= 3 \times$ coeff. of linear expansion. Therefore,

$$\text{Increase in volume} = 3\alpha \times V \times (\theta_2 - \theta_1).$$

Note that *linear* refers to any linear dimension such as length, breadth, thickness, diameter, circumference, and so on, and the expression for linear expansion can be applied to any of these dimensions. It is true for internal dimensions as well as external.

The expression for superficial expansion covers any area of the solid, cross-sectional area, surface area, etc., and holds good for internal areas as well as external.

The expression for cubical expansion is also applicable for internal volumes of a hollow vessel.

Example. A metal sphere is exactly 25 mm diameter at 20°C. Find the increase in diameter, increase in surface area, and increase in volume, when heated to 260°C, if the coefficient of linear expansion of the metal is $1\cdot8 \times 10^{-5}/°C$.

$$\text{Increase in temperature} = 260 - 20 = 240°C$$
$$\text{Increase in diameter} = \alpha \times d \times (\theta_2 - \theta_1)$$
$$= 1\cdot8 \times 10^{-5} \times 25 \times 240$$
$$= 0\cdot108 \text{ mm} \quad \text{Ans. (i)}$$
$$\text{Surface area of sphere} = \pi d^2$$
$$\text{Increase in area} = 2\alpha \times A \times (\theta_2 - \theta_1)$$
$$= 2 \times 1\cdot8 \times 10^{-5} \times \pi \times 25^2 \times 240$$
$$= 16\cdot96 \text{ mm}^2 \quad \text{Ans. (ii)}$$
$$\text{Volume of sphere} = \frac{\pi}{6} d^3$$
$$\text{Increase in volume} = 3\alpha \times V \times (\theta_2 - \theta_1)$$
$$= 3 \times 1\cdot8 \times 10^{-5} \times \frac{\pi}{6} \times 25^3 \times 240$$
$$= 106 \text{ mm}^3 \quad \text{Ans. (iii)}$$

EXPANSION OF LIQUIDS

Liquids have no definite shape of their own, therefore no linear dimensions, hence the coefficient of cubical expansion of a liquid is an independent quantity. The coefficient of cubical expansion is usually represented by β, the Greek letter beta, thus,

$$\text{Increase in volume} = \beta \times V \times (\theta_2 - \theta_1)$$

Example. 2500 litres of oil are heated through 50°C. If the coefficient of cubical expansion of this oil is 0·0008/°C, find the increase in volume in cubic metres.

$$2500 \text{ litres} = 2 \cdot 5 \text{ m}^3$$
$$\text{Increase in volume} = \beta \times V \times (\theta_2 - \theta_1)$$
$$= 0 \cdot 0008 \times 2 \cdot 5 \times 50$$
$$= 0 \cdot 1 \text{ m}^3 \quad \text{Ans.}$$

APPARENT CUBICAL EXPANSION. The tank or vessel which contains a liquid will also expand when heated. It is therefore useful to know the expansion of the liquid relative to its container so that the correct allowance can be made for changes of temperature.

The apparent or relative increase in volume of a liquid is the difference between the volumetric expansion of the liquid and the volumetric expansion of its container. If both have the same initial volume and are raised through the same range of temperature, then, letting suffix L represent the liquid and suffix C the container:

Apparent increase in volume of the liquid
$$= \text{vol. increase of liquid} - \text{vol. increase of container}$$
$$= \beta_L \times V \times (\theta_2 - \theta_1) - \beta_C \times V \times (\theta_2 - \theta_1)$$
$$= (\beta_L - \beta_C) \times V \times (\theta_2 - \theta_1)$$

The difference between the coefficients of cubical expansion of the liquid and its container can therefore be termed the *apparent* coefficient of cubical expansion of the liquid.

RESTRICTED THERMAL EXPANSION

If the natural thermal expansion of a metal is restricted, the metal will be strained from its natural length and the material will be stressed. To enable the effects of restricted expansion to be seen here, it is necessary to briefly revise the relationships between stress, strain and modulus of elasticity, which are explained in detail in Volume 2 (Applied Mechanics).

STRESS is the internal resistance set up in a material when an external force is applied. It is expressed as the force carried by the material per unit area of its cross-section.

The unit of force is the newton [N] and the unit of area is the square metre [m²], therefore the unit of stress is newton per square metre [N/m²], thus,

$$\text{Stress [N/m}^2\text{]} = \frac{\text{force [N]}}{\text{area [m}^2\text{]}}$$

High values of stress are expressed in multiples of the force unit, such as kN/m² (= 10^3 N/m²) and MN/m² (= 10^6 N/m²). In some industries, stress may be expressed in hectobars [hbar], one hectobar = 10^7 N/m².

STRAIN is the change of shape that takes place in a material due to it being stressed. Linear strain is the change of length per unit length, thus,

$$\text{Linear strain} = \frac{\text{change of length}}{\text{original length}}$$

Strain is therefore a pure number because the units cancel.

The MODULUS OF ELASTICITY of a material is the constant obtained by dividing the stress set up in it by the strain it endures under that stress, provided the elastic limit is not exceeded. Modulus of elasticity is represented by E and its units are the same as for stress.

$$\text{Modulus of elasticity } (E) = \frac{\text{stress}}{\text{strain}}$$

The following should now be readily understood.

Example. A solid steel stay 50 mm diameter and 300 mm long at a temperature of 25°C is firmly secured at each end so that expansion is fully restricted. Find the stress set up in the stay and the equivalent total axial force when it is heated to 150°C, taking the coefficient of linear expansion of steel as 1.2×10^{-5}/°C and its modulus of elasticity as 206 GN/m².

If the stay was perfectly free to expand without restriction,
$$\text{Free expansion} = \alpha \times l \times (\theta_2 - \theta_1)$$
$$= 1.2 \times 10^{-5} \times 300 \times (150 - 25)$$
$$= 0.45 \text{ mm}$$
Free and unrestrained length would then be
$$300 + 0.45 = 300.45 \text{ mm}$$

If prevented from expanding the effect is that the stay is compressed from its natural unstrained length of 300·45 mm to 300 mm,

$$\text{Strain} = \frac{\text{change of length}}{\text{original length}}$$

$$= \frac{0·45}{300·45} = 0·001\,498$$

It is usual however, to make a slight approximation at this stage by taking the original cold length of 300 mm instead of the heated length of 300·45 mm. The difference in the final result is negligible and it proves much more convenient in solving more complicated problems. It will also be seen that by making this slight approximation the length of the material will not be required because the strain can be obtained direct from:

$$\text{Strain} = \frac{\text{change of length}}{\text{original length}}$$

$$= \frac{\alpha \times l \times (\theta_2 - \theta_1)}{l} = \alpha(\theta_2 - \theta_1)$$

Hence the strain for this stay can be taken as:
$$\text{Strain} = \alpha(\theta_2 - \theta_1)$$
$$= 1·2 \times 10^{-5} \times 125 = 0·0015$$

which is the same as $\dfrac{0·45}{300} = 0·0015$

$$E \,[\text{N/m}^2] = \frac{\text{stress} \,[\text{N/m}^2]}{\text{strain}}$$

$$\therefore \text{Stress} = \text{strain} \times E$$
$$= 0·0015 \times 206 \times 10^9$$
$$= 3·09 \times 10^8 \,\text{N/m}^2$$
$$= 309 \,\text{MN/m}^2 \text{ or } 30·9 \text{ hbar} \quad \text{Ans. (i)}$$

$$\text{Stress} \,[\text{N/m}^2] = \frac{\text{force} \,[\text{N}]}{\text{area} \,[\text{m}^2]}$$

$$\therefore \text{Total axial force} = \text{stress} \times \text{area}$$
$$= 3·09 \times 10^8 \times 0·7854 \times 0·05^2$$
$$= 6·068 \times 10^5 \,\text{N}$$
$$= 606·8 \,\text{kN} \quad \text{Ans. (ii)}$$

Further examples including stresses in compound bars due to thermal expansion are given in Volume 2 of this series, Applied Mechanics.

TEST EXAMPLES 3

1. A steam pipe is 3·85 m long when fitted at a temperature of 18°C. Find the increase in length if free to expand, when carrying steam at a temperature of 260°C, taking the coefficient of linear expansion of the pipe material as $1·25 \times 10^{-5}$/°C.

2. A solid cast iron sphere is 150 mm diameter. If 2110 kJ of heat energy is transferred to it, find the increase in diameter, taking the following values for cast iron: density = 7·21 g/cm³, specific heat = 0·54 kJ/kg K, coefficient of linear expansion = $1·12 \times 10^{-5}$/°C.

3. A bi-metal control device is made up of a thin flat strip of aluminium and a thin flat strip of steel of the same dimensions, connected together in parallel and separated from each other by two brass distance pieces 2·5 mm long, their centres being 50 mm apart. Find the radius of curvature of the strips when heated through 200°C, taking the following values for the coefficients of linear expansion:

$$\text{Aluminium } \alpha = 2·5 \times 10^{-5}/°C$$
$$\text{Steel } \alpha = 1·2 \times 10^{-5}/°C$$
$$\text{Brass } \alpha = 2·0 \times 10^{-5}/°C$$

4. The pipe line of a hydraulic system consists of a total length of steel pipe of 13·7 m and internal diameter 30 mm. If the coefficient of linear expansion of the steel is $1·2 \times 10^{-5}$/°C and coefficient of cubical expansion of the oil in the pipe is 9×10^{-4}/°C, calculate the volumetric allowance in litres to be made for oil overflow from the pipe when the temperature rises by 27°C.

5. In an experiment to determine the coefficient of cubical expansion of an oil, a glass vessel of 40 ml capacity was filled with the oil at a temperature of 15°C. When the vessel and oil was heated to 65°C the quantity of oil spilt over from the vessel was 1·57 ml. Taking the coefficient of linear expansion of the glass as $8·5 \times 10^{-6}$/°C calculate the coefficient of cubical expansion of the oil.

6. A length of steel wire 2·5 mm diameter is heated from 16°C to 113°C and its ends are securely fastened while the wire is hot. Find the stress in the wire when it cools down to its original temperature assuming the elastic limit of the wire is not exceeded.

Take the coefficient of linear expansion of the steel as $1.2 \times 10^{-5}/°C$ and the modulus of elasticity as 207 GN/m². What total force applied to the ends of the wire would cause the same stress?

7. A straight length of steam pipe is to be fitted between two fixed points with no allowance for expansion. If the compressive stress in the pipe is to be limited to 35 hbar (= 350 MN/m²) calculate the initial tensile stress to be exerted on the pipe when fitted cold at 17°C to allow for a steam temperature of 220°C. Take the coefficient of linear expansion of the pipe material as $1.12 \times 10^{-5}/°C$ and the modulus of elasticity as 206 GN/m².

CHAPTER 4

HEAT TRANSFER

Heat is transferred from one system to another by one of the three methods known as *Conduction*, *Convection*, and *Radiation*, or by a combination of these.

CONDUCTION

Conduction is the flow of heat energy through a body, or from one body to another in contact with each other, due to difference in temperature. The natural flow of heat takes place from a region of high temperature to a region of lower temperature.

Generally speaking, metals are good conductors of heat. Air, and some materials such as asbestos, cork, glass wool, are very bad conductors, these are called insulators and are used to minimise heat loss. Asbestos and special compositions are used to lag steam boilers, pipes, turbine casings, etc. to reduce loss of heat energy from the steam to the colder outside surroundings. Cork and glass wool are common insulating materials to pack the hollow walls of refrigerating chambers to reduce heat flow into the cold chambers from the warmer outside surroundings.

Water is a poor conductor. A simple experiment commonly demonstrated to show this is illustrated in Fig. 8 which represents a glass flask of water containing a piece of ice at the bottom

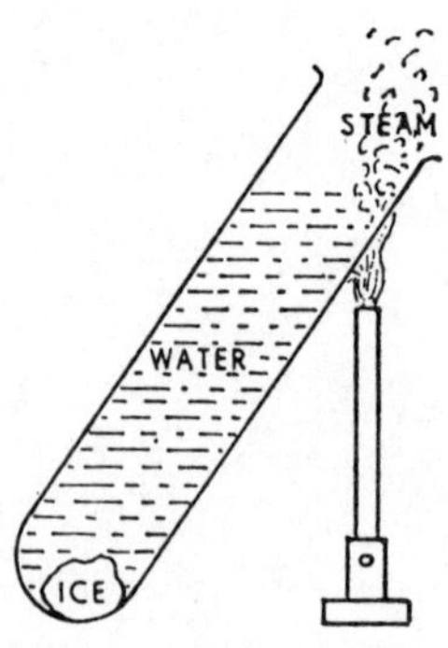

Fig. 8

(loaded to keep it down, otherwise it would float to the top), with a flame applied near the top of the water. The water at the top begins to boil at the temperature of 100°C while the ice remains unmelted and the water at the bottom remains cold, showing a poor conduction of heat through the water.

The quantity of heat conducted through a material in a given time depends upon the thermal conductivity of the material, is proportional to the surface area exposed to the source of heat, is proportional to the temperature difference between the hot and cold ends, and is inversely proportional to the distance or thickness through which the heat is conducted, thus,

$$\text{Quantity of heat varies as } \frac{\text{area} \times \text{time} \times \text{temp. difference}}{\text{thickness}}$$

The thermal conductivity depends upon the nature of the material and its ability to conduct heat. This varies for different materials and sometimes varies slightly for the same material depending upon the temperature range.

The *coefficient of thermal conductivity* of a material, represented by the Greek letter lambda λ, expresses the quantity of heat energy conducted through unit area in unit time for unit temperature difference between two opposite faces of a material of unit distance apart.

Considering the heat flow through a flat wall, taking the following symbols and basic units, and referring to Fig. 9:

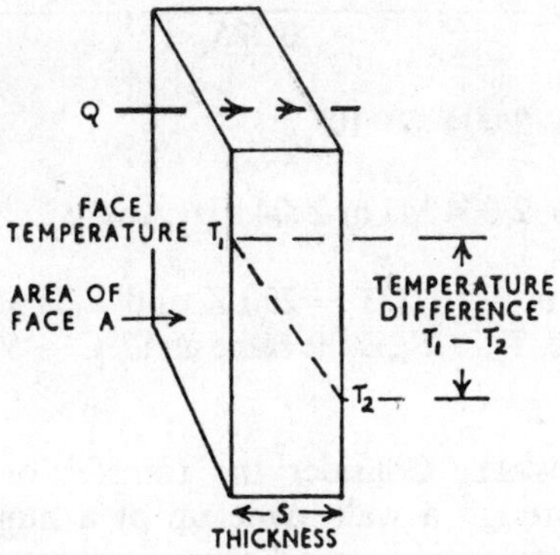

Fig. 9

Q = quantity of heat energy conducted, in joules [J].

A = area through which heat flows, in square metres [m²].

t = time of heat flow, in seconds [s].

$T_1 - T_2$ = temperature difference between the two faces [K].

S = thickness of wall, in metres [m].

Then, the units of λ the coefficient of thermal conductivity are $Jm/m^2 s K$. For convenience this is usually shortened by cancelling m into m² and substituting W watts in the place of J/s joules per second:

$$\frac{Jm}{m^2 s K} = \frac{J}{ms K} = \frac{W}{m K} = W/m K$$

Hence, the quantity of heat energy transferred by conduction is:

$$Q = \frac{\lambda A t (T_1 - T_2)}{S}$$

Example. Calculate the heat transfer per hour through a solid brick wall 6 m long, 2·9 m high, and 225 mm thick, when the outer surface is at 5°C and the inner surface 17°C, the coefficient of thermal conductivity of the brick being 0·6 W/m K.

$$Q[J] = \frac{\lambda [W/m K] \times A[m^2] \times t[s] \times (T_1 - T_2)[K]}{S[m]}$$

$$= \frac{0 \cdot 6 \times 6 \times 2 \cdot 9 \times 3600 \times (17 - 5)}{0 \cdot 225}$$

$$= 2 \cdot 004 \times 10^6 \text{ J}$$

$$= 2 \cdot 004 \text{ MJ or } 2004 \text{ kJ} \quad \text{Ans.}$$

Note: that T_1 is $17 + 273 = 290$ K and T_2 is $5 + 273 = 278$ K, their difference, $T_1 - T_2$, is the same as $17°C - 5°C$.

f COMPOSITE WALL. Consider the transfer of heat energy by conduction through a wall made up of a number of layers of different materials, take as an example three slabs of different thicknesses as shown in Fig. 10.

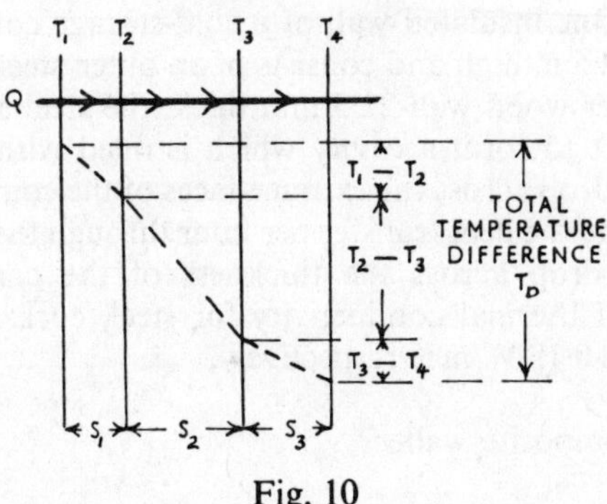

Fig. 10

For each thickness:

$$Q = \frac{\lambda A t \times \text{temp. diff.}}{S} \qquad \therefore \ \text{temp. diff.} = \frac{QS}{\lambda A t}$$

Total drop in temperature across the three thicknesses:

$$T_1 - T_4 = \frac{Q_1 S_1}{\lambda_1 A_1 t_1} + \frac{Q_2 S_2}{\lambda_2 A_2 t_2} + \frac{Q_3 S_3}{\lambda_3 A_3 t_3}$$

The same quantity of heat energy is transferred across each layer through the same area in the same time, therefore Q, A and t are common:

$$T_1 - T_4 = \frac{Q}{At}\left\{ \frac{S_1}{\lambda_1} + \frac{S_2}{\lambda_2} + \frac{S_3}{\lambda_3} \right\}$$

For any number of layers, let T_D represent the total temperature drop, that is,

$$T_D = (T_1 - T_2) + (T_2 - T_3) + (T_3 - T_4) + \text{etc.}$$

and using the summation sign Σ for the sum of the quantities inside the brackets, that is,

$$\Sigma\left\{\frac{S}{\lambda}\right\} = \frac{S_1}{\lambda_1} + \frac{S_2}{\lambda_2} + \frac{S_3}{\lambda_3} + \text{etc.},$$

then,

$$T_D = \frac{Q}{At}\Sigma\left\{\frac{S}{\lambda}\right\} \quad \text{or} \quad Q = \frac{At T_D}{\Sigma\left\{\frac{S}{\lambda}\right\}}$$

f Example. One insulated wall of a cold-storage compartment is 8 m long by 2·5 m high and consists of an outer steel plate 18 mm thick, an inner wood wall 22·5 mm thick, the steel and wood are 90 mm apart to form a cavity which is filled with cork. If the temperature drop across the extreme faces of the composite wall is 15°C, calculate the heat transfer per hour through the wall and the temperature drop across the thickness of the cork. Take the coefficients of thermal conductivity for steel, cork and wood as 45, 0·045, and 0·18 W/m K respectively.

For the composite wall,

$$T_1 - T_4 = \frac{Q}{At}\left\{\frac{S_1}{\lambda_1} + \frac{S_2}{\lambda_2} + \frac{S_3}{\lambda_3}\right\}$$

where
$$T_1 - T_4 = 15 \text{ K}$$
$$A = 8 \times 2\cdot5 = 20 \text{ m}^2$$
$$t = 3600 \text{ seconds}$$

$$\Sigma\left\{\frac{S}{\lambda}\right\} = \frac{0\cdot018}{45} + \frac{0\cdot09}{0\cdot045} + \frac{0\cdot0225}{0\cdot18}$$

$$= 0\cdot0004 + 2 + 0\cdot125$$
$$= 2\cdot1254$$

$$\therefore \ 15 = \frac{Q}{20 \times 3600} \times 2\cdot1254$$

$$Q = \frac{15 \times 20 \times 3600}{2\cdot1254}$$

$$= 5\cdot082 \times 10^5 \text{ J}$$
$$= 508\cdot2 \text{ kJ} \quad \text{Ans. (i)}$$

Temperature drop across the cork:

$$= \frac{QS}{At\lambda}$$

$$= \frac{5\cdot082 \times 10^5 \times 0\cdot09}{20 \times 3600 \times 0\cdot045} = 14\cdot11 \text{ K}$$

$$= 14\cdot11°\text{C} \quad \text{Ans. (ii)}$$

f TRANSFER OF HEAT FROM ONE FLUID TO ANOTHER THROUGH A
DIVIDING WALL. This is a practical application in many engineering
appliances. Consider transfer of heat from a fluid to a flat plate,
through the plate, and from the plate to another fluid, as illustrated
in Fig. 11. On each side of the plate, a thin film of almost stagnant
fluid clings to the surface, the heat transfer through the film is
practically by conduction only.

The quantity of heat conducted through the film of fluid per
unit area of surface, in unit time, for unit temperature drop across
the thickness of the film, is expressed by the *coefficient of heat
transfer*, and this depends to a large extent upon the velocity of the
fluid (such as the velocity of the cooling water through condenser
tubes, etc.) and the condition of the surface.

This heat transfer coefficient can be represented by *h* and, since
the quantity of heat conducted in a given time is proportional to
the surface area and temperature drop, then the units of *h* are
joules per metre² of area per second per degree $= \mathrm{J/m^2 s\,K} =$
$\mathrm{W/m^2\,K}$, hence:

$$Q[\mathrm{J}] = h[\mathrm{W/m^2\,K}] \times A[\mathrm{m^2}] \times t[\mathrm{s}] \times (T_1 - T_2)\,[\mathrm{K}]$$

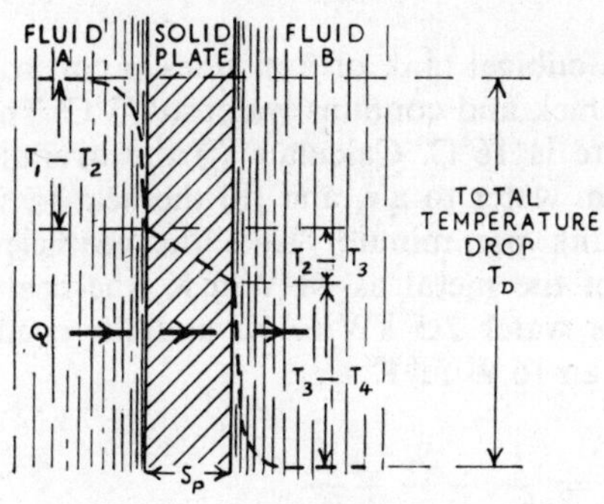

Fig. 11

Referring to Fig. 11,

Heat conducted through film of fluid A:

$$Q = h_A A t(T_1 - T_2) \qquad \therefore T_1 - T_2 = \frac{Q}{h_A A t}$$

Heat conducted through solid plate:

$$Q = \frac{\lambda_P A t(T_2 - T_3)}{S_P} \qquad \therefore T_2 - T_3 = \frac{Q S_P}{\lambda_P A t}$$

Heat conducted through film of fluid B:

$$Q = h_B At(T_3 - T_4) \qquad \therefore T_3 - T_4 = \frac{Q}{h_B At}$$

Total drop in temperature:

$$T_D = (T_1 - T_2) + (T_2 - T_3) + (T_3 - T_4)$$

$$= \frac{Q}{At} \left\{ \frac{1}{h_A} + \frac{S_P}{\lambda_P} + \frac{1}{h_B} \right\}$$

The quantity inside the brackets may be represented by $1/U$, where U is called the overall heat transfer coefficient, then,

$$\frac{1}{U} = \frac{1}{h_A} + \frac{S_P}{\lambda_P} + \frac{1}{h_B}$$

hence,

$$T_D = \frac{Q}{UAt} \qquad \text{or, } Q = UAtT_D$$

Example. A cubical tank of 2 m sides is constructed of metal plate 12 mm thick and contains water at 75°C. The surrounding air temperature is 16°C. Calculate (i) the overall heat transfer coefficient from water to air, and (ii) the heat loss through each side of the tank per minute. Take the coefficient of thermal conductivity of the metal as 48 W/m K, the coefficient of heat transfer of the water 2·5 kW/m²K, and the coefficient of heat transfer of the air 16 W/m²K.

$$\frac{1}{U} = \frac{1}{h_w} + \frac{S_P}{\lambda_P} + \frac{1}{h_A}$$

$$= \frac{1}{2·5 \times 10^3} + \frac{0·012}{48} + \frac{1}{16}$$

$$= 0·0004 + 0·00025 + 0·0625$$
$$= 0·06315$$
$$U = 1/0·06315 = 15·84 \text{ W/m}^2\text{K} \quad \text{Ans. (i)}$$
$$Q = UAtT_D$$
$$= 15·84 \times 2^2 \times 60 \times (75 - 16)$$
$$= 2·243 \times 10^5$$
$$= 224·3 \text{ kJ} \quad \text{Ans. (ii)}$$

CONVECTION

Convection is the method of transferring heat through a fluid by the movement of heated particles of the fluid.

Fig. 12 shows a vessel with an inclined tube connected at the bottom. When this contains water, and heat is applied to the tube, the heated particles of the water become less dense and rise, denser particles move to take their place and thus convection currents are set moving resulting in all the water in the vessel and tube becoming heated almost uniformly due to the continuous circulation of the water. This is the principle of the water-tube boiler.

Fig. 13 illustrates air in a room heated by convection, the fire, radiator or other heat source being placed at the bottom of the room.

Fig. 14 illustrates the air in a room cooled by convection, the coolers (such as refrigerator pipes) being situated near the top of the room.

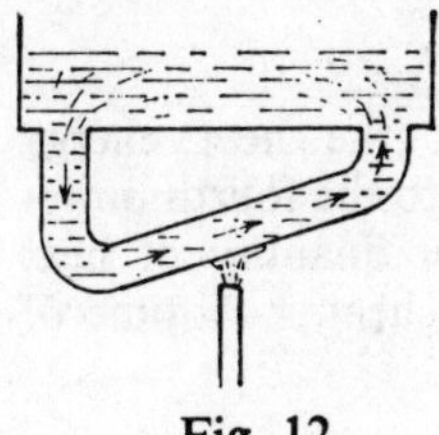

Fig. 12

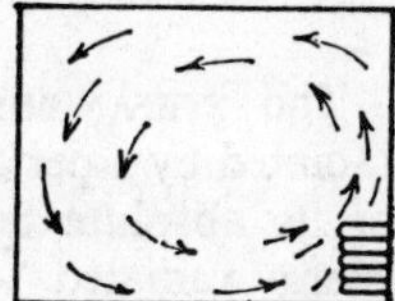

Fig. 13

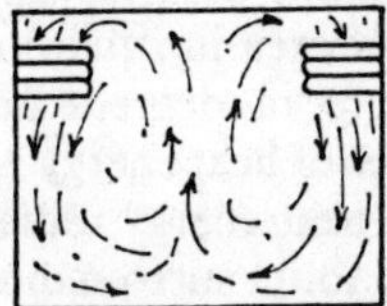

Fig. 14

The above are examples of natural circulation of the fluid and is referred to as *free convection*. When the motion of the fluid is produced mechanically, such as by means of a pump or a fan, it is referred to as *forced convection*.

RADIATION

Radiation is the transfer of heat energy from one body to another through space by rays of electro-magnetic waves. The rays of heat travel in straight lines in all directions at the same velocity as light, that is, at nearly 300 000 kilometres per second, which, for all practical purposes in engineering may be considered as instantaneous.

Some of the radiant heat falling upon a body is reflected in the same manner as light is reflected, the remainder is absorbed except for a negligible amount which is transmitted through the body. Dark and rough surfaces are good absorbers of radiant heat whereas bright and polished surfaces reflect most of the heat and therefore the amount of absorption is small. A body which is a poor absorber is also a poor radiator, and a good absorber is a good radiator. A perfect absorber and radiator of heat energy is termed a perfect "black" body.

The *emissivity* of a radiating body is the ratio of the heat emitted by that body compared with the heat emitted by a perfect black body of the same surface area and temperature in the same time. Emissivity may be represented by ε, the Greek letter epsilon, and its value for the ideal radiator is therefore unity.

The STEFAN-BOLTZMANN LAW states that the heat energy radiated by a perfect radiator is proportional to the fourth power of its absolute temperature. Hence, if Q = quantity of heat energy radiated, A = surface area radiating heat, t = time of radiation, T = absolute temperature, then:

$$Q = AtT^4 \times \text{a constant}$$

The value of the constant depends upon the units employed, and for the basic units of area in square metres, time in seconds, and absolute temperature in degrees kelvin, the value of the constant to give kilojoules of heat energy is $5 \cdot 67 \times 10^{-11} \text{kJ/m}^2\text{s K}^4$.

Thus, the quantity of heat energy radiated from a hot body of absolute temperature T_1 to its surroundings at absolute temperature T_2 is therefore:

$$Q = 5 \cdot 67 \times 10^{-11} \times \varepsilon At(T_1^4 - T_2^4)$$

Example. The temperature of the flame in a furnace is 1277°C and the temperature of its surrounds is 277°C. Calculate the maximum theoretical quantity of heat energy radiated per minute per square metre to the surrounding surface area.

$$\begin{aligned}
T_1 &= 1277°C + 273 = 1550 \text{ K} \\
T_2 &= 277°C + 273 = 550 \text{ K} \\
Q &= 5 \cdot 67 \times 10^{-11} \times At(T_1^4 - T_2^4) \\
&= 5 \cdot 67 \times 10^{-11} \times 1 \times 60 \times (1550^4 - 550^4)
\end{aligned}$$

Note the convenience of factorising the term inside the brackets, it also avoids excessively large numbers:

$$
\begin{aligned}
1550^4 - 550^4 &= (1550^2 + 550^2)(1550^2 - 550^2) \\
&= (2\,402\,500 + 302\,500)(2\,402\,500 - 302\,500) \\
&= 2\,705\,000 \times 2\,100\,000 \\
&= 2.705 \times 2.1 \times 10^{12} \\
Q &= 5.67 \times 10^{-11} \times 1 \times 60 \times 2.705 \times 2.1 \times 10^{12} \\
&= 1.933 \times 10^4 \text{ kJ or } 19.33 \text{ MJ} \quad \text{Ans.}
\end{aligned}
$$

TEST EXAMPLES 4

1. Calculate the quantity of heat conducted per minute through a duralumin circular disc 127 mm diameter and 19 mm thick when the temperature drop across the thickness of the plate is 5°C. Take the coefficient of thermal conductivity of duralumin as 150 W/m K.

2. A cold storage compartment is 4·5 m long by 4 m wide by 2·5 m high. The four walls, ceiling and floor are covered to a thickness of 150 mm with insulating material which has a coefficient of thermal conductivity of $5·8 \times 10^{-2}$ W/m K. Calculate the quantity of heat leaking through the insulation per hour when the outside and inside face temperatures of the material is 15°C and —5°C respectively.

ƒ 3. One side of a refrigerated cold chamber is 6 m long by 3·7 m high and consists of 168 mm thickness of cork between outer and inner walls of wood. The outer wood wall is 30 mm thick and its outside face temperature is 20°C, the inner wood wall is 35 mm thick and its inside face temperature is —3°C. Taking the coefficients of thermal conductivity of cork and wood as 0·042 and 0·2 W/m K respectively, calculate (i) the heat transfer per second per square metre of surface area, (ii) the total heat transfer through the chamber side per hour, (iii) the interface temperatures.

ƒ 4. Hot gases at 280°C flow on one side of a metal plate of 10 mm thickness and air at 35°C flows on the other side. The heat transfer coefficient of the gases is 31·5 W/m²K and that of the air is 32 W/m²K. The coefficient of thermal conductivity of the metal plate is 50 W/m K. Calculate (i) the overall heat transfer coefficient, and (ii) the heat transfer from gases to air per minute per square metre of plate area.

ƒ 5. The wall of a cold room consists of a layer of cork sandwiched between outer and inner walls of wood, the wood walls being each 30 mm thick. The inside atmosphere of the room is maintained at —20°C when the external atmospheric temperature is 25°C, and the heat loss through the wall is 42 W/m². Taking the thermal conductivity of wood and cork as 0·2 W/m K and 0·05 W/m K respectively, and the rate of heat transfer between each exposed wood surface and their respective atmospheres as 15 W/m²K, calculate (i) the temperatures of the exposed surfaces, (ii) the temperatures of the interfaces, and (iii) the thickness of the cork.

6. A flat circular plate is 500 mm diameter. Calculate the theoretical quantity of heat radiated per hour when its temperature is 215°C and the temperature of its surrounds is 45°C. Take the value of the radiation constant as 5.67×10^{-11} kJ/m²s K⁴.

f 7. The steam drum of a water-tube boiler has hemispherical ends, the diameter is 1·22 m and the overall length is 6 m. Under steaming conditions the temperature of the shell before lagging was 230°C and the temperature of the surrounds was 51°C. The temperature of the cleading after lagging was 69°C and the surrounds 27°C. Assuming 75% of the total shell area to be lagged and taking the radiation constant as 5.67×10^{-11} kJ/m²s K⁴, estimate the saving in heat energy per hour due to lagging.

CHAPTER 5

LAWS OF PERFECT GASES

When a substance has been evaporated it can exist as a gas or vapour and one of its most important characteristics is its elastic property. For instance, if a certain volume of a liquid is put into a vessel of larger volume, the liquid will only partially fill the vessel, taking up no more nor less volume than it did before, but when a gas enters a vessel it immediately fills up every part of that vessel no matter how large it is. Practically speaking, liquids cannot be compressed nor expanded, but gases can be compressed into smaller volumes and expanded to larger volumes.

A perfect gas is a theoretically ideal gas which strictly follows Boyle's and Charles' laws of gases.

Consider a given mass of a perfect gas enclosed in a cylinder by a gas-tight moveable piston. When the piston is pushed inward, the gas is compressed to a smaller volume, when pulled outward the gas is expanded to a larger volume. However, not only is there a change in volume but the pressure and temperature also change. These three quantities, pressure, volume and temperature, are related to each other, and to determine their relationship it is usual to perform experiments with each one of these quantities in turn kept constant while observing the relationship between the other two.

In such basic laws, the pressure, volume and temperature must all be the absolute values, that is, measured from absolute zero, and not measured from some artificial level. Absolute values were explained in Chapter 1, a brief reminder will suffice:

ABSOLUTE PRESSURE (p) is the pressure measured above a perfect vacuum. Ordinary pressure gauges indicate the pressure from the level of atmospheric pressure and the reading of a pressure gauge is termed the *gauge pressure*. The absolute pressure is therefore obtained by adding the atmospheric pressure to the gauge pressure. For many calculations in engineering the atmospheric pressure may be assumed to be 100 kN/m² which is 1 bar. For more accurate results the atmospheric pressure would be measured probably by a barometer graduated in millimetres of mercury [mm Hg]:

$$\text{Atmospheric pressure [N/m}^2\text{]} = \text{mm Hg} \times 133 \cdot 3$$

For example, if the barometer reads 756 mm Hg, this is

$$\text{Atmospheric pressure} = 756 \times 133 \cdot 3$$
$$= 1 \cdot 008 \times 10^5 \, \text{N/m}^2$$
$$= 100 \cdot 8 \, \text{kN/m}^2 \text{ or } 1 \cdot 008 \text{ bar}$$

The student is reminded that for absolute pressures, the word "absolute" need not follow the given pressure, it is to be taken as such unless the pressure is distinctly marked "gauge" to indicate that it is a pressure gauge reading.

VOLUME (V). The volume of the gas is equal to the full volume of the vessel containing it. Note that in the case of the cylinder of an air compressor or reciprocating engine, the total volume of air or gas includes the volume of the clearance space between the piston at its top dead centre and the cylinder cover, as well as the piston swept volume.

ABSOLUTE TEMPERATURE (T). This is the temperature in degrees kelvin measured above the absolute zero of temperature. Since the absolute zero is 273° below 0°C, then, as explained in Chapter 1:

$$T[\text{K}] = \theta[°\text{C}] + 273$$

BOYLE'S LAW

Boyle's law states that the absolute pressure of a fixed mass of a perfect gas varies inversely as its volume if the temperature remains unchanged.

$$p \propto \frac{1}{V} \qquad \text{therefore } p \times V = \text{a constant.}$$

Hence,
$$p_1 \times V_1 = p_2 \times V_2$$

To illustrate this, imagine 2 m³ of gas at a pressure of 100 kN/m² ($= 10^5$N/m² $= 1$ bar) contained in a cylinder with a gas-tight moveable piston as illustrated in Fig. 15. When the piston is pushed inward the pressure will increase as the gas is compressed to a smaller volume and, provided the temperature remains unchanged, the product of pressure and volume will be a constant quantity for all positions of the piston. From the known initial conditions the constant is calculated:

$$p_1 \times V_1 = \text{constant}$$
$$100 \times 2 = 200$$

and the pressure at any other volume can be determined:

When the volume is 1·5 m³,
$$p_2 \times 1·5 = 200$$
$$p_2 = 133·3 \text{ kN/m}^2$$
When the volume is 1 m³,
$$p_3 \times 1 = 200$$
$$p_3 = 200 \text{ kN/m}^2$$
When the volume is 0·5 m³,
$$p_4 \times 0·5 = 200$$
$$p_4 = 400 \text{ kN/m}^2$$
And so on.

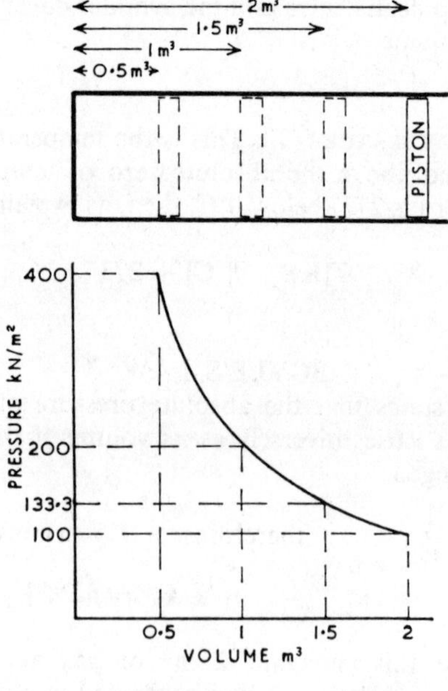

Fig. 15

The variation of pressure with change of volume is shown in the graph below the cylinder in Fig. 15. The graph produced by joining up the plotted points is a rectangular hyperbola, consequently we refer to compression or expansion where $pV = $ constant as hyperbolic compression or hyperbolic expansion. When the temperature is constant as in this example, the operation may also be termed "isothermal".

Note that as the ordinates (vertical measurements) represent pressure, and the abscissae (horizontal measurements) represent volume, and since the product of pressure and volume is constant, then all rectangles drawn from the axes with their corners touching the curve, will be of equal area.

Example. 3·5 m³ of air at a pressure of 20 kN/m² gauge is compressed at constant temperature to a pressure of 425 kN/m² gauge. Taking the atmospheric pressure as 100 kN/m² calculate the final volume of the air.

$$\text{Initial absolute pressure} = \ 20 + 100 = 120 \text{ kN/m}^2$$
$$\text{Final absolute pressure} = 425 + 100 = 525 \text{ kN/m}^2$$
$$p_1 V_1 = p_2 V_2$$
$$120 \times 3{\cdot}5 = 525 \times V_2$$

$$V_2 = \frac{120 \times 3{\cdot}5}{525} = 0{\cdot}8 \text{ m}^3 \quad \text{Ans.}$$

CHARLES' LAW

Charles' law states that the volume of a fixed mass of a perfect gas varies directly as its absolute temperature if the pressure remains unchanged, also, the absolute pressure varies directly as the absolute temperature if the volume remains unchanged.

From the above statement we have:

For constant pressure, $V \propto T$ $\qquad \therefore \dfrac{V}{T} = \text{constant}$

hence, $\qquad \dfrac{V_1}{T_1} = \dfrac{V_2}{T_2} \qquad$ or $\qquad \dfrac{V_1}{V_2} = \dfrac{T_1}{T_2}$

For constant volume, $p \propto T$ $\qquad \therefore \dfrac{p}{T} = \text{constant}$

hence, $\qquad \dfrac{p_1}{T_1} = \dfrac{p_2}{T_2} \qquad$ or $\qquad \dfrac{p_1}{p_2} = \dfrac{T_1}{T_2}$

Example. The pressure of the air in a starting air vessel is 40 bar ($= 40 \times 10^5$ N/m^2 or 4 MN/m^2) and the temperature is 24°C. If a fire in the vicinity causes the temperature to rise to 65°C, find the pressure of the air. Neglect any increase in volume of the vessel.

As the term "gauge" does not follow the given pressure, it is assumed that this is the initial absolute pressure.

$$\text{Initial absolute temperature} = 24°C + 273 = 297 \text{ K}$$
$$\text{Final absolute temperature} = 65°C + 273 = 338 \text{ K}$$

$$\frac{p_1}{T_1} = \frac{p_2}{T_2} \qquad \therefore \ p_2 = \frac{p_1 T_2}{T_1}$$

$$p_2 = \frac{40 \times 338}{297} = 45 \cdot 52 \text{ bar}$$

$$= 45 \cdot 52 \times 10^5 \text{ N/m}^2 \text{ or } 4 \cdot 552 \text{ MN/m}^2 \quad \text{Ans.}$$

COMBINATION OF BOYLE'S AND CHARLES' LAWS

Each one of these laws states how one quantity varies with another if the third quantity remains unchanged, but if the three quantities change simultaneously, it is necessary to combine these laws in order to determine the final conditions of the gas.

Referring to Fig. 16 which again represents a cylinder with a piston, gas-tight so that the mass of gas within the cylinder is always the same. Let the gas be compressed from its initial state of pressure p_1 volume V_1 and temperature T_1 to its final state of $p_2 V_2$ and T_2, but to arrive at the final state let it pass through two stages, the first to satisfy Boyle's law and the second to satisfy Charles' law.

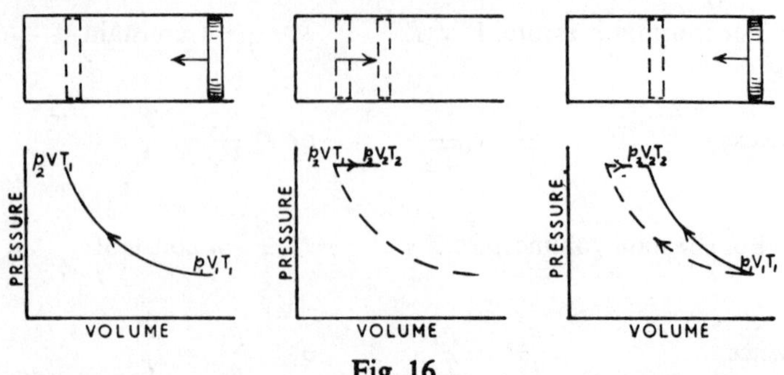

Fig. 16

Imagine the piston pushed inward to compress the gas until it reaches the final pressure of p_2 and let its volume then be represented by V. Normally the temperature would tend to increase due to the work done in compressing the gas, but any heat so generated must be taken away from it during compression so that its temperature remains unchanged at T_1 hence following Boyle's law:

$$p_1 V_1 = p_2 V \quad \dots \quad \dots \quad \dots \quad \text{(i)}$$

Now apply heat to raise the temperature from T_1 to T_2 and at the same time draw the piston outward to prevent a rise of pressure and keep it constant at p_2. The volume will increase in direct proportion to the increase in absolute temperature according to Charles' law:

$$\frac{V_2}{V} = \frac{T_2}{T_1} \quad \dots \quad \dots \quad \dots \quad \text{(ii)}$$

By substituting the value of V from (ii) into (i) this quantity will be eliminated:

From (ii) $\qquad\qquad V = \dfrac{V_2 T_1}{T_2}$

Substituting into (i)

$$p_1 V_1 = p_2 \times \frac{V_2 T_1}{T_2}$$

$$\therefore \frac{p_1 V_1}{T_1} = \frac{p_2 V_2}{T_2}$$

This combined law of Boyle's and Charles' is true for a given mass of any perfect gas subjected to any form of compression or expansion.

Example. 0·5 m³ of a perfect gas at a pressure of 0·95 bar (= 95 kN/m²) and temperature 17°C are compressed to a volume of 0·125 m³ and the final pressure is 5·6 bar (= 560 kN/m²). Calculate the final temperature.

$$\text{Initial absolute temperature} = 17°C + 273 = 290 \text{ K}$$

$$\frac{p_1 V_1}{T_1} = \frac{p_2 V_2}{T_2}$$

$$\frac{95 \times 0{\cdot}5}{290} = \frac{560 \times 0{\cdot}125}{T_2}$$

$$T_2 = \frac{560 \times 0{\cdot}125 \times 290}{95 \times 0{\cdot}5}$$

$$= 427{\cdot}4 \text{ K}$$

$$427{\cdot}4 - 273 = 154{\cdot}4°C \quad \text{Ans.}$$

CHARACTERISTIC EQUATION OF A PERFECT GAS

Since pV/T is a constant, its value can be determined for a given mass of any perfect gas. To form a basis on which to work, the constant is calculated on the specific volume (v) of the gas, that is, the volume in cubic metres occupied by a mass of one kilogramme. The constant so obtained is termed the *gas constant*, it is represented by R and is different for all different gases:

$$\frac{pv}{T} = R \quad \text{or} \quad pv = RT$$

The volume occupied by m kg of mass being represented by V then:

$$pV = mRT$$

This expression is called the *characteristic equation of a perfect gas*.

Taking air as an example, experiments show that at standard atmospheric pressure and temperature, that is, at $101{\cdot}325$ kN/m^2 and $0°C$, the specific volume is $0{\cdot}7734$ m^3/kg. Inserting these values to find R:

$$R = \frac{pv}{T}$$

$$= \frac{101 \cdot 325 \ [\text{kN/m}^2] \times 0 \cdot 7734 \ [\text{m}^3/\text{kg}]}{273 \ [\text{K}]}$$

$$= 0 \cdot 287 \ \text{kJ/kgK}$$

Note the units for R. Repeating the above with the units only:

$$R = \frac{\text{kN/m}^2 \times \text{m}^3/\text{kg}}{\text{K}} = \frac{\text{kN}}{\text{m}^2} \times \frac{\text{m}^3}{\text{kg}} \times \frac{1}{\text{K}}$$

$$= \frac{\text{kN}\,\text{m}}{\text{kg}\,\text{K}} = \text{kN}\,\text{m/kgK} = \text{kJ/kgK}$$

Example. An air compressor delivers $0 \cdot 2$ m^3 of air at a pressure of 850 kN/m^2 and 31°C into an air reservoir. Taking the gas constant for air as $0 \cdot 287$ kJ/kgK, calculate the mass of air delivered.

$$pV = mRT$$

$$m = \frac{pV}{RT} = \frac{850 \times 0 \cdot 2}{0 \cdot 287 \times (31 + 273)}$$

$$= \frac{850 \times 0 \cdot 2}{0 \cdot 287 \times 304} = 1 \cdot 948 \ \text{kg} \quad \text{Ans.}$$

f UNIVERSAL GAS CONSTANT

The *kilogramme-mol* of a substance is a mass of that substance numerically equal to its molecular weight. The kilogramme-mol may be simply abbreviated to *mol* and represented by the symbol kg-mol.

As examples, the molecular weight of oxygen is 32, therefore the mass of one kg-mol of oxygen is 32 kg. The molecular weight of hydrogen is 2 hence one kg-mol of hydrogen is a mass of 2 kg. And so on.

f AVOGADRO'S LAW states that under equal conditions of temperature and pressure, equal volumes of all gases contain the same number of molecules.

Consider two gases, one of mass m_1 of molecular weight M_1 and containing n_1 molecules, the other of mass m_2 of molecular weight M_2 and containing n_2 molecules:

$$\text{Ratio of masses} = \frac{m_1}{m_2} = \frac{n_1 M_1}{n_2 M_2}$$

For equal volumes of the gases at the same temperature and pressure, the gases contain an equal number of molecules, therefore $n_1 = n_2$ and cancel:

$$\frac{m_1}{m_2} = \frac{M_1}{M_2}$$

From the characteristic gas equation $pV = mRT$, substituting for $m = pV/RT$:

$$\frac{p_1 V_1 / R_1 T_1}{p_2 V_2 / R_2 T_2} = \frac{M_1}{M_2} \qquad \therefore \frac{p_1 V_1 R_2 T_2}{p_2 V_2 R_1 T_1} = \frac{M_1}{M_2}$$

pV and T cancel because they are equal, therefore,

$$\frac{R_2}{R_1} = \frac{M_1}{M_2} \quad \text{or} \quad R_1 M_1 = R_2 M_2$$

Hence the product of the gas constant R and the molecular weight M is the same for all gases. This product is termed the *Universal Gas Constant*, it is represented by R_0 and its value has been found by experiment to be $8 \cdot 314$ kJ/kg-mol K.

$$R_0 = RM$$

Any particular gas constant R can therefore be found if its molecular weight is known.

Thus, the molecular weight of nitrogen is 28, therefore the gas constant for nitrogen is:

$$R = \frac{R_0}{M} = \frac{8 \cdot 314}{28} = 0 \cdot 2969 \text{ kJ/kg K}$$

f DALTON'S LAW OF PARTIAL PRESSURES

This law states that the pressure exerted in a vessel by a mixture of gases is equal to the sum of the pressures that each separate gas would exert if it alone occupied the whole volume of the vessel.

The pressure exerted by each gas is termed a partial pressure:

Total pressure of the mixture
$$= \frac{\text{partial press.}}{\text{due to gas}_1} + \frac{\text{partial press.}}{\text{due to gas}_2} + \frac{\text{partial press.}}{\text{due to gas}_3} + \text{etc.}$$

$$p = p_1 + p_2 + p_3 + \text{etc.}$$

From the characteristic gas equation $pV = mRT$, substituting for the pressure of each gas, and also for the mixture:

$$\frac{mRT}{V} = \frac{m_1 R_1 T_1}{V_1} + \frac{m_2 R_2 T_2}{V_2} + \frac{m_3 R_3 T_3}{V_3} + \text{etc.}$$

Since it can be considered as though each gas on its own occupied the whole space then the volume V is common throughout. The temperature is also common, therefore V and T being common to all terms will cancel:

$$\frac{mRT}{V} = \frac{T}{V} \{m_1 R_1 + m_2 R_2 + m_3 R_3 + \text{etc}\}.$$

$$mR = m_1 R_1 + m_2 R_2 + m_3 R_3 + \text{etc.}$$

where m is the total mass of the mixture and R is the gas constant of the mixture.

f Example. The analysis by mass of a sample of air is $23 \cdot 14\%$ oxygen, $75 \cdot 53\%$ nitrogen, $1 \cdot 28\%$ argon and $0 \cdot 05\%$ carbon dioxide. Estimate the gas constant for air (to the nearest four figures) taking the molecular weights of O_2, N_2, Ar and CO_2 as 32, 28, 40 and 44 respectively, and the universal gas constant $R_0 = 8 \cdot 314$ kJ/kg-mol K.

$$R = \frac{R_0}{M} \quad \text{Considering 1 kg of air:}$$

$$\text{Oxygen} \quad R_1 = \frac{8\cdot314}{32} = 0\cdot2598$$

$$m_1R_1 = 0\cdot2314 \times 0\cdot2598 = 0\cdot060\ 12$$

$$\text{Nitrogen} \quad R_2 = \frac{8\cdot314}{28} = 0\cdot2969$$

$$m_2R_2 = 0\cdot7553 \times 0\cdot2969 = 0\cdot224\ 2$$

$$\text{Argon} \quad R_3 = \frac{8\cdot314}{40} = 0\cdot2078$$

$$m_3R_3 = 0\cdot0128 \times 0\cdot2078 = 0\cdot002\ 661$$

$$\text{Carbon dioxide} \quad R_4 = \frac{8\cdot314}{44} = 0\cdot1889$$

$$m_4R_4 = 0\cdot0005 \times 0\cdot1889 = 0\cdot000\ 094\ 45$$
$$\Sigma mR = \overline{0\cdot287\ 075\ 45}$$

$$R\,\text{air} = \frac{0\cdot2871}{1} = 0\cdot2871\ \text{kJ/kg\,K} \quad \text{Ans.}$$

f PARTIAL VOLUMES. If each gas of a mixture in a closed vessel at the full volume of V and partial pressure $p_1\,(p_2$, etc.), is considered as being compressed until its pressure is the full pressure p and its volume is its partial volume $V_1\,(V_2$, etc.), then, since its temperature remains the same:

$$p_1 \times V = p \times V_1 \quad \text{and} \quad p_2 \times V = p \times V_2 \quad \text{etc.}$$
$$\therefore (p_1 + p_2 + \text{etc.}) \times V = p \times (V_1 + V_2 + \text{etc.})$$

hence, the total volume of the mixture is equal to the sum of the partial volumes of the gases at the same total pressure and temperature.

$$\text{Also,}\ \frac{p_1}{p_2} = \frac{V_1}{V_2}\ \text{therefore:}$$

the ratio of the partial volumes is equal to the ratio of the partial pressures.

Dalton's law is also applied for the determination of the quantity of air-leakage into steam condensers. An example is given in Chapter 10 after the study of properties of steam.

SPECIFIC HEATS OF GASES

The specific heat (c) of a substance is defined (see Chapter 2) as the quantity of heat energy required to be transferred to unit mass of that substance to raise its temperature by one degree. Hence, the quantity (Q) of heat in kilojoules required to be given to a mass of m kilogrammes of the substance to raise its temperature from T_1 to T_2 is:

$$Q[\text{kJ}] = m[\text{kg}] \times c[\text{kJ/kgK}] \times (T_2 - T_1)\ [\text{K}]$$

Since the characteristics of gases vary considerably at different temperatures and pressures, heat energy may be transferred under an infinite number of conditions, therefore the specific heat can have an infinite number of different values. Consider, however, two important conditions under which heat may be transferred, (i) while the *volume* of the gas remains constant, (ii) while the *pressure* of the gas remains constant.

The specific heat of a gas at constant volume is represented by c_V.

The specific heat of a gas at constant pressure is represented by c_P. This is a higher value than c_V because, when the gas is receiving heat it must be allowed to expand in volume to prevent a rise in pressure and, whilst expanding, the gas is expending energy in doing external work, hence extra heat energy must be supplied equivalent to the external work done.

Example. A quantity of air of mass 0·23 kg, pressure 100 kN/m², volume 0·1934 m³ and temperature 20°C is enclosed in a cylinder with a gas-tight moveable piston, and heat energy is transferred to the air to raise its temperature to 142°C.

(a) If the piston is prevented from moving during heat transfer so that the volume of the air remains unchanged, calculate (i) the heat supplied, taking the specific heat at constant volume $c_V = 0·718$ kJ/kgK, and (ii) the final pressure.

(b) If the piston moves to allow the air to expand in volume at such a rate as to keep the pressure constant, calculate (i) the heat supplied, taking the specific heat at constant pressure $c_P = 1·005$ kJ/kgK, and (ii) the final volume.

$$\text{Initial absolute temperature} = 20 + 273 = 293\,\text{K}$$
$$\text{Final absolute temperature} = 142 + 273 = 415\,\text{K}$$
$$\text{Temperature rise} = 142 - 20 = 122°\text{C} = 122\,\text{K}$$

(a)
$$Q = m \times c \times (T_2 - T_1)$$
$$= 0.23 \times 0.718 \times 122$$
$$= 20.15 \text{ kJ} \quad \text{Ans. (a i)}$$

$$\frac{p_1 V_1}{T_1} = \frac{p_2 V_2}{T_2}$$

The volume is constant $\therefore V_2 = V_1$ and cancels,

$$p_2 = \frac{p_1 T_2}{T_1} \text{ (which is Charles' law)}$$

$$= \frac{100 \times 415}{293} = 141.6 \text{ kN/m}^2 \quad \text{Ans. (a ii)}$$

(b)
$$Q = m \times c \times (T_2 - T_1)$$
$$= 0.23 \times 1.005 \times 122$$
$$= 28.19 \text{ kJ} \quad \text{Ans. (b i)}$$

$$\frac{p_1 V_1}{T_1} = \frac{p_2 V_2}{T_2}$$

The pressure is constant $\therefore p_1 = p_2$ and cancels

$$V_2 = \frac{V_1 T_2}{T_1} \text{ (which is Charles' law)}$$

$$= \frac{0.1934 \times 415}{293} = 0.2738 \text{ m}^3 \quad \text{Ans. (b ii)}$$

Note the difference between the quantity of heat energy supplied to the air in the two cases:

$$28.19 - 20.15 = 8.04 \text{ kJ}$$

This amount of heat energy was expended in moving the piston:

Referring to Fig. 17, the piston is pushed forward by the gas at a constant pressure of, say, $p[\text{kN/m}^2]$. Let $A[\text{m}^2]$ represent the area of the piston, then the total force on the piston is $p \times A[\text{kN}]$. If the piston moves $S[\text{metres}]$, the work done, being the product of force

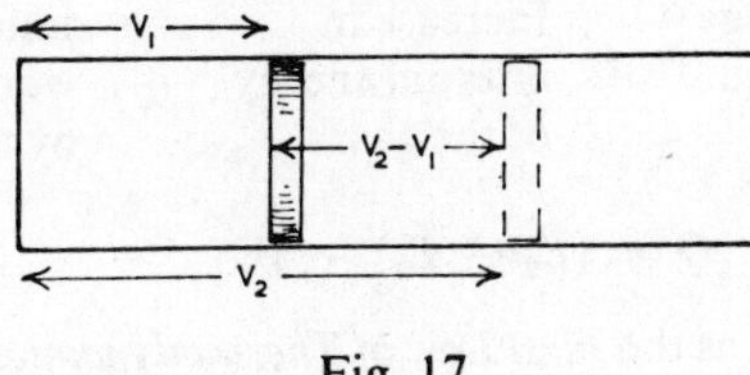

Fig. 17

and distance, is $p \times A \times S$ [kN m = kJ]. The product of the area A and the distance moved S is the volume swept through by the piston, hence,

$$\text{Work done} = p(V_2 - V_1)$$

In the last example, when the pressure was constant at 100 kN/m² the volume of the air increased from 0·1934 m³ to 0·2738 m³ so that the piston swept volume is the difference between these.

$$\begin{aligned}
\text{Work done} &= p(V_2 - V_1) \\
&= 100(0\cdot2738 - 0\cdot1934) \\
&= 100 \times 0\cdot0804 \\
&= 8\cdot04 \text{ kJ}
\end{aligned}$$

Showing that the extra heat energy given to the air at constant pressure compared with that at constant volume is the heat energy expended in doing external work by pushing the piston forward.

ENERGY EQUATION

The *internal energy* (E) of a gas is the energy contained in it as stored up work by virtue of the movement of its molecules.

JOULE'S LAW states that the internal energy of a gas depends only upon its temperature and is independent of changes in pressure and volume.

By the principle of the conservation of energy, that is, that energy can neither be created nor destroyed, it follows that the total heat energy (Q) transferred to a gas will be the sum of the increase in its internal energy $(E_2 - E_1)$ and any work (W) that is done by the gas during the transfer of heat energy to it, thus:

| Heat energy transferred to the gas | $=$ | Increase in internal energy of the gas | $+$ | External work done by the gas |

$$Q = (E_2 - E_1) + W$$

This is known as the *First Law of Thermodynamics.*

RELATIONSHIP BETWEEN SPECIFIC HEATS. Referring again to the last example, with Fig. 17, and applying the energy equation:

In the first case, heat energy is supplied to the gas to raise its temperature from T_1 to T_2 at *constant volume*, the heat supplied is $mc_V(T_2 - T_1)$ but no work is done because, as the volume is constant, the piston does not move.

| Heat supplied | $=$ | Increase in internal energy | $+$ | External work done |

$$mc_V(T_2 - T_1) = (E_2 - E_1) + 0$$
$$E_2 - E_1 = mc_V(T_2 - T_1) \quad \text{....} \quad \text{....} \quad \text{....} \quad \text{(i)}$$

In the second case, heat energy is supplied to raise the temperature through the same range, from T_1 to T_2 at *constant pressure*. The heat supplied is $mc_P(T_2 - T_1)$ and external work is done equal to $p(V_2 - V_1)$. Expressing the work done in terms of temperature by substituting the value of V from the characteristic gas equation, and inserting into the energy equation,

$$p(V_2 - V_1) = p\left\{ \frac{mRT_2}{p} - \frac{mRT_1}{p} \right\} = mR(T_2 - T_1)$$

| Heat supplied | $=$ | Increase in internal energy | $+$ | External work done |

$$mc_P(T_2 - T_1) = (E_2 - E_1) + mR(T_2 - T_1)$$
$$E_2 - E_1 = mc_P(T_2 - T_1) - mR(T_2 - T_1) \quad \text{....} \quad \text{(ii)}$$

Since the temperature change is the same in each case then the change in internal energy is the same, hence, from (i) and (ii):

$$mc_V(T_2 - T_1) = mc_P(T_2 - T_1) - mR(T_2 - T_1)$$
m and $(T_2 - T_1)$ are common to all terms and cancel:

$$c_V = c_P - R$$
$$\therefore\ R = c_P - c_V$$

Inserting the values for air as previously given,

$$R = 1{\cdot}005 - 0{\cdot}718 = 0{\cdot}287\ \text{kJ/kg K}$$

RATIO OF SPECIFIC HEATS. Another important relationship between the specific heats is the ratio c_P/c_V, the symbol denoting this ratio is the Greek letter gamma γ:

$$\gamma = \frac{c_P}{c_V}$$

and for air this ratio is:

$$\gamma = \frac{1{\cdot}005}{0{\cdot}718} = 1{\cdot}4$$

Taking this further:

$$\gamma = \frac{c_P}{c_V} \qquad \therefore\ c_P = \gamma c_V$$

substituting this into the relationship:

$$R = c_P - c_V$$
$$= \gamma c_V - c_V$$
$$\therefore\ R = c_V(\gamma - 1)$$

$$\text{or}\ \ c_V = \frac{R}{\gamma - 1}$$

These expressions will be found useful in later calculations.

TEST EXAMPLES 5

1. Express the atmospheric pressure in kN/m² and bar when the barometer stands at 760 mm Hg.

2. 0·14 m³ of air at 1500 kN/m² expands to a volume of 0·35 m³ at constant temperature. Find the final pressure.

3. Assuming compression according to the law $pV = $ constant:
(i) Calculate the final volume when 1 m³ of gas at 120 kN/m² is compressed to a pressure of 960 kN/m².
(ii) Calculate the initial volume of gas at a pressure of 1·05 bar which will occupy a volume of 5·6 m³ when it is compressed to a pressure of 42 bar.

4. An air storage system consists of two cylindrical vessels with hemispherical ends, each being 300 mm diameter and 1·5 m long overall. Calculate the volume of "free" air (that is, air at atmospheric pressure, say 100 kN/m²) to be taken from the atmosphere and pumped into the two vessels at 2800 kN/m² gauge. Assume that the vessels initially contain air at atmospheric pressure and that the temperature of the compressed air delivered is the same as the temperature of the atmosphere.

5. A single-acting air pump has an effective swept stroke volume of 0·08 m³ and the total volume of the vessel and pipe-line to which it is connected is 14 m³. Commencing with a pressure of 101·3 kN/m² in the line, find the number of suction strokes required to reduce the pressure to 35 kN/m² assuming the temperature remains constant throughout.

6. 0·25 m³ of a gas at 21°C is heated at constant pressure to a temperature of 315°C. Calculate the final volume.

7. A closed vessel contains air at a pressure of 140 kN/m² gauge and temperature 20°C. Find the final gauge pressure if the air is heated at constant volume to 40°C. Take the atmospheric pressure as 759 mm Hg.

8. 0·2 m³ of gas at a pressure of 1350 kN/m² and temperature 177°C is expanded in a cylinder to a volume of 0·9 m³ and pressure 250 kN/m². Calculate the final temperature.

9. Taking the characteristic gas constant R for nitrogen as 0.297 kJ/kg K, calculate (i) the mass of 0.05 m^3 of nitrogen at 550 kN/m^2 and 28°C, (ii) the volume of 1 kg of nitrogen at 1 MN/m^2 and 0°C.

10. A cylindrical oxygen bottle has one flat end and one hemispherical end, the diameter is 240 mm and the overall length is 1.24 m. Calculate (i) the mass of oxygen contained in the bottle when the pressure is 700 kN/m^2 gauge and temperature 20°C. Calculate (ii) the mass of oxygen drawn off when the pressure falls to 500 kN/m^2 gauge and temperature 15°C. Take the atmospheric pressure as 1.013 bar and R for oxygen 0.26 kJ/kg K.

11. An air reservoir contains 20 kg of air at 3200 kN/m^2 gauge and 16°C. Calculate the new pressure and heat energy transfer if the air is heated to 35°C. Neglect any expansion of the reservoir, take R for air $= 0.287$ kJ/kg K, specific heat at constant volume $c_V = 0.718$ kJ/kg K, and atmospheric pressure $= 100$ kN/m^2.

12. The dimensions of a ship's saloon are 12 m by 16.5 m by 4 m. The air is completely changed once every 30 minutes and the temperature is maintained at 21°C. If the temperature of the outside atmosphere is 30°C, calculate the quantity of heat required to be extracted from the supply air per hour, and the equivalent power, taking the density of air at atmospheric pressure and 0°C as 1.293 kg/m^3 and the specific heat at constant pressure as 1.005 kJ/kg K.

13. The burning period of the fuel in a diesel engine cylinder takes place at constant pressure while the piston moves a fraction of its power stroke, calculate the external work done during the burning period if (i) the pressure is constant at 34.5 bar and the volume of the air increases from 0.0425 m^3 to 0.0935 m^3, (ii) the cylinder contains 0.63 kg of air and the temperature increases from 538°C to 1511°C. Take R for air $= 0.287$ kJ/kg K.

f 14. Heat energy is transferred to 1.36 kg of air which causes its temperature to increase from 40°C to 468°C. Calculate, for the two separate cases of heat transfer at (a) constant volume, (b) constant pressure:
(i)   the quantity of heat energy transferred,
(ii) the external work done,
(iii) the increase in internal energy.
Take c_V and c_P as 0.718 and 1.005 kJ/kg K respectively.

f 15. A closed vessel of 500 cm³ capacity contains a sample of flue gas at 1·015 bar and 20°C. If the analysis of the gas by volume is 10% carbon dioxide, 8% oxygen, and 82% nitrogen, calculate the partial pressure and mass of each constituent in the sample.

R for CO_2, O_2 and N_2 = 0·189, 0·26 and 0·297 kJ/kgK respectively.

CHAPTER 6

EXPANSION AND COMPRESSION OF PERFECT GASES

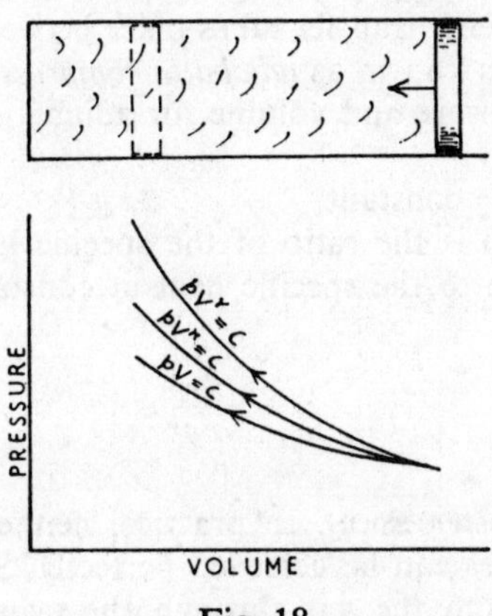

Fig. 18

COMPRESSION OF A GAS IN A CLOSED SYSTEM

When a gas is compressed in a cylinder by the inward movement of a gas-tight piston (Fig. 18), the pressure of the gas increases as the volume decreases. The work done *on* the gas to compress it appears as heat energy in the gas and the temperature tends to rise. This effect can readily be seen with a tyre inflator; in pumping up the tyre the discharge end of the inflator gets hot due to compressing the air.

ISOTHERMAL COMPRESSION. Imagine the piston pushed inward slowly to compress the gas and, at the same time, let heat be taken away via the cylinder walls (by a water-jacket or other means) to avoid any rise in temperature. If the gas could be compressed in this manner, *at constant temperature*, the process would be referred to as *isothermal compression* and the relationship between pressure and volume would follow Boyle's law as stated in the previous chapter:

$$pV = \text{constant} \qquad \therefore \ p_1 V_1 = p_2 V_2$$

ADIABATIC COMPRESSION. Now imagine the piston pushed inward quickly so that there is insufficient time for any heat energy to be transferred from the gas to the cylinder walls. All the work done in compressing the gas appears as stored up heat energy. The temperature at the end of compression will therefore be high and, for the same ratio of compression as the first case, the pressure will consequently be higher. This form of compression, where no heat energy transfer takes place between the gas and an external source, is known as *adiabatic compression*. The relationship between pressure and volume for adiabatic compression is:

$$pV^\gamma = \text{constant} \qquad \therefore p_1V_1^\gamma = p_2V_2^\gamma$$

where γ (gamma) is the ratio of the specific heat of the gas at constant pressure to the specific heat at constant volume, thus,

$$\gamma = \frac{c_P}{c_V}$$

POLYTROPIC COMPRESSION. In practice, neither isothermal nor adiabatic processes can be achieved perfectly. Some heat energy is always lost from the gas through the cylinder walls, more especially if the cylinder is water cooled, but this is never as much as the whole amount of the generated heat of compression. Consequently, the compression curve representing the relationship between pressure and volume lies somewhere between the two theoretical cases of isothermal and adiabatic. Such compression, where a partial amount of heat energy exchange takes place between the gas and an outside source during the process, is termed *polytropic compression* and the compression curve follows the law:

$$pV^n = \text{constant} \qquad \therefore p_1V_1^n = p_2V_2^n$$

Thus, the law $pV^n = $ constant may be taken as the general case to cover all forms of compression from isothermal to adiabatic wherein the value of n for isothermal compression is unity, for adiabatic compression $n = \gamma$, and for polytropic compression n generally lies somewhere between 1 and γ.

EXPANSION OF A GAS IN A CLOSED SYSTEM

When a gas is expanded in a cylinder (Fig. 19) the pressure

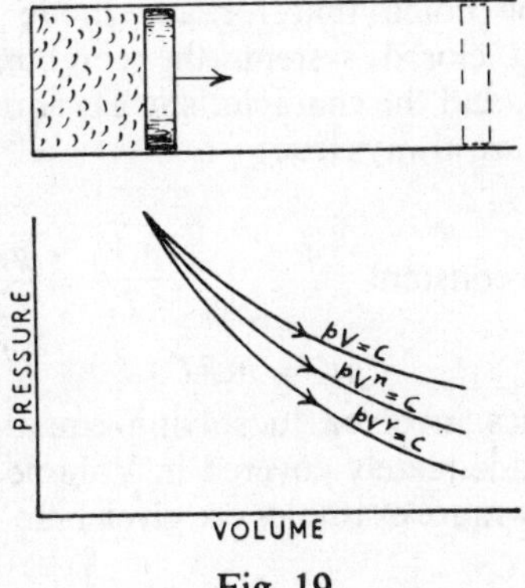

Fig. 19

falls and the volume increases as the piston is pushed outward by the energy in the gas.

This is exactly the opposite to compression. Work is done *by* the gas in pushing the piston outward and there is a tendency for the temperature to fall due to the heat energy in the gas being converted into mechanical energy. Therefore to expand the gas *isothermally*, heat energy must be transferred to the gas from an external source during the expansion in order to maintain its temperature constant. The expansion would then follow Boyle's law, $pV = $ constant.

The gas would expand *adiabatically* if no heat energy transfer, to or from the gas, occurs during the expansion, the external work done in pushing the piston forward being entirely at the expense of the stored up heat energy. Therefore the temperature of the gas will fall during the expansion. As for adiabatic compression, the law for adiabatic expansion is $pV^\gamma = $ constant.

During a *polytropic* expansion, a partial amount of heat energy will be transferred to the gas from an outside source but not sufficient to maintain a uniform temperature during the expansion. The law for polytropic expansion is $pV^n = $ constant as it is for polytropic compression.

With reference to Figs. 18 and 19 the student should note that the adiabatic curve is the steepest, the isothermal curve is the least steep, and the polytropic curve lies between the two. Thus, the higher the index of the law of expansion or compression, the steeper will be the curve.

It must also be noted that for any mode of expansion or compression in a closed system, the combination of Boyle's and Charles' laws, and the characteristic gas equation given in the previous chapter, are always true:

$$\frac{pV}{T} = \text{constant} \qquad \therefore \frac{p_1V_1}{T_1} = \frac{p_2V_2}{T_2}$$

$$pV = mRT$$

The mathematics involved in solving equations of the form $p_1V_1{}^n = p_2V_2{}^n$ is adequately covered in Volume 1 (Mathematics) of this series, a few more examples are given here.

Example. 0.25 m³ of air at 90 kN/m² and $10°$C are compressed in an engine cylinder to a volume of 0.05 m³, the law of compression being $pV^{1.4} = $ constant. Calculate (i) the final pressure, (ii) the final temperature, (iii) the mass of air in the cylinder, taking the characteristic gas constant for air $R = 0.287$ kJ/kg K.

$$p_1V_1{}^{1.4} = p_2V_2{}^{1.4}$$
$$90 \times 0.25^{1.4} = p_2 \times 0.05^{1.4}$$

$$p_2 = \frac{90 \times 0.25^{1.4}}{0.05^{1.4}} = 90 \times \left(\frac{0.25}{0.05}\right)^{1.4}$$

$$= 90 \times 5^{1.4} = 856.7 \text{ kN/m}^2 \quad \text{Ans. (i)}$$

$$\frac{p_1V_1}{T_1} = \frac{p_2V_2}{T_2}$$

$$\frac{90 \times 0.25}{(10 + 273)} = \frac{856.7 \times 0.05}{T_2}$$

$$T_2 = \frac{283 \times 856.7 \times 0.05}{90 \times 0.25} = 538.8 \text{ K}$$

$$538.8 - 273 = 265.8°\text{C} \quad \text{Ans. (ii)}$$

$$p_1V_1 = mRT_1$$

$$m = \frac{90 \times 0.25}{0.287 \times 283} = 0.277 \text{ kg} \quad \text{Ans. (iii)}$$

Example. 0·07 m³ of gas at 4·14 MN/m² is expanded in an engine cylinder and the pressure at the end of expansion is 310 kN/m². If expansion follows the law $pV^{1.35} = $ constant, find the final volume.

$$p_1 V_1^{1.35} = p_2 V_2^{1.35}$$

$$4140 \times 0.07^{1.35} = 310 \times V_2^{1.35}$$

$$V_2^{1.35} = \frac{4140 \times 0.07^{1.35}}{310}$$

$$V_2 = 0.07 \times \sqrt[1.35]{\frac{4140}{310}} = 0.4774 \text{ m}^3 \quad \text{Ans.}$$

Example. 0·014 m³ of gas at 3·15 MN/m² is expanded in a closed system to a volume of 0·154 m³ and the final pressure is 120 kN/m². If the expansion takes place according to the law $pV^n = $ constant, find the value of n.

$$p_1 V_1^n = p_2 V_2^n$$
$$3150 \times 0.014^n = 120 \times 0.154^n$$

$$\frac{3150}{120} = \left\{\frac{0.154}{0.014}\right\}^n$$

$$26.25 = 11^n$$

$$\log 26.25 = n \times \log 11$$
$$1.4191 = n \times 1.0414$$

$$n = \frac{1.4191}{1.0414} = 1.363 \quad \text{Ans.}$$

DETERMINATION OF n FROM GRAPH

It will be appreciated that it would be most difficult to obtain two pairs of sufficiently accurate values of pressure and volume from a running engine to enable the law of expansion or compression to be determined as in the previous example. One practical method of finding a fairly close approximation of the law is as follows:

(i) measure a series of connected values of p and V from the curve of an indicator diagram, (ii) reduce the equation $pV^n = C$ to a straight line logarithmic equation, (iii) draw a straight line graph as near as possible through the plotted points of $\log p$ and $\log V$ to eliminate slight errors of measurement, (iv) determine the law of this graph to obtain the value of n. Thus:

$$p \times V^n = C$$
$$\log p + n \log V = \log C$$
$$\log p = \log C - n \log V$$

This is the same form of equation as,

$$y = a - bx$$

which represents a simple straight-line graph.

The terms $\log p$ and $\log V$ are the two variables comparable with y and x respectively, and $\log C$ and n are constants comparable with a and b respectively. The constant n (like constant b) represents the slope of the straight-line graph and, being a negative value, the line will slope downwards from left to right.

Example. The following related values of the pressure p in kN/m^2 and the volume V in m^3 were measured from the compression curve of an internal combustion engine indicator diagram. Assuming that p and V are connected by the law $pV^n = C$, find the value of n.

p	3450	2350	1725	680	270	130
V	0·0085	0·0113	0·0142	0·0283	0·0566	0·0991

The solution to a similar type of problem is shown in Chapter 7 on Graphs in Volume 1 (Mathematics). It was there explained that when the value of V is less than unity, its logarithm has a negative characteristic, therefore to plot such a quantity on a graph the true numerical value of the log must be obtained by combining the mixture of negative characteristic with positive mantissa which produces an all-negative quantity. However, it will be appreciated that the scales of the graph can be of any convenient choice. In this case both pressure and volume can be expressed in more convenient units, the pressure in bars (1 bar $= 10^5$ N/m^2 $= 10^2$ kN/m^2) to proportionally reduce the high figures, and the volume in litres (10^3 litres $= 1$ m^3) to express all volumes above unity, thereby avoiding negative logs and so reducing labour. Hence the following tabulated values of p and V are in bars and litres respectively with their corresponding logarithms to obtain graph plotting points from the respective pairs of $\log p$ and $\log V$.

p [bar]	$\log p$	V [litre]	$\log V$
34·5	1·5378	8·5	0·9294
23·5	1·3711	11·3	1·0531
17·25	1·2368	14·2	1·1523
6·8	0·8325	28·3	1·4518
2·7	0·4314	56·6	1·7528
1·3	0·1139	99·1	1·9961

The graph is then plotted as shown in Fig. 20. Note that it is not necessary to commence at zero origin when only the slope of the line (value of n) is required, a larger graph on the available squared paper can be drawn by starting and finishing to suit the minimum and maximum values to be plotted.

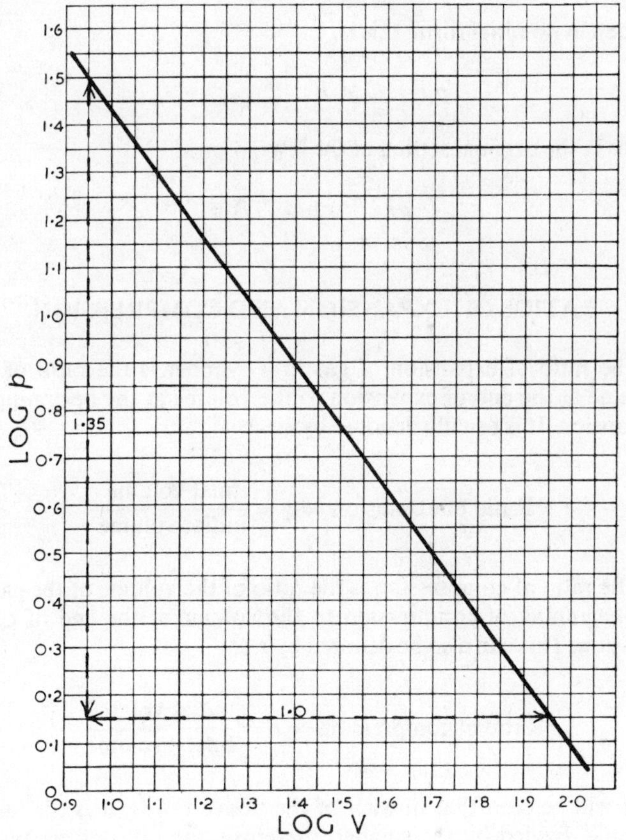

Fig. 20

Choosing any two points on the line such as those shown:

$$n = \frac{\text{decrease of } \log p}{\text{increase of } \log V}$$

$$= \frac{1 \cdot 5 - 0 \cdot 15}{1 \cdot 95 - 0 \cdot 95}$$

$$= \frac{1 \cdot 35}{1} = 1 \cdot 35$$

Thus, $\log p = \log C - 1 \cdot 35 \log V$

written in nominal form, this is,

$$p = C \times V^{-1 \cdot 35}$$

or, as in the original setting of the law:

$$pV^{1 \cdot 35} = C \quad \text{Ans.}$$

RATIOS OF EXPANSION AND COMPRESSION

The ratio of expansion of gas in a cylinder is the ratio of the volume at the end of expansion to the volume at the beginning of expansion. It is usually denoted by r.

$$\text{Ratio of expansion} = r = \frac{\text{final volume}}{\text{initial volume}}$$

The ratio of compression is the ratio of the volume of the gas at the beginning of compression to the volume at the end of compression. This can also be denoted by r.

$$\text{Ratio of compression} = r = \frac{\text{initial volume}}{\text{final volume}}$$

It will be seen that in each of the above ratios, it is the larger volume divided by the smaller, therefore, the ratio of expansion and ratio of compression is always greater than unity.

RELATIONSHIPS
BETWEEN TEMPERATURE AND VOLUME,
AND TEMPERATURE AND PRESSURE,
when $pV^n = C$

As stated previously, the equations,

$$p_1 V_1{}^n = p_2 V_2{}^n \quad \text{and} \quad \frac{p_1 V_1}{T_1} = \frac{p_2 V_2}{T_2}$$

are always true for any kind of expansion or compression of a perfect gas in a closed system. Some problems arise, however, where neither p_1 nor p_2 are given, and the unknown temperature or volume has to be solved by substituting the value of one of the pressures from one equation into the other. Similarly, where neither volume is given, substitution has to be made for one of the volumes to obtain the unknown temperature or pressure.

Substitution can be made in one of the above equations to eliminate either pressure or volume and so derive relationships for direct solution, as follows.

$$p_1 V_1{}^n = p_2 V_2{}^n \qquad \therefore p_1 = \frac{p_2 V_2{}^n}{V_1{}^n}$$

Substituting this value of p_1 into the combined law equation:

$$\frac{p_1 V_1}{T_1} = \frac{p_2 V_2}{T_2}$$

$$\frac{p_2 \times V_2{}^n \times V_1}{T_1 \times V_1{}^n} = \frac{p_2 V_2}{T_2}$$

$$T_1 \times V_1{}^n \times p_2 \times V_2 = T_2 \times p_2 \times V_2{}^n \times V_1$$

$$\frac{T_1}{T_2} = \frac{p_2 \times V_2{}^n \times V_1}{p_2 \times V_2 \times V_1{}^n}$$

p_2 cancels,

$$\text{dividing } V_2{}^n \text{ by } V_2 = V_2{}^n \div V_2 = V_2{}^{n-1}$$
$$\text{dividing } V_1{}^n \text{ by } V_1 = V_1{}^n \div V_1 = V_1{}^{n-1}$$

$$\frac{T_1}{T_2} = \frac{V_2{}^{n-1}}{V_1{}^{n-1}}$$

$$\frac{T_1}{T_2} = \left\{\frac{V_2}{V_1}\right\}^{n-1} \qquad \cdots \qquad \cdots \qquad \text{(i)}$$

Again from $p_1 V_1{}^n = p_2 V_2{}^n$ $\qquad V_1{}^n = \dfrac{p_2 V_2{}^n}{p_1}$

$$\therefore \ V_1 = \frac{p_2{}^{1/n} \times V_2}{p_1{}^{1/n}}$$

Substituting this value of V_1 into the combined law equation:

$$\frac{p_1 V_1}{T_1} = \frac{p_2 V_2}{T_2}$$

$$\frac{p_1 \times p_2{}^{1/n} \times V_2}{T_1 \times p_1{}^{1/n}} = \frac{p_2 \times V_2}{T_2}$$

$$T_1 \ \times \ p_1{}^{1/n} \ \times \ p_2 \ \times \ V_2 = T_2 \times p_1 \times p_2{}^{1/n} \times V_2$$

$$\frac{T_1}{T_2} = \frac{p_1 \times p_2{}^{1/n} \times V_2}{p_1{}^{1/n} \times p_2 \times V_2}$$

V_2 cancels,
$\qquad$ dividing p_1 by $p_1{}^{1/n} = p_1 \div p_1{}^{1/n} = p_1{}^{1-1/n}$
$\qquad$ dividing p_2 by $p_2{}^{1/n} = p_2 \div p_2{}^{1/n} = p_2{}^{1-1/n}$

$$\frac{T_1}{T_2} = \frac{p_1{}^{1-1/n}}{p_2{}^{1-1/n}}$$

$$\frac{T_1}{T_2} = \left\{\frac{p_1}{p_2}\right\}^{1-1/n}$$

$$\frac{T_1}{T_2} = \left\{\frac{p_1}{p_2}\right\}^{\frac{n-1}{n}} \qquad \cdots \qquad \cdots \qquad \text{(ii)}$$

From (i) and (ii) we have the very useful relationship:

$$\frac{T_1}{T_2} = \left\{\frac{V_2}{V_1}\right\}^{n-1} = \left\{\frac{p_1}{p_2}\right\}^{\frac{n-1}{n}}$$

For an adiabatic process, the adiabatic index γ is substituted for the polytropic index n.

Example. Air is expanded adiabatically from a pressure of 800 kN/m² to 128 kN/m². If the final temperature is 57°C, calculate the temperature at the beginning of expansion, taking $\gamma = 1\cdot4$.

$$\frac{T_1}{T_2} = \left\{\frac{p_1}{p_2}\right\}^{\frac{\gamma-1}{\gamma}}$$

$$\text{Note } \frac{\gamma-1}{\gamma} = \frac{1\cdot4-1}{1\cdot4} = \frac{0\cdot4}{1\cdot4} = \frac{2}{7}$$

$$\frac{T_1}{(57+273)} = \left\{\frac{800}{128}\right\}^{2/7}$$

$$T_1 = 330 \times 6\cdot25^{2/7}$$
$$= 557\cdot1 \text{ K}$$
$$557\cdot1 - 273 = 284\cdot1°C \quad \text{Ans.}$$

Example. The ratio of compression in a petrol engine is 9 to 1. Find the temperature of the gas at the end of compression if the temperature at the beginning is 24°C, assuming compression to follow the law $pV^n = $ constant, where $n = 1\cdot36$.

$$\frac{T_1}{T_2} = \left\{\frac{V_2}{V_1}\right\}^{n-1} \qquad \text{or} \quad \frac{T_2}{T_1} = \left\{\frac{V_1}{V_2}\right\}^{n-1}$$

$$n-1 = 1\cdot36-1 = 0\cdot36$$
$$\frac{T_2}{(24+273)} = \left\{\frac{9}{1}\right\}^{0\cdot36}$$

$$T_2 = 297 \times 9^{0\cdot36} = 655\cdot1 \text{ K}$$
$$655\cdot1 - 273 = 382\cdot1°C \quad \text{Ans.}$$

Example. The volume and temperature of a gas at the beginning of expansion are 0·0056 m³ and 183°C, at the end of expansion the values are 0·0238 m³ and 22°C respectively. Assuming expansion follows the law $pV^n = C$, find the value of n.

$$\frac{T_1}{T_2} = \left\{\frac{V_2}{V_1}\right\}^{n-1}$$

$$\frac{183 + 273}{22 + 273} = \left\{\frac{0·0238}{0·0056}\right\}^{n-1}$$

$$
\begin{aligned}
1·546 &= 4·25^{n-1} \\
\log 1·546 &= (\log 4·25) \times (n-1) \\
0·1892 &= 0·6284\,(n-1) \\
0·1892 &= 0·6284n - 0·6284 \\
0·8176 &= 0·6284n
\end{aligned}
$$

$$n = \frac{0·8176}{0·6284} = 1·301 \quad \text{Ans.}$$

WORK TRANSFER

Firstly, consider a gas expanding at *constant pressure* in a cylinder fitted with a gas-tight piston (also refer to Fig. 17 with notes in Chapter 5).

Work done [kJ = kN m] = force [kN] × distance [m]

The total force [kN] on the piston is the product of the pressure p [kN/m²] of the gas and the area A [m²] of the piston and, if the piston moves through a distance of S metres as the gas expands, then,

$$\text{Work done} = p \times A \times S$$

The product of the piston area A and the distance it moves S, is the volume swept through by the piston, this is also the increase in volume of the gas in the cylinder. If V_1 is the volume of the gas at the beginning of expansion, and V_2 is the volume at the end of expansion, then $A \times S$ is equal to $V_2 - V_1$ cubic metres, hence,

$$\text{Work done} = p(V_2 - V_1)$$

Fig. 21 shows the pressure-volume diagram representing work done at constant pressure. The graph is a straight horizontal line

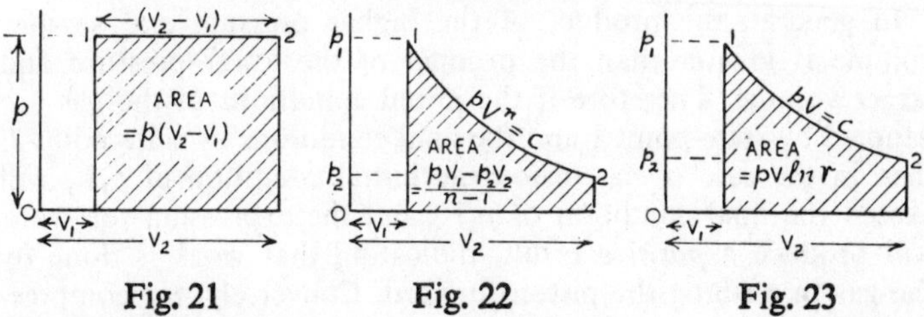

Fig. 21   Fig. 22   Fig. 23

and the area under it is a rectangle. The area of a rectangle is height $\times$ length which, in this case, is $p \times (V_2 - V_1)$. Hence *the area under the pressure-volume line represents work done.*

Now consider cases where the pressure falls during the expansion of the gas. The formula giving the area under the polytropic curve representing the general relationship between pressure and volume, i.e. $pV^n = C$ can only be derived satisfactorily by the use of the calculus. The expression is therefore given here without proof and illustrated in Fig. 22.

$$\text{Work done during polytropic expansion} = \frac{p_1 V_1 - p_2 V_2}{n - 1}$$

The above is the general expression for work done. For pure adiabatic expansion, n is replaced by γ. For isothermal expansion however, since the value of n is 1, and as $p_1 V_1 = p_2 V_2$ then substitution in this expression for work will produce $0 \div 0$ which is indeterminate. A different expression is therefore necessary to obtain work during isothermal expansion:

$$\text{Work done during isothermal expansion} = pV \ln r$$

where r is the ratio of expansion.

ln represents the logarithm to base 2·7183. These are called natural logarithms as distinct from common logs which are to base 10, an alternative symbol to ln is $\log_e$. These tables are under the heading "Natural Logarithms" in some sets of mathematical tables, and "Hyperbolic or Naperian Logarithms" in others. If these are not readily to hand, the natural logarithm of a number can be obtained by multiplying the common log by 2·3026.

The above expressions give the work done *by* the gas during expansion. The same expressions give the work done *on* the gas during compression.

In general, the product of the higher pressure and smaller volume is greater than the product of the lower pressure and larger volume. Therefore if the initial conditions of the gas are denoted by state-point 1 and the final conditions by state-point 2, then in the case of expansion the initial condition of $p_1 V_1$ will exceed the final condition of $p_2 V_2$ and the expression for work will produce a *positive* result, indicating that work is done *by* the gas in pushing the piston forward. Conversely, for compression, the initial condition of $p_1 V_1$ will be less than the final condition of $p_2 V_2$ and therefore a *negative* result will be obtained, indicating that work is done *on* the gas by the piston.

It is important to bear in mind that in calculating work, the units must be consistent. For example, to express work in kilojoules, the pressure must be in kN/m^2 and the volume in m^3, thus,

$$kN/m^2 \times m^3 = kN\,m = kJ$$

Work is transfer of energy, therefore the above can be referred to as *work transfer* from the closed system within the boundary of the cylinder to the external mechanism, or vice-versa. In the first case, when the gas expands, work is being transferred from the energy in the gas to the piston, which, in turn, transmits the work through connecting mechanism to the crank shaft, the work transfer in this case is referred to as being positive. In the second case, when the gas is compressed, work is being transferred from the crank shaft, through the connecting mechanism and piston to the gas, thereby increasing the energy in the gas, and this work transfer is called negative.

Further, since $pV = mRT$, the expressions for work may be stated in terms of mRT instead of pV:

Polytropic expansion

$$\text{Work} = \frac{p_1 V_1 - p_2 V_2}{n-1} = \frac{mR(T_1 - T_2)}{n-1}$$

Isothermal expansion
$$\text{Work} = pV \ln r = mRT \ln r$$

Example. $0{\cdot}04\ m^3$ of gas at a pressure of $1482\ kN/m^2$ is expanded isothermally until the volume is $0{\cdot}09\ m^3$. Calculate the work done during the expansion.

$$\text{Work done} = pV \ln r$$

$$r = \frac{\text{final volume}}{\text{initial volume}} = \frac{0.09}{0.04} = 2.25$$

$$\text{from natural log tables, } \ln 2.25 = 0.8109$$
$$\text{Work done} = 1482 \times 0.04 \times 0.8109$$
$$= 48.08 \text{ kJ} \quad \text{Ans.}$$

Example. 7.08 litres of air at a pressure of 13.79 bar and temperature 335°C are expanded according to the law $pV^{1.32}$ = constant, and the final pressure is 1.206 bar. Calculate (i) the volume at the end of expansion, (ii) the work transfer from the air, (iii) the temperature at the end of expansion, (iv) the mass of air in the system, taking $R = 0.287$ kJ/kg K.

$$p_1 V_1^{1.32} = p_2 V_2^{1.32}$$
$$1379 \times 0.00708^{1.32} = 120.6 \times V_2^{1.32}$$

$$V_2^{1.32} = \frac{1379 \times 0.00708^{1.32}}{120.6}$$

$$V_2 = 0.00708 \times \sqrt[1.32]{\frac{1379}{120.6}}$$

$$= 0.04484 \text{ m}^3 \text{ or } 44.84 \text{ litres}$$
$$\text{Ans. (i)}$$

Note that if the units of pressure and volume are of the same kind on each side of the equation, the units cancel each other out and hence any convenient units can be used. The above could therefore be worked in bars of pressure and litres of volume in which the question data is given. However, as pointed out, it is essential to work in fundamental units in such expressions as used in parts (ii) and (iv) of this problem, therefore the student might prefer to use fundamental units throughout by expressing the pressure in kN/m² and the volume in m³.

$$\text{Work} = \frac{p_1 V_1 - p_2 V_2}{n - 1}$$

$$= \frac{1379 \times 0.00708 - 120.6 \times 0.04484}{1.32 - 1}$$

$$= \frac{9 \cdot 763 - 5 \cdot 408}{0 \cdot 32}$$

$$= \frac{4 \cdot 355}{0 \cdot 32} = 13 \cdot 61 \text{ kJ} \quad \text{Ans. (ii)}$$

$$\frac{p_1 V_1}{T_1} = \frac{p_2 V_2}{T_2}$$

$$\frac{1379 \times 0 \cdot 00708}{(335 + 273)} = \frac{120 \cdot 6 \times 0 \cdot 04484}{T_2}$$

$$T_2 = \frac{608 \times 120 \cdot 6 \times 0 \cdot 04484}{1379 \times 0 \cdot 00708}$$

$$= 336 \cdot 6 \text{ K}$$
$$336 \cdot 6 - 273 = 63 \cdot 6°\text{C} \quad \text{Ans. (iii)}$$

$$p_1 V_1 = mRT_1$$

$$m = \frac{1379 \times 0 \cdot 00708}{0 \cdot 287 \times 608}$$

$$= 0 \cdot 05595 \text{ kg} \quad \text{Ans. (iv)}$$

Example. A perfect gas is compressed in a cylinder according to the law $pV^{1 \cdot 3} = $ constant. The initial condition of the gas is 1·05 bar, 0·34 m³ and 17°C. If the final pressure is 6·32 bar, calculate (i) the mass of gas in the cylinder, (ii) the final volume, (iii) the final temperature, (iv) the work done to compress the gas, (v) the change in internal energy, (vi) the transfer of heat between the gas and cylinder walls.

Take $c_V = 0 \cdot 7175$ kJ/kg K and $R = 0 \cdot 287$ kJ/kg K.

$$p_1 V_1 = mRT_1$$

$$m = \frac{105 \times 0 \cdot 34}{0 \cdot 287 \times (17 + 273)}$$

$$= \frac{105 \times 0.34}{0.287 \times 290} = 0.4289 \text{ kg Ans. (i)}$$

$$p_1 V_1^{1.3} = p_2 V_2^{1.3}$$
$$1.05 \times 0.34^{1.3} = 6.32 \times V_2^{1.3}$$
$$V_2^{1.3} = \frac{0.34^{1.3} \times 1.05}{6.32}$$

$$V_2 = 0.34 \times \sqrt[1.3]{\frac{1.05}{6.32}}$$

$$\text{or } 0.34 \div \sqrt[1.3]{\frac{6.32}{1.05}}$$
$$= 0.08549 \text{ m}^3 \quad \text{Ans. (ii)}$$

$$\frac{p_1 V_1}{T_1} = \frac{p_2 V_2}{T_2}$$

$$\frac{1.05 \times 0.34}{(17 + 273)} = \frac{6.32 \times 0.08549}{T_2}$$

$$T_2 = \frac{290 \times 6.32 \times 0.08549}{1.05 \times 0.34} = 438.8 \text{ K}$$

$$438.8 - 273 = 165.8°\text{C} \quad \text{Ans. (iii)}$$

Alternatively, the final temperature could be obtained from

$$\frac{T_1}{T_2} = \left\{ \frac{p_1}{p_2} \right\}^{\frac{n-1}{n}}$$

The student should do this as an additional exercise.

$$\text{Work done} = \frac{p_1 V_1 - p_2 V_2}{n - 1}$$

$$= \frac{105 \times 0.34 - 632 \times 0.08549}{1.3 - 1}$$

$$= \frac{35.71 - 54.04}{0.3}$$

$$= \frac{-18.33}{0.3} = -61.1 \text{ kJ}$$

Alternatively, the work done could be obtained from:

$$\frac{mR(T_1 - T_2)}{n - 1}$$

This is a further useful exercise for the student.
Note that the minus sign indicates that work is done **ON** the gas.

Work to compress gas $= 61\cdot1$ kJ   Ans. (iv)

Increase in internal energy:

$$
\begin{aligned}
E_2 - E_1 &= mc_\mathrm{v}(T_2 - T_1) \\
&= 0\cdot4289 \times 0\cdot7175 \times (438\cdot8 - 290) \\
&= 0\cdot4289 \times 0\cdot7175 \times 148\cdot8 \\
&= 45\cdot78 \text{ kJ}   \text{Ans. (v)}
\end{aligned}
$$

$$
\begin{array}{ccc}
\text{Heat supplied} & = & \text{Increase in} & + & \text{Work done} \\
\text{to the gas} & & \text{internal energy} & & \text{by the gas}
\end{array}
$$

$$
\begin{aligned}
&= 45\cdot78 - 61\cdot1 \\
&= -15\cdot32 \text{ kJ}
\end{aligned}
$$

The minus sign means that heat is rejected by the gas during compression, that is, this amount of heat energy is transferred *from* the gas *to* the cylinder wall surrounds.

Transfer of heat $= 15\cdot32$ kJ   Ans. (vi)

RELATIONSHIP BETWEEN HEAT ENERGY SUPPLIED AND WORK DONE

Consider polytropic expansion of a gas in a cylinder from initial conditions represented by state-point 1 to the final conditions of state-point 2, and apply the energy equation:

$$
\begin{array}{ccc}
\text{Heat supplied} & = & \text{Increase in} & + & \text{Work done} \\
\text{to the gas} & & \text{internal energy} & & \text{by the gas}
\end{array}
$$

$$= mc_\mathrm{v}(T_2 - T_1) + \frac{mR(T_1 - T_2)}{n - 1}$$

Substituting $c_\mathrm{v} = \dfrac{R}{\gamma - 1}$

$$\text{Heat supplied} = \frac{mR(T_2 - T_1)}{\gamma - 1} + \frac{mR(T_1 - T_2)}{n - 1}$$

$$= \frac{mR(T_1 - T_2)}{n-1} - \frac{mR(T_1 - T_2)}{\gamma - 1}$$

$$= \frac{mR(T_1 - T_2)}{n-1}\left\{ 1 - \frac{n-1}{\gamma - 1}\right\}$$

$$= \frac{mR(T_1 - T_2)}{n-1}\left\{ \frac{\gamma - 1 - n + 1}{\gamma - 1}\right\}$$

$$= \frac{mR(T_1 - T_2)}{n-1} \times \frac{\gamma - n}{\gamma - 1}$$

$$= \text{Work} \times \frac{\gamma - n}{\gamma - 1}$$

Note that during expansion of gas, work is done *by* the gas and is a positive quantity, therefore a positive quantity of heat energy is supplied from the cylinder walls to the gas.

In compression of a gas, work is done *on* the gas which is negative work done by the gas, the result of the above expression is negative heat supplied, meaning that heat energy is transferred from the gas to the cylinder walls.

TEST EXAMPLES 6

1. A gas expands isothermally in a closed system from a volume of 0.015 m^3 to a volume of 0.075 m^3. If the initial pressure was 600 kN/m^2, find the final pressure.

2. 0.012 m^3 of air at 101.5 kN/m^2 is compressed adiabatically to a volume of 0.002 m^3. Taking $\gamma = 1.4$, find the pressure at the end of compression.

3. Gas is expanded in an engine cylinder, following the law $pV^n = $ constant where the value of n is 1.3. The initial pressure is 2550 kN/m^2 and the final pressure is 210 kN/m^2. If the volume at the end of expansion is 0.75 m^3, calculate the volume at the beginning of expansion.

4. The ratio of compression in a petrol engine is 8.6 to 1. At the beginning of compression the pressure of the gas is 98 kN/m^2 and the temperature is $28°$C. Find the pressure and temperature at the end of compression, assuming it follows the law $pV^{1.36} = $ constant.

5. Gas is expanded in an engine cylinder according to the law $pV^n = C$. At the beginning of expansion the pressure and volume are 1750 kN/m^2 and 0.05 m^3 respectively, and at the end of expansion the respective values are 122.5 kN/m^2 and 0.375 m^3. Calculate the value of n.

6. The following data of pressure in bars and volume in litres were taken from the expansion curve of an indicator diagram off an internal combustion engine. Assuming that the expansion follows the law $pV^n = $ constant, plot the values of $\log p$ on a base of $\log V$ and determine the value of n:

p	30	17.7	12.2	9.1	7.2	5.9
V	20	30	40	50	60	70

7. The temperature and pressure of the air at the beginning of compression in a compressor cylinder are $20°$C and 101.3 kN/m^2, and the pressure at the end of compression is 1420 kN/m^2. If the law of compression is $pV^{1.35}$ find the temperature at the end of compression.

8. 0·014 m³ of a gas at 66°C is expanded adiabatically in a closed system and the temperature at the end of expansion is 2°C. Taking the specific heats of the gas at constant pressure and constant volume as 1·005 and 0·718 kJ/kg K respectively, calculate the volume at the end of expansion.

9. A perfect gas is expanded polytropically behind a piston in a cylinder. The initial volume and temperature is 0·06 m³ and 147°C, the final volume and temperature is 0·21 m³ and 21°C, calculate the index of expansion.

10. Air is compressed in a diesel engine from 1·17 bar to 36·55 bar. If the temperatures at the beginning and end of compression are 32°C and 500°C respectively, find the law of compression assuming it is polytropic.

11. 0·113 m³ of air at 8·25 bar is expanded in a cylinder until the volume is 0·331 m³. Calculate the final pressure and work done if the expansion is (i) isothermal, (ii) adiabatic, taking $\gamma = 1·4$.

12. A gas is expanded in a cylinder behind a gas-tight piston. At the beginning of expansion the pressure is 36 bar, volume 0·125 m³, and temperature 510°C. At the end of expansion the volume is 1·5 m³ and temperature 40°C. Taking $R = 0·284$ kJ/kg K and $c_V = 0·71$ kJ/kg K, calculate (i) the pressure at the end of expansion, (ii) the index of expansion, (iii) the mass of gas in the cylinder, (iv) change of internal energy, (v) work done by the gas, (vi) heat transfer during expansion.

CHAPTER 7

I.C. ENGINES — ELEMENTARY PRINCIPLES

Internal combustion engines are so named because combustion of the fuel takes place *inside* the engine. When the fuel burns inside the engine cylinder, it gives out heat which is absorbed by the air previously taken into the cylinder, the temperature of the air is therefore increased with a consequent increase in pressure and/or volume, thus energy is imparted to the piston. The reciprocating motion of the piston is converted into a rotary motion at the crank shaft by connecting rod and crank.

The method of igniting the fuel varies. In diesel engines the air in the cylinder is compressed to a high pressure so that it attains a high temperature, and when oil fuel is injected into this high temperature air the fuel immediately ignites. When the ignition of the fuel is caused solely by the heat of compression, the engine is classed as a *compression-ignition* engine. In petrol and paraffin engines the fuel is usually taken in with the charge of air, compressed and then ignited by an electric spark.

THE FOUR-STROKE DIESEL ENGINE

In this type of engine it takes four strokes of the piston (*i.e.* two revolutions of the crank) to complete one working cycle of operations, hence the name *four-stroke* cycle.

Fig. 24 illustrates each of these four strokes in one cylinder. On the cylinder head is shown the fuel valve which lifts to admit oil fuel (under pressure) into the cylinder, the air-induction valve through which air is drawn in, and the exhaust valve through which the exhaust gases are expelled from the cylinder. There are two more valves which are not shown here because they do not operate during the normal working cycle; one is the relief valve which opens against the compression of its spring when the pressure in the cylinder rises too high, the other is the air-starting valve which is cam-operated and opens to admit high pressure air into the cylinder to move the piston and start the engine.

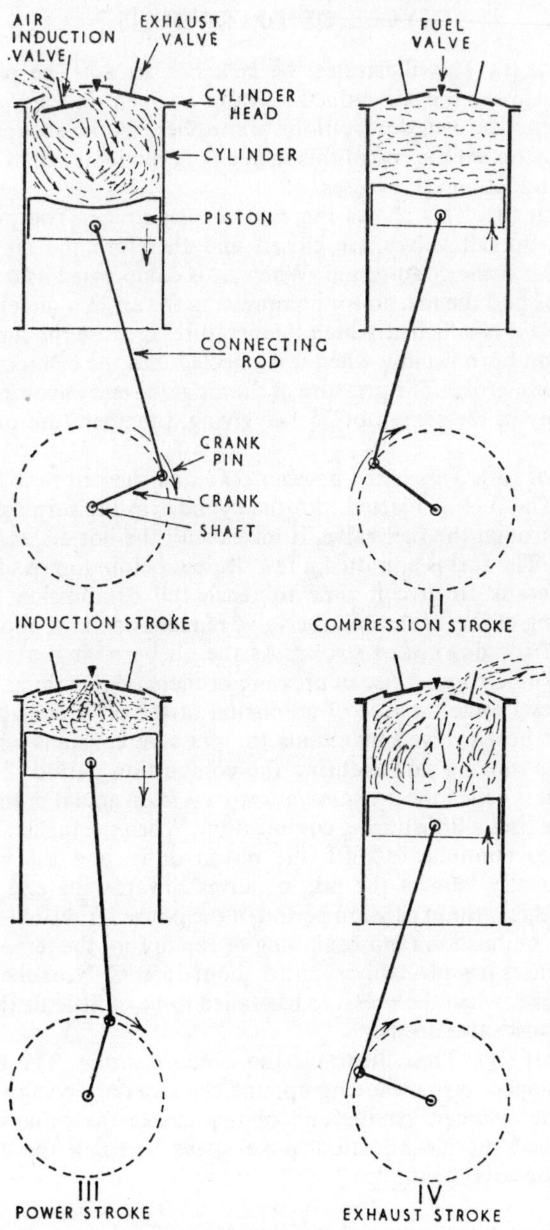

FOUR-STROKE DIESEL ENGINE   Fig. 24

CYCLE OF OPERATIONS

SKETCH (i). This illustrates the *induction stroke*. The piston is moving down, the air induction valve is open and air is being drawn into the cylinder from the atmosphere by the suction effect of the piston. At the end of this stroke the cylinder is full of air and the air-induction valve closes.

SKETCH (ii). This shows the *compression stroke*. The piston is moving up, all valves are closed and therefore the air in the cylinder is being compressed. When air is compressed its temperature rises and the reason for compressing the air in a diesel engine is to obtain a sufficiently high temperature to cause the fuel oil to ignite and burn rapidly when it is injected into the cylinder at the end of this stroke. The pressure of the air at the end of compression is usually in the region of 35 bar giving a temperature of about 540°C.

SKETCH (iii). This is the *power stroke* and the piston is moving down. The fuel is injected into the cylinder in the form of a fine spray through the fuel valve, it mixes with the hot air and burns rapidly. The fuel is admitted a few degrees before top dead centre of the crank to give it time to reach full combustion for the beginning of the stroke and the valve remains open for about one-tenth of the downward stroke. As the oil burns it heats the air which would cause a rise in pressure or increase in volume; in the pure diesel cycle the rate of admission of the oil is controlled so that the heat evolved maintains the pressure constant while the piston is moving down during the combustion period. For this reason it is called the *constant pressure cycle* (in actual practice the pressure rises a little during combustion). When the fuel is shut off, the gases continue to push the piston down and the pressure consequently falls as the piston moves towards the end of this stroke, this is the expansion period of the power stroke. At the end of fuel combustion (and beginning of expansion) the temperature of the gases has probably risen to about 1650°C. Near the end of this stroke, when the pressure has fallen to be of little further use, the exhaust valve opens.

SKETCH (iv). This illustrates the *exhaust stroke*. The exhaust valve is open, piston moving up, and the gases are being expelled from the cylinder. At the end of this stroke the exhaust valve closes and the air-induction valve opens to begin the cycle of operations over again.

The valves and ports of an internal combustion engine do not open and close when the crank is exactly on top or bottom centre (*i.e.* when the piston is at the top or bottom of its stroke). For instance, the induction valve must begin to open well before the crank is on top centre so that it is fully open when the piston begins to move down on its induction stroke. As the fast-moving piston moves down, the air rushes into the cylinder at a high velocity and the momentum gained causes it to continue to "pour" into the cylinder after the piston completes its downward stroke. Thus the induction valve is kept open until the crank has passed bottom centre and the momentum of the inrush of air is reduced to practically nil. At the end of compression the temperature of the air is sufficiently high to ignite the fuel when it is injected into the cylinder and cause it to burn rapidly. However, since it takes a fraction of a second to get the full burning effect, the oil is injected before top dead centre of the crank, and it follows that the faster the speed of the engine the earlier the beginning of the fuel injection. The fuel valve is kept open for just sufficient oil to be injected to produce the required power during the power stroke, and

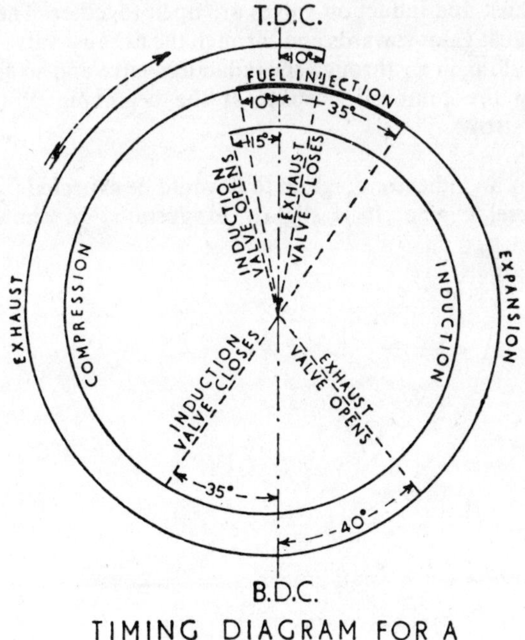

TIMING DIAGRAM FOR A
FOUR-STROKE DIESEL ENGINE

Fig. 25

to allow for a reasonable drop in pressure before exhaust begins. The exhaust valve opens a little before the piston reaches the end of the power stroke, to allow any remaining pressure after expansion to fall to about atmospheric pressure, before the piston begins to move up again, and it remains open for a short while after the piston completes its upward exhaust stroke because exhaust continues for a brief period due to the momentum of the gases being entrained towards the exhaust opening.

A diagram representing two revolutions of the crank (four strokes of the piston) showing the timing of the opening and closing of the valves with respect to the position of the crank, is shown on Fig. 25. This is called a timing diagram and the crank angles shown are typical values for a four-stroke diesel engine. These, however, vary considerably for different engines and depend partly on the designed speed of the engine and the grade of fuel burned.

Note the exhaust-induction overlap, that is the short period when exhaust and induction valves are open together. The sweep of the exhaust gases towards and through the exhaust valve has the effect of pulling in air through the induction valve and so assists in scavenging the combustion space at the beginning of the air-induction stroke.

Fig. 26 is an indicator diagram that would be expected from a 4-stroke diesel engine. It is slightly exaggerated to show more

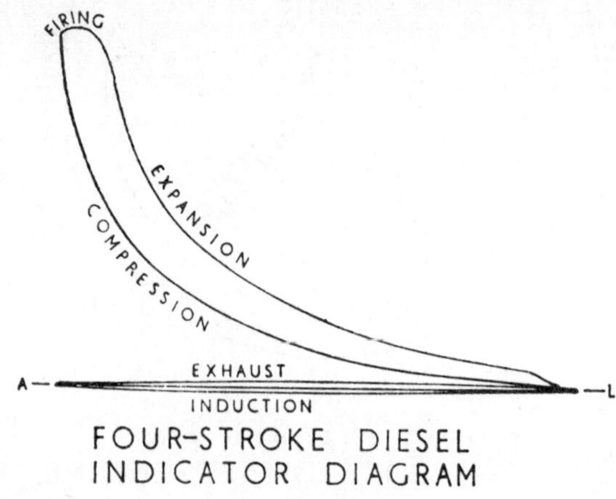

FOUR-STROKE DIESEL
INDICATOR DIAGRAM

Fig. 26

distinctly the suction pressure a little below atmospheric during the induction stroke, and the pressure a little above atmospheric during the exhaust stroke.

VALVE MECHANISM

The air-induction, exhaust, and starting-air valves are opened by means of rocking levers fulcrummed about their centre and actuated by cams fixed to the cam shaft. Each cam has a peak, which, when it comes around to contact the roller on the end of the rocking lever, pushes this end up, and the other end is depressed to open the valve against a spring in the valve housing (see Fig. 27).

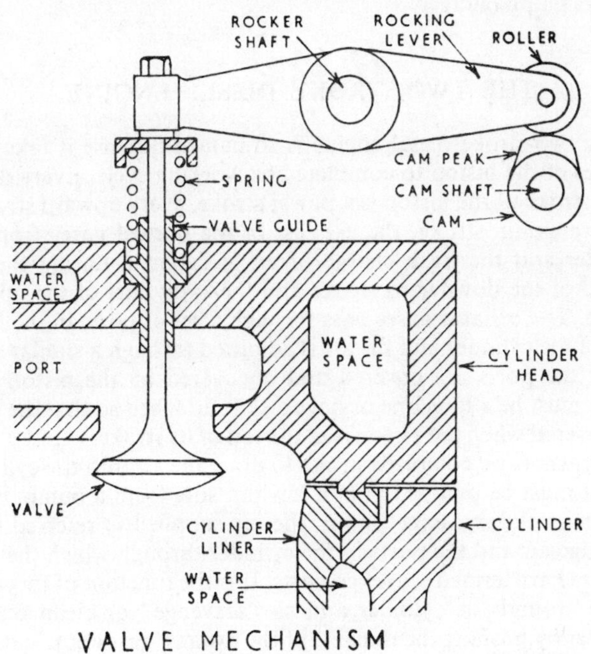

Fig. 27

Push rods may be used as distance pieces between cams and rocking levers. The cams are set in the correct position relative to the crank so that the valves open and close at the exact moment and for the required period in the working cycle. As the timing of opening and closing of the valves is relative to the crank position and direction of movement, separate cams fixed at different relative positions are required to actuate the rocking levers to run the engine in the reverse direction.

The working cycle constitutes four strokes of the piston, which is two revolutions of the crank shaft. During one cycle each valve is opened only once, hence in the four-stroke engine *the cam shaft is driven at half the speed of the crank shaft.*

The fuel valve is usually opened by the pressure of the oil discharged by the fuel pump and closes under the action of a spring when the pressure is released. The timing of the beginning of opening, period of opening, and closing of the valve is varied by the fuel pump plunger.

THE TWO-STROKE DIESEL ENGINE

The two-stroke diesel engine is so named because it takes two strokes of the piston to complete one working cycle. Every downward stroke of the piston is a power stroke, every upward stroke is a compression stroke, the exhaust of the burned gases from the cylinder and the fresh charge of air is taken in during the late period of the downward stroke and the early part of the upward stroke. The exhaust gases pass through a set of ports in the lower part of the cylinder and the air is admitted through a similar set of ports, the ports are covered and uncovered by the piston itself which must be a long one or have a skirt attached so that the ports are covered when the piston is at the top of its stroke.

As there is no complete stroke to draw the air into the cylinder, the air must be pumped in at a low pressure from a pump, this is known as the scavenge pump, the air supplied is referred to as scavenge-air and the ports in the cylinder through which the air is admitted are termed scavenge ports. It is the function of this air to sweep around the cylinder and so "scavenge" or clean out the cylinder by pushing the remains of the exhaust gases out, leaving a clean charge of air to be compressed.

There is therefore no air-induction valve or exhaust valve in the cylinder head as there are in a four-stroke engine and the cylinder head is consequently a much simpler and stronger casting; there are of course the fuel valve, air-starting valve and relief valve.

As the cycle of operations takes place in two-strokes of the piston, which is one revolution of the crank shaft, *the cam shaft is driven at the same speed as the crank shaft.*

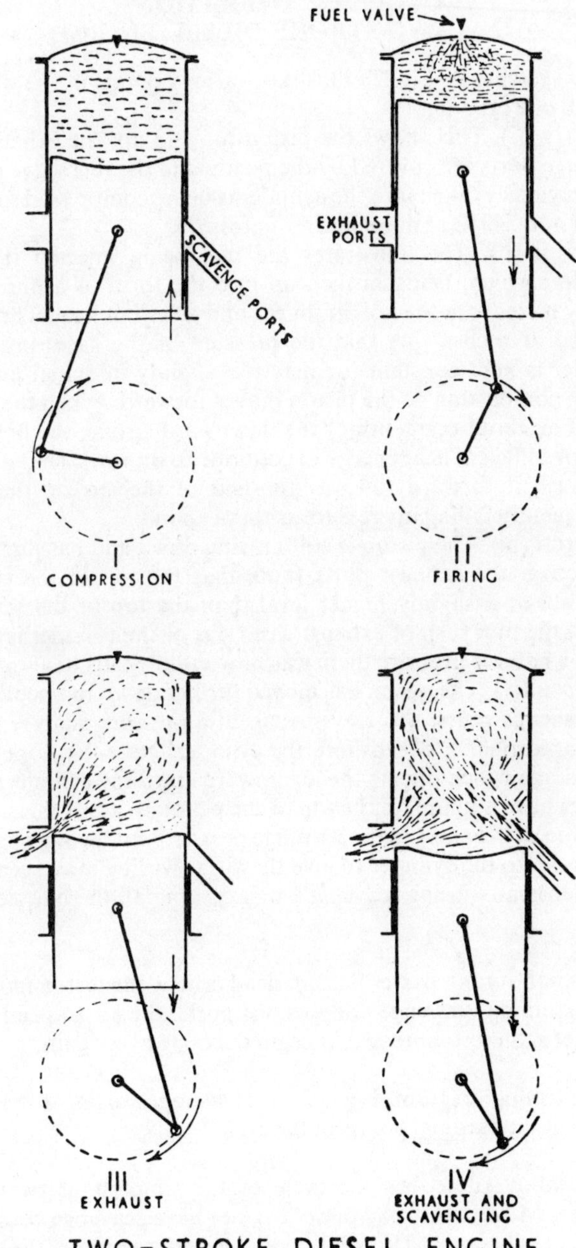

TWO-STROKE DIESEL ENGINE

CYCLE OF OPERATIONS
IN A TWO-STROKE DIESEL ENGINE

The sketches in Fig. 28 illustrate various points in the working cycle in one cylinder.

SKETCH (i). This shows the piston moving up, the exhaust and scavenge ports are covered by the piston and the fuel valve is shut. Air previously taken into the cylinder is being compressed to about 35 bar and 540°C at the end of compression.

SKETCH (ii). This illustrates the fuel being injected into the cylinder, which, being broken up into the form of a fine spray, readily mixes with the hot air, burns and gives out heat. The fuel is injected at such a rate that the pressure of the gases inside the cylinder is kept constant (or may rise slightly in actual practice) during combustion as the piston moves forward. When the fuel is cut off at about one-tenth of the downward stroke, the hot gases contain sufficient heat energy to continue to do work on the piston and push it forward towards the end of the stroke, the gases consequently falling in pressure as they expand.

SKETCH (iii). The piston is still moving down and has just begun to uncover the exhaust ports (note that the top of the exhaust ports are at a slightly higher level than the top of the scavenge ports), the first rush of exhaust gases out of the cylinder is taking place, whatever pressure there was now rapidly falls to about zero.

SKETCH (iv). The piston has moved further down to uncover also the scavenge ports, the scavenge air at a pressure slightly higher than atmospheric, sweeps into the cylinder. Note the slope of the air passage which directs the air upwards into the cylinder, many engines have a hump on the top of the piston to assist this upward direction of the air; and the scavenge ports are cut through partially tangential to the cylinder to give the air a swirling movement. The cylinder is now being scavenged by expelling all the burned gases out.

When the crank passes bottom dead centre, the piston moves up and covers the scavenge and exhaust ports, the air trapped in the cylinder is then compressed to begin the cycle over again.

The timing diagram, Fig. 29, gives average values of the crank angles at the cardinal points of the cycle.

The above describes the cycle of the simplest of two-stroke engines. Most two-stroke diesel engines have scavenge control by valves or other means in addition to the sleeve action of the piston,

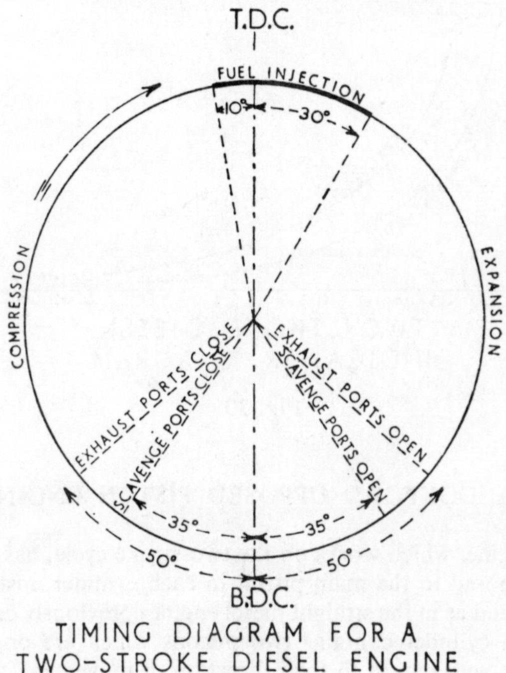

TIMING DIAGRAM FOR A
TWO-STROKE DIESEL ENGINE

Fig. 29

an extra set of scavenge ports are arranged above the main scavenge ports which are closed by the valves on the downward stroke of the piston to allow exhaust to take place first, and opened when the pressure in the cylinder falls below scavenge pressure. By this means the scavenge air remains open on the upward stroke after the exhaust ports are closed and the pressure in the cylinder at the beginning of compression can be equal to the pressure of the scavenge air. A slightly higher pressure at the beginning of compression means a greater mass of air in the cylinder and therefore more fuel can be burned to produce more power. Engines in which there is an extra pressure of air at the beginning of compression are said to be supercharged.

Fig. 30 is an indicator diagram, slightly exaggerated from a two-stroke diesel engine.

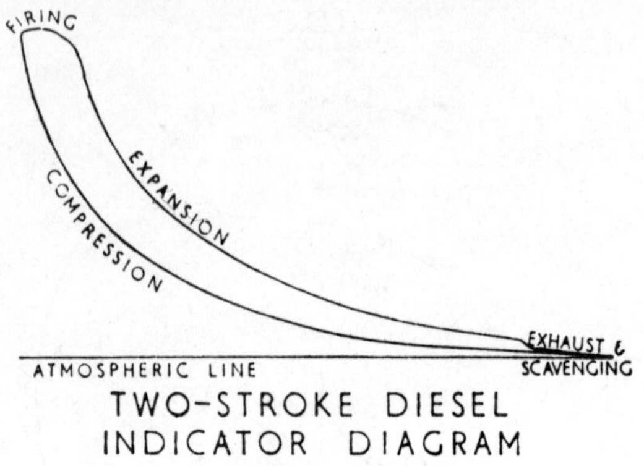

TWO-STROKE DIESEL
INDICATOR DIAGRAM

Fig. 30

THE DOXFORD OPPOSED PISTON ENGINE

This engine, which works on the two-stroke cycle, has an extra piston opposed to the main piston in each cylinder, instead of a cylinder head as in the straight diesel engines previously described. Thus each cylinder contains two pistons which are opposed to each other with regard to their direction of movement, when the bottom piston is moving up the top piston is moving down, and vice-versa. The bottom piston is connected by piston rod, cross-head and connecting rod to a centre crank; the top piston is coupled by a yoke to two side rods and connected by connecting rods to side cranks, one on either side of the centre crank. The side cranks are at 180 degrees to the centre crank which gives their respective pistons opposite directions of motion. The fuel valves are positioned at the centre of length of the cylinder, there is a complete ring of exhaust ports near the top of the cylinder which are covered and uncovered by the top piston, and a complete ring of scavenge ports near the bottom which are covered and uncovered by the bottom piston.

CYCLE OF OPERATIONS IN A
DOXFORD ENGINE

Although this engine works on a two-stroke cycle, there are some important differences from an ordinary diesel engine. Firstly, the high temperature of air required to ignite and burn the fuel is

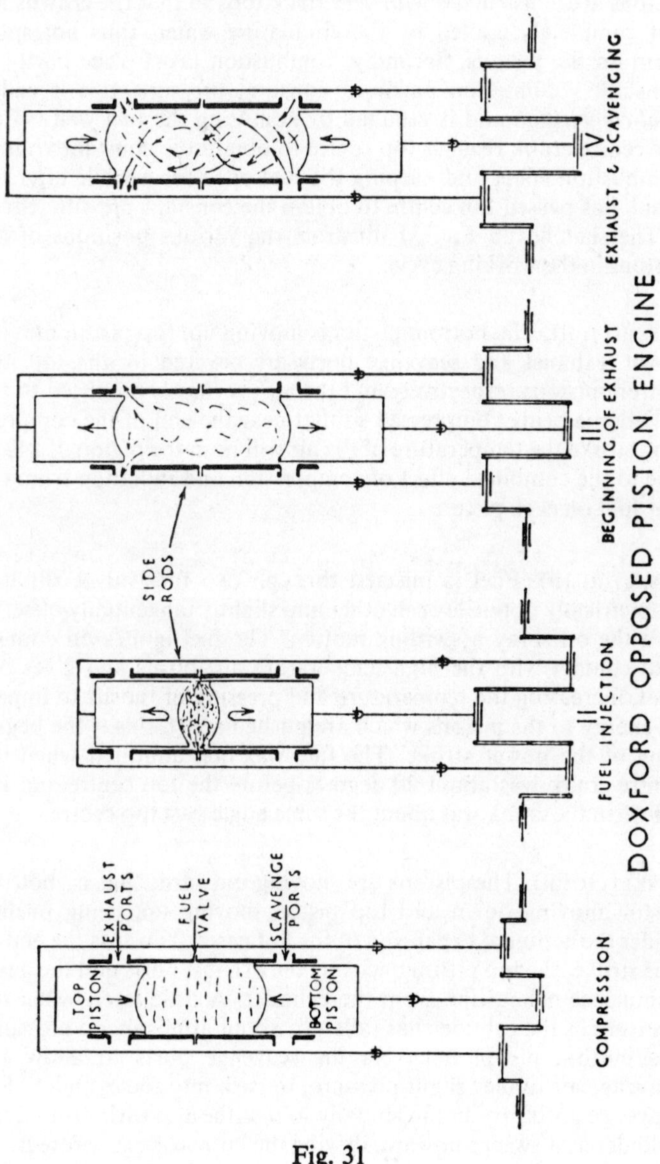

Fig. 31

not derived solely by compression, but partly by compression and
partly by heat radiated from the hot crowns of the pistons. The

pistons are constructed with very thick tops so that the crowns are not completely cooled by the circulating water, thus hot spots form on the pistons. Secondly, combustion takes place partly at constant volume and partly at constant pressure, this is called *dual-combustion* and is attained by admitting the fuel well before the centre crank reaches top centre to obtain the constant volume combustion effect and keeping the fuel open for a while after the crank has passed top centre to obtain the constant pressure effect.

The sketches in Fig. 31 illustrate the various positions of the pistons in the working cycle.

SKETCH (i). The bottom piston is moving up, top piston moving down, exhaust and scavenge ports are covered by the top and bottom pistons respectively and the air previously admitted to the cylinder is being compressed so that near the end of the compression stroke the temperature of the air will be in the region of 540°C due to the combined effect of compression and radiation from the hot spot on each piston.

SKETCH (ii). Fuel is injected through two fuel valves situated diametrically opposite each other and slightly tangentially offset to give the oil spray a swirling motion. The fuel ignites on coming into contact with the high temperature air, burns and gives out heat, increasing the temperature and pressure of the air to impart its energy to the pistons which are pushed apart, this is the beginning of the power stroke. The fuel was first admitted when the centre crank was about 20 degrees before the top centre and cut off when the crank was about the same angle past top centre.

SKETCH (iii). The pistons are moving outward, that is, bottom piston moving down and top piston moving up, being pushed under the action of expansion of the hot gases. Towards the end of this stroke, the top piston uncovers the exhaust ports and the gases commence exhausting from the cylinder. A little later, when the pressure in the cylinder has fallen to about atmospheric pressure, the bottom piston uncovers the scavenge ports to allow the scavenge air (under slight pressure) to rush into the cylinder. The scavenge ports are cut tangentially so that the air swirls around the cylinder as it sweeps upward, driving the burned gases before it.

SKETCH (iv). The pistons have completed their outward stroke and are beginning to move inward during the scavenging process.

The scavenging effect is practically 100% under the ideal arrangement of admitting the scavenge-air in the bottom of the cylinder to drive the exhaust gases out through the top.

STARTING

As previously explained, the ignition of the fuel in diesel engines is caused by the heat of compression of the air previously admitted into the cylinders, thus, for the engine to begin firing, air must first be drawn or pumped into the cylinder and this air must be compressed by the upward movement of the piston to obtain the high temperature necessary to burn the fuel when it is injected. Hence the engine must be driven for a few revolutions by some outside source before allowing the fuel into the cylinders. In heavy marine engines the practice is to drive the engine on compressed air which has previously been stored up (at pressures ranging from 25 to 40 bar depending upon the type of engine) in starting-air reservoirs.

The compressed air is admitted to each cylinder through a cam-operated starting-air valve when the piston has just passed its top centre and commencing what will be its power stroke and remaining open until the piston has travelled part of that stroke. The period of opening depends upon the number of cylinders and whether it is a four-stroke or two-stroke engine. When the starting-air valve closes on one cylinder, another starting-air valve has already opened on another cylinder whose piston has just commenced its downward stroke, when this valve closes another valve on another cylinder has already opened and thus, no matter in what position the engine stops there will always be at least one of the cylinders with its starting-air valve open to admit compressed air to start the engine. When the engine attains sufficient speed, the fuel pumps and valves are brought into operation and the starting-air valves put out of commission.

The reservoirs are pumped up by two-stage or three-stage compressors. In a two-stage compressor, atmospheric air is taken into the low pressure cylinder where it is compressed to about 5·5 bar and discharged through an intercooler into the high pressure cylinder; it is now compressed to its final pressure and passed through an aftercooler into the air reservoirs. Sufficient air is stored to enable a minimum of about fourteen starts of the engine without replenishing from the compressor. In some diesel installations there is an air compressor driven off the main engines in addition to the independently driven air compressors.

SPEED CONTROL

The speed of the engine is controlled by varying the quantity of fuel injected into the cylinders, from zero quantity for stop to maximum quantity for full speed. This is commonly effected by varying the discharge period of the oil from the fuel pump.

REVERSING

There are two cams for each valve, one is set so that its peak will lift the valve rocking lever to open the valve and keep it open for the correct period of the cycle when the engine is running in one direction; the other cam alongside is set for running in the opposite direction. The reversing mechanism is therefore arranged to bring either ahead or astern cams into line with the valve rocking lever as required.

The action of one type of reversing gear is first to lift the rocking levers clear of the cams, slide the cam shaft along so that the opposite cam comes into line and then return the rocking lever to its working position.

PETROL ENGINES

Engines which run with petrol (or paraffin) as the fuel are often termed "light oil" engines. The main difference between the majority of petrol engines and the diesel engine is that the petrol engine takes in a charge of air and petrol vapour, this explosive mixture is compressed and ignited by an electric spark; whereas in the diesel engine the cylinder is charged with air only so that only pure air is compressed and the fuel is injected at the moment ignition and burning of the fuel is required, ignition being caused solely by the heat of the compressed air.

When the air is compressed in a diesel engine there is no possibility of firing before the fuel is injected. In a petrol engine, an explosive mixture of petrol and air is compressed and there is danger of the mixture firing spontaneously due to the heat of compression alone and before the electric spark occurs, therefore the ratio of compression must be limited to prevent this. The ratio of compression in diesel engines can be high, such as twelve to one and upwards whereas the ratio of compression in petrol engines is much less, in the region of eight or nine to one.

Petrol engines usually work on the *constant volume* cycle, that is, when combustion takes place there is theoretically no change in volume but a considerable increase in pressure.

CYCLE OF OPERATIONS
IN THE FOUR-STROKE PETROL ENGINE

Fig. 32 illustrates the four strokes which constitute the cycle of operations in a four-stroke petrol engine.

SKETCH (i). This shows the *Induction Stroke*. The piston is moving down, the induction valve is open and a mixture of air and petrol-vapour is being drawn into the cylinder through the induction valve, from the carburettor.

SKETCH (ii). This is the *Compression Stroke*. The induction valve has closed, no valves are open, the piston is moving up and the petrol-air mixture is being compressed. The pressure at the end of compression will be about 17 bar and the temperature in the region of 370°C. The temperature should be as high as possible without causing spontaneous ignition so that the petrol will burn very rapidly when ignited by the electric spark towards the end of this stroke.

SKETCH (iii). This is the *Power Stroke*. An electric spark across the points of a sparking plug is timed to ignite the fuel before the crank reaches its top centre so that full burning effect takes place almost instantaneously while the piston is at the top of its stroke. The heat given out by the fuel causes a rapid rise of pressure and temperature, combustion takes place in about one three-hundredth of a second and is therefore often referred to as an "explosion". The hot gases in the cylinder drive the piston down, falling in pressure and expanding in volume as the piston travels towards the bottom of the cylinder.

SKETCH (iv). This shows the *Exhaust Stroke*. The exhaust valve has opened, the piston is moving up, and the burned gases are being pushed out of the cylinder leaving the cylinder empty at the end of this stroke to start the cycle of operations over again.

Timing diagrams of petrol engines vary considerably depending upon the quality of the petrol the engine is designed to run on and the range of speeds. Some average values of the opening and

INDUCTION VALVE
EXHAUST VALVE
SPARKING PLUG
I
INDUCTION STROKE
II
COMPRESSION STROKE
III
POWER STROKE
IV
EXHAUST STROKE
FOUR-STROKE PETROL ENGINE

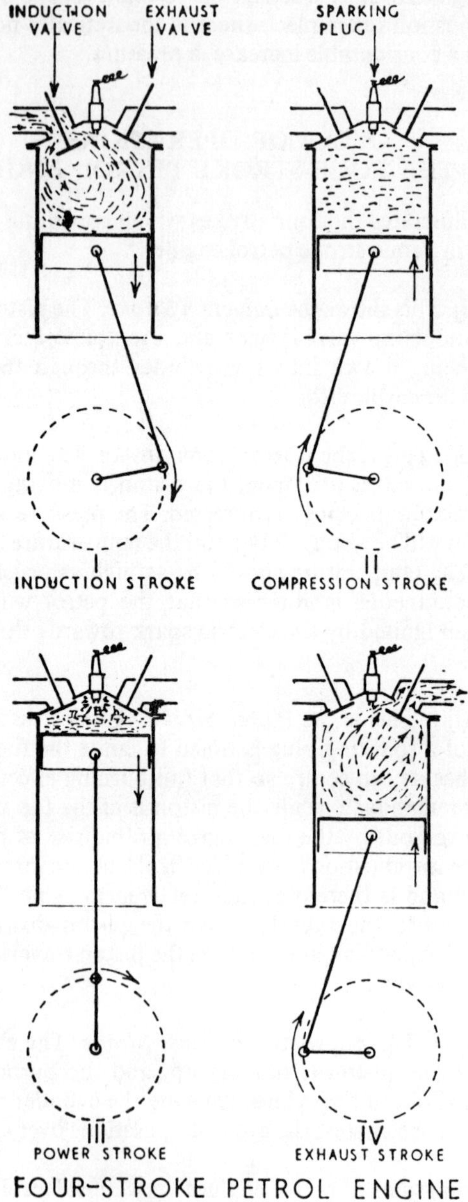

Fig. 32

closing of the valves and the timing of the spark are shown in Fig. 33. As in all internal combustion engines, the valves do not open and close when the crank is exactly on top and bottom dead centres; modifications are necessary to give time for the various events to come fully into operation. For instance, the induction valve must begin to open well before the crank is on top centre so that it is full open when the piston begins to move down on its induction stroke, as the fast-moving piston moves down, the mixture rushes into the cylinder at a high velocity, the momentum gained causes the gas to continue to "pour" into the cylinder even after the piston completes its downward stroke hence the reason for not closing the induction valve until after the crank has passed bottom centre.

The spark sets the fuel alight but it takes a fraction of a second to get the full burning effect, therefore the spark occurs before top centre, the faster the speed of the engine the earlier the spark and most engines have an automatic device to advance or retard the spark as the engine speed increases or decreases.

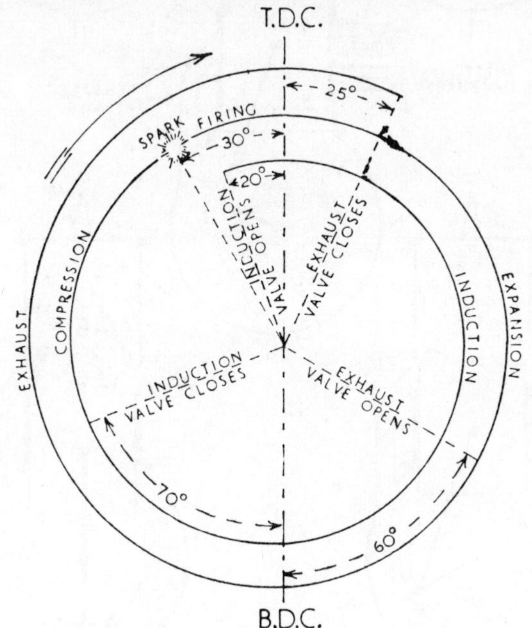

TIMING DIAGRAM FOR A
FOUR-STROKE PETROL ENGINE

Fig. 33

CYCLE OF OPERATIONS
IN THE TWO-STROKE PETROL ENGINE

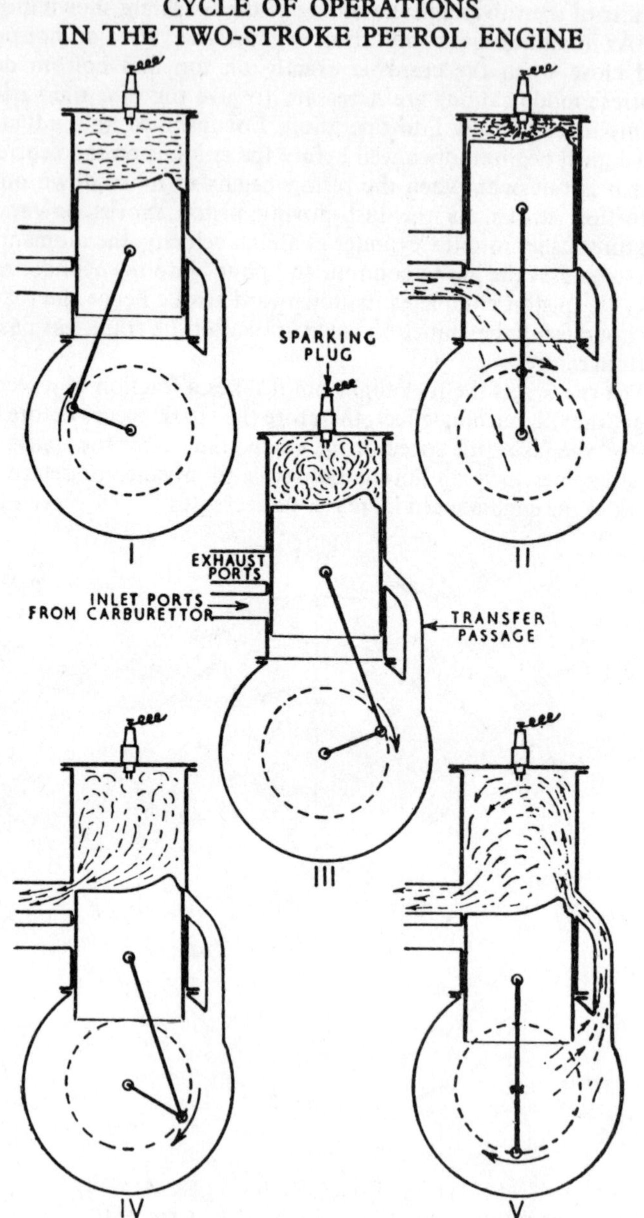

TWO-STROKE PETROL ENGINE   Fig. 34

The exhaust valve opens before the piston gets to the end of its power stroke to allow any remaining pressure after expansion to fall to atmospheric pressure before the piston starts to move up again, and the valve remains open for a short while after the piston completes its exhaust stroke. Note the exhaust-induction valve overlap when both exhaust and induction valves are open together for a short period. The momentum of the exhaust gases rushing out through the exhaust valve induces the fresh charge to begin entering the cylinder through the induction valve, and thus scavenges the combustion space.

Fig. 34 illustrates a typical two-stroke petrol engine with an enclosed crankcase to act as the scavenge pump. As in the two-stroke diesel engine, exhaust and scavenging takes place through ports in the cylinder wall and the ports are covered and uncovered by the piston.

SKETCH (i). Air and petrol vapour, previously admitted into the cylinder, is being compressed by the upward movement of the piston, at the same time a partial vacuum is formed in the crankcase.

SKETCH (ii). As the piston nears the top of its stroke, the mixture of petrol and air, now at a high temperature, is ignited by an electric spark across the gap of the points of a sparking plug, the heat of combustion causes a sudden rise of pressure which imparts its energy to the piston. The bottom of the piston has uncovered the inlet ports into the crankcase and a fresh charge of petrol-air mixture rushes in from the carburettor.

SKETCH (iii). The piston is being pushed down by the expansion of the gases in the cylinder. The mixture in the crankcase is being slightly compressed.

SKETCH (iv). End of expansion and beginning of exhaust. The piston has moved further down and uncovered the exhaust ports to allow the burned gases to leave the cylinder. The mixture in the crankcase has now been compressed to a little above atmospheric pressure.

SKETCH (v). The piston is at the bottom of its stroke, exhaust ports are full open. The scavenge ports have been uncovered by the piston and the charge of air and petrol rushes up from the crankcase, through the transfer passage and into the cylinder.

As the piston moves up, scavenge and exhaust ports are closed and the mixture in the cylinder is compressed as in sketch (i) to begin the cycle over again.

MEAN EFFECTIVE PRESSURE AND POWER

We have seen in the last chapter that the area of a pV diagram represents work. The indicator diagrams shown in Figs. 26 and 30 are examples of practical pV diagrams taken off engines by means of an engine indicator, the areas of these indicator diagrams represent the work done per cycle.

ENGINE INDICATOR. An engine indicator consists of a small bore cylinder containing a short stroke piston which is subjected to the same varying pressure that takes place inside the engine cylinder during one cycle of operations. This is done by connecting the indicator cylinder to the top of the engine cylinder in the case of single-acting engines, or through change-over cocks and pipes leading to the top and bottom ends of the engine cylinder in the case of double-acting engines. The gas pressure pushes the indicator piston up against the resistance of a spring, a choice of specially scaled springs of different stiffness being available to suit the operating pressures within the cylinder and a reasonable height of diagram.

A spindle connects the indicator piston to a system of small levers designed to produce a vertical straight-line motion at the pencil on the end of the pencil lever, parallel (but magnified about six times) to the motion of the indicator piston. The "pencil" is often a brass point, or stylus, this is brought to press lightly on specially prepared indicator paper which is wrapped around a cylindrical drum and clipped to it. The drum, which has a built-in recoil spring, is actuated in a semi-rotary manner by a cord wrapped around a groove in the bottom of it; the cord, passing over a guide pulley, is attached by a hook at its lower end to a reduction lever system from the engine crosshead. Instead of the lever system from the crosshead, many engines are fitted with a special cam and tappet gear to reproduce the stroke of the engine piston to a small scale. The drum therefore turns part of a revolution when the engine piston moves down, and turns back again when the engine piston moves up, thus the pencil or stylus on the end of the indicator lever draws a diagram which is a record of the pressure in the engine cylinder during one complete cycle.

Fig. 35 shows a Maihak indicator which is suitable for taking indicator diagrams off steam reciprocating engines and internal combustion engines up to rotational speeds of about 300 rev/min. In this type, the pressure scale spring is anchored at its bottom end to the framework, and the top of the piston spindle bears upwards on the top coil of the spring, the upward motion of the indicator piston thus stretches the spring.

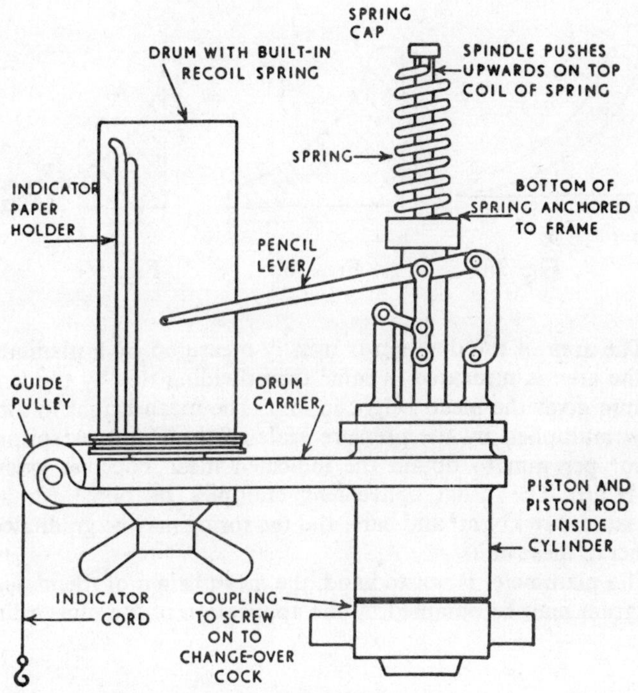

ENGINE INDICATOR

Fig. 35

MEAN EFFECTIVE PRESSURE. Consider the two-stroke diesel engine indicator diagram shown in Fig. 36. The positive work done in one cycle of operations *by* the gas during the burning period and expansion of the gas is shown by the shaded area of Fig. 36b. The work done *on* the air during the compression period, representing negative work done by the engine is shown by the shaded area of Fig. 36c. Hence the net useful work done in one cycle is the difference between positive and negative work and represented by the actual diagram of Fig. 36a. Therefore, if the area of the indicator diagram is divided by its length, the average height is obtained which, to scale, is the average or mean pressure effectively pushing the piston forward and transmitting useful energy to the crank during one cycle. This, expressed in N/m² or a suitable multiple of the basic pressure unit, is termed the *indicated mean effective pressure*.

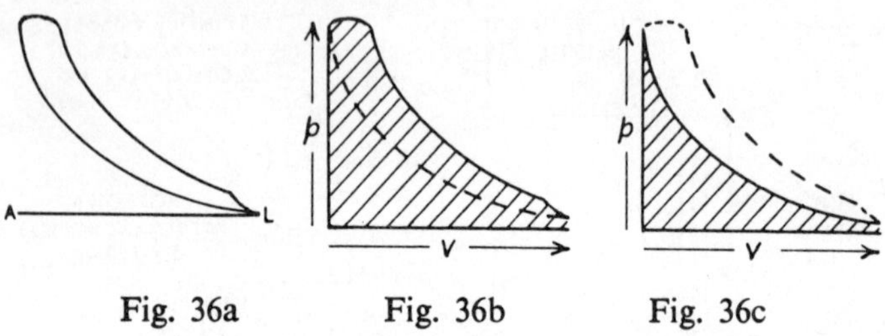

Fig. 36a   Fig. 36b   Fig. 36c

The area of the diagram is usually measured by a planimeter. If the area is measured in mm² then dividing this by the length in mm gives the mean height in mm. The mean height in mm is now multiplied by the pressure scale of the indicator spring in N/m² per mm to obtain the indicated mean effective pressure in N/m². The usual convenient multiples of N/m² for such pressures are kN/m² and bars, and the spring may be graduated in either of these units.

If a planimeter is not to hand, the mean height of the indicator diagram may be obtained by the application of the mid-ordinate rule.

INDICATED POWER. Power is the rate of doing work, that is, the quantity of work done in a given time. The basic unit of power is the *watt* [W] which is equal to the rate of one joule of work being done every second. In symbols:

$$1 \text{ W} = 1 \text{ J/s} = 1 \text{N m/s}$$

The watt is a small unit and only suitable for expressing the power of small machines. For normal powers in marine engineering, mechanical, electrical or hydraulic, the kilowatt [kW] is usually a more convenient size, and heavy powers may be expressed in megawatts [MW]. The familiar unit of horse-power is now gradually becoming obsolete, powers of all engines, pumps, motors, compressors, etc., will now be measured in multiples of the watt. One horse-power is equal to 745·7 W or 0·7457 kW, hence, as an example, an engine previously rated as 14 000 hp is now expressed as 10 440 kW or 10·44 MW.

$$\text{Let } p_m = \text{mean effective pressure [N/m}^2\text{]}$$
$$A = \text{area of piston [m}^2\text{]}$$
$$L = \text{length of stroke [m]}$$
$$n = \text{number of power strokes per second}$$

then,

Average force [N] on piston
$$= p_m \times A \text{ newtons}$$

Work done [J] in one power stroke
$$= p_m \times A \times L \text{ newton-metres} = \text{joules}$$

Work per second [J/s = W]
$$= p_m \times A \times L \times n \text{ watts of power}$$

therefore,

Indicated power $= p_m ALn$

This is the power indicated in one cylinder. The total power of a multi-cylinder engine is that multiplied by the number of cylinders, if the mean effective pressure is the same for all cylinders.

Note that when the mean effective pressure is in N/m^2 the power obtained by the above expression is in watts. If the mean effective pressure in kN/m^2 is inserted, the result will be the power in kW, and this is usually more convenient.

The value of n, the number of power strokes per second, depends upon the working cycle of the engine (two-stroke or four-stroke), its rotational speed, and whether it is a single-acting or double-acting engine.

Referring to single-acting engines, wherein the cycle of operations takes place only on the top side of the piston:

In the four-stroke cycle, there is one power stroke in every four strokes, that is, one power stroke in every two revolutions, hence,
$$n = \text{rev/s} \div 2$$

In the two-stroke cycle, there is one power stroke in every two strokes, that is, one power stroke in every revolution, thus,
$$n = \text{rev/s}$$

In double-acting engines the bottom of the cylinders is closed by a cylinder cover which is fitted with a full set of operating valves, the piston rod passing through a gas tight gland in the cover, and a similar cycle of operations takes place on the bottom side of the piston. If the mean effective pressure on the bottom side of the piston is the same as on the top, and, for rough calculations, if the space taken up in the cylinder by the piston rod is neglected, then the power developed in a double-acting engine would be twice as much as if it were single-acting, and the value of n twice that for a single-acting engine. For more accurate calculations, the power developed on the top and bottom of the piston is treated as separate single-acting cycles, by obtaining each mean effective pressure separately and taking the correct effective piston area on the bottom side as the difference between the cross-sectional areas of the cylinder and piston rod.

In the steam reciprocating engine (dealt with later) there is one power stroke in every revolution, and these engines are always double-acting, therefore it is a double-acting two-stroke engine.

In reciprocating engines run on town's gas, governing to run at constant speed is often on the "hit and miss" method. In this case, when the engine governor increases in speed it causes the fuel cam peak to miss the gas intake valve tappet and for one or more suction strokes, air but no fuel gas is drawn into the cylinder, and so there is no explosion and no power strokes during those periods. The valve of n is then obtained by counting the number of explosions over a considerable period (per minute or more) and obtaining the average number of power strokes per second.

Example. The area of an indicator diagram taken off one cylinder of a four-cylinder, four-stroke, single-acting internal combustion engine is 378 mm², the length is 70 mm, and the indicator spring scale is 1 mm = 1 bar. The diameter of the cylinders is 250 mm, stroke 300 mm, and rotational speed 5 rev/s. Calculate the indicated power of the engine assuming all cylinders develop equal power.

$$\text{Mean height of diagram} = \text{area} \div \text{length}$$
$$= 378 \div 70 = 5.4 \text{ mm}$$
$$\text{Indicated } p_m = \text{mean height} \times \text{spring scale}$$
$$= 5.4 \times 1 = 5.4 \text{ bar}$$
$$5.4 \text{ bar} \times 10^2 = 540 \text{ kN/m}^2$$
$$n = \text{rev/s} \div 2$$
$$= 5 \div 2 = 2.5$$
$$\text{Indicated power} = p_m A L n$$
$$= 540 \times 0.7854 \times 0.25^2 \times 0.3 \times 2.5$$
$$= 19.87 \text{ kW per cylinder}$$
$$\text{Total power for four cylinders}$$
$$= 4 \times 19.87 = 79.48 \text{ kW} \quad \text{Ans.}$$

Example. The diameter of the cylinders of a six-cylinder, single-acting, two-stroke diesel engine, is 635 mm and the stroke is 1010 mm. Indicator diagrams taken off the engine when running at 132 rev/min give an average area of 563 mm², the length of the diagram being 80 mm and the scale of the indicator spring 1 mm $= 80$ kN/m². Calculate the indicated power.

Indicated mean effective pressure

$$= \text{mean height of diagram} \times \text{spring scale}$$

$$= \frac{\text{area of diagram}}{\text{length of diagram}} \times \text{spring scale}$$

$$= \frac{563}{80} \times 80 = 563 \text{ kN/m}^2$$

For a single-acting two-stroke

$$n = \text{rev/s} = \frac{132}{60} = 2 \cdot 2$$

For a six-cylinder engine,

$$\text{Indicated power} = p_m A L n \times 6$$
$$= 563 \times 0 \cdot 7854 \times 0 \cdot 635^2 \times 1 \cdot 01 \times 2 \cdot 2 \times 6$$
$$= 2378 \text{ kW} \quad \text{Ans.}$$

BRAKE POWER AND MECHANICAL EFFICIENCY

Power is absorbed in overcoming frictional resistances at the various rubbing surfaces of the engine, such as at the piston rings, crosshead, crank and shaft bearings, therefore only part of the *indicated power* (ip) developed in the cylinders is transmitted as useful power at the engine shaft. The power absorbed in overcoming friction is termed the *friction power* (fp). The power available at the shaft is termed *shaft power* (sp) or, as this is measured by means of a brake it is also called *brake power* (bp).

Brake power = indicated power — friction power

The *mechanical efficiency* is the ratio of the brake power to the indicated power:

$$\text{Mechanical efficiency} = \frac{\text{brake power}}{\text{indicated power}}$$

Since the brake power is always less than the indicated power, the above expresses the mechanical efficiency as a fraction less than unity. It is common practice to state the efficiency as a percentage, by multiplying the fraction by 100.

Brake power is measured by applying a resisting torque as a brake on the shaft, the heat generated by the friction at the brake being transferred to and carried away by circulating water.

Let F = resisting force of brake, in newtons, applied at a radius of R metres when the rotational speed is in revolutions per second, then:

$$\text{Work absorbed per revolution [N m = J]}$$
$$= \text{force [N]} \times \text{circumference [m]}$$
$$= F \times 2\pi R$$
$$\text{Work absorbed per second [J/s]} = \text{power absorbed [W]}$$
$$= F \times 2\pi R \times \text{rev/s}$$
$$F \times R = \text{torque in N m} = T,$$
$$\therefore \text{brake power} = T \times 2\pi \times \text{rev/s}$$
$$2\pi \times \text{rev/s} = \text{angular velocity in radians/second}$$
$$= \omega$$
$$\therefore \text{brake power} = T\omega$$

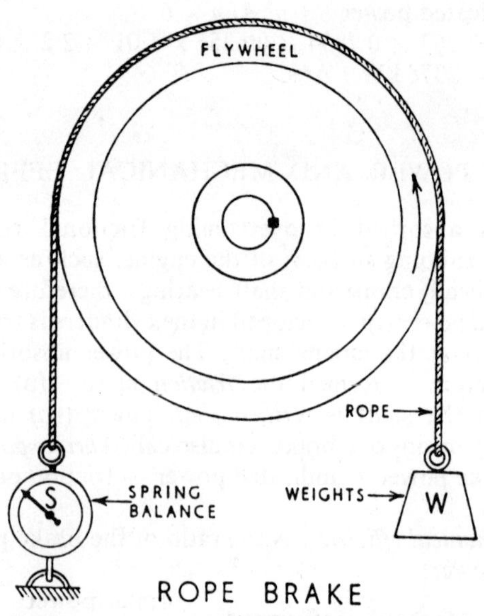

ROPE BRAKE

Fig. 37

Common types of brakes for measuring brake power are, for small engines, a loaded rope or steel band around a flywheel on the shaft, and, for large engines, a hydraulic dynamometer.

Fig. 37 illustrates a simple rope brake. The rope passes over the flywheel with one end of the rope anchored to the engine base and the other end hanging freely and loaded against the direction of rotation of the flywheel, the amount of loading depending upon the desired speed.

If W = weight of the load in newtons, and S = reading of spring balance in newtons, then the effective tangential braking force on the flywheel rim is $(W - S)$ newtons. If R = effective radius in metres from centre of shaft to centre of rope, then the braking torque in newton-metres is:

$$T = (W - S) \times R$$

Example. In a single-cylinder four-stroke single-acting gas engine, the cylinder diameter is 180 mm and the stroke 350 mm. When running at 250 rev/min the mean area of the indicator diagrams taken off the engine is 355 mm², length of diagram 75 mm, scale of the indicator spring 90 kN/m² per mm, and the number of explosions was counted to be 114 per minute. Calculate (i) the indicated power. If the effective radius of the rope brake on the flywheel is 600 mm, load on free end of rope 425 N, and reading of spring balance 72 N, calculate (ii) the brake power, and (iii) the mechanical efficiency.

Indicated mean effective pressure
$$= \text{mean height of diagram} \times \text{spring scale}$$

$$= \frac{\text{area of diagram}}{\text{length of diagram}} \times \text{spring scale}$$

$$= \frac{355}{75} \times 90 = 426 \text{ kN/m}^2$$

Indicated power $= p_m A L n$

$$= 426 \times 0{\cdot}7854 \times 0{\cdot}18^2 \times 0{\cdot}35 \times \frac{114}{60}$$

$$= 7{\cdot}21 \text{ kW} \quad \text{Ans. (i)}$$

$$\text{Braking torque} = (W - S) \times R$$
$$= (425 - 72) \times 0{\cdot}6 = 211{\cdot}8 \text{ N m}$$
$$\text{Brake power} = T\omega$$
$$= 211{\cdot}8 \times \frac{250 \times 2\pi}{60}$$
$$= 5546 \text{ W} = 5{\cdot}546 \text{ kW} \quad \text{Ans. (ii)}$$

$$\text{Mech. efficiency} = \frac{\text{brake power}}{\text{indicated power}}$$
$$= \frac{5{\cdot}546}{7{\cdot}21}$$
$$= 0{\cdot}7693 \text{ or } 76{\cdot}93\% \quad \text{Ans. (iii)}$$

MORSE TEST

In multi-cylinder internal combustion engines wherein all cylinders are of the same cubic capacity, a reasonable estimate of the indicated power developed in each cylinder can be made by the *morse test*. This is most useful in small high speed engines where indicator diagrams cannot be taken satisfactorily by the standard mechanical indicator.

The test consists of measuring the brake power at the shaft when all cylinders are firing and then measuring the brake power of the remaining cylinders when each one is "cut out" in turn. Cutting out the power of each cylinder is done in petrol engines by shorting the sparking plug, and in diesel engines by by-passing the cylinder fuel supply. The speed of the engine and the petrol throttle or fuel pump setting is kept constant during the test so that friction and pumping losses are approximately constant.

Taking a four-cylinder engine as an example:

With all four cylinders working,

$$\text{Total bp} = \text{total ip} - \text{total fp}$$
$$= \text{sum of ip's of the four cylinders} - \text{sum of the fp's of the four cylinders}$$

When the power of one cylinder is cut out,

$$\text{Total bp} = \text{sum of ip's of the three cylinders} - \text{sum of the fp's of the four cylinders}$$

Hence, we can see from the above that, when one cylinder is cut out, the loss of *brake* power at the shaft is the loss of the *indicated* power of that cylinder which is not firing.

Example. During a morse test on a four-cylinder four-stroke petrol engine, the throttle was set in a fixed position and the speed maintained constant at 35 rev/s by adjusting the brake, and the following powers in kW were measured at the brake,

With all cylinders working, bp developed $= 57$
With sparking plug of no. 1 cyl. shorted, bp $= 38.5$
   ,,   ,,   ,,   ,, no. 2 cyl.   ,,   bp $= 37$
   ,,   ,,   ,,   ,, no. 3 cyl.   ,,   bp $= 37.5$
   ,,   ,,   ,,   ,, no. 4 cyl.   ,,   bp $= 38$

Estimate the indicated power of the engine and the mechanical efficiency.

$$\text{ip of no. 1 cyl.} = 57 - 38.5 = 18.5$$
$$\text{,, ,, no. 2 cyl.} = 57 - 37 = 20$$
$$\text{,, ,, no. 3 cyl.} = 57 - 37.5 = 19.5$$
$$\text{,, ,, no. 4 cyl.} = 57 - 38 = 19$$
$$\text{total ip} = \overline{77.0} \text{ kW} \quad \text{Ans. (i)}$$

$$\text{Mechanical effic.} = \frac{\text{brake power}}{\text{indicated power}}$$

$$= \frac{57}{77} = 0.74 \text{ or } 74\% \quad \text{Ans. (ii)}$$

THERMAL EFFICIENCY

In engine trials it is usually most convenient to base calculations on a running time of one hour. Also, to enable comparisons to be made on the quantity of fuel oil to run the engine under different conditions, or comparisons of one engine with another, the fuel consumption is expressed per unit power developed. The fuel consumed in unit time per unit power developed is termed the *specific fuel consumption* and commonly stated in the units kilogrammes of fuel per kilowatt-hour [kg/kWh].

The thermal efficiency of an engine is the relationship between the quantity of heat energy converted into work and the quantity of heat energy supplied:

$$\text{Thermal efficiency} = \frac{\text{heat energy converted into work}}{\text{heat energy supplied}}$$

In internal combustion engines the heat is supplied directly into the cylinders by the burning of the injected fuel. The heat energy given off during complete combustion of unit mass of the fuel is termed the *calorific value* and may be expressed in kilojoules of heat energy given off during the burning of one kilogramme of fuel [kJ/kg]. However, the calorific value of fuel oil ranges from about 40000 to 44000 kJ/kg and is therefore more conveniently expressed in megajoules per kilogramme [MJ/kg], that is, 40 to 44 MJ/kg.

Hence the heat supplied in megajoules is the product of the mass of fuel burned in kilogrammes and its calorific value in megajoules per kilogramme. Therefore, on a basis of one kilowatt-hour:

$$\text{Thermal effic.} = \frac{\text{heat energy equivalent of 1 kWh [MJ/kWh]}}{\text{spec. fuel cons. [kg/kWh]} \times \text{cal. value [MJ/kg]}}$$

It was shown in Chapter 1 that the heat energy equivalent of one kilowatt-hour is:

$$
\begin{aligned}
\text{Energy} &= \text{power} \times \text{time} \\
&= 1000\,[\text{W}] \times 3600\,[\text{s}] \\
&= 3{\cdot}6 \times 10^6\,\text{J or } 3{\cdot}6 \times 10^3\,\text{kJ or } 3{\cdot}6\,\text{MJ}
\end{aligned}
$$

Thermal efficiency may be based on the heat energy supplied to develop 1 kW of indicated power in the cylinders, or the heat energy supplied to obtain 1 kW of brake power at the shaft. In the former, the specific fuel consumption (indicated) is expressed as the kilogrammes of fuel per indicated kilowatt-hour [kg/ind. kWh] and the efficiency is the *indicated thermal efficiency*. In the latter, the specific fuel consumption (brake) is expressed as the kilogrammes of fuel per brake kilowatt-hour [kg/brake kWh] and the efficiency is the *brake thermal efficiency*.

Thus, on the basis of one kilowatt-hour:

$$\text{Indicated thermal effic.} = \frac{3\cdot6\,[\text{MJ/kWh}]}{\text{kg fuel/ind.kWh} \times \text{cal. value [MJ/kg]}}$$

$$\text{Brake thermal effic.} = \frac{3\cdot6\,[\text{MJ/kWh}]}{\text{kg fuel/brake kWh} \times \text{cal. value [MJ/kg]}}$$

The brake thermal efficiency is also the product of the indicated thermal efficiency and the mechanical efficiency, and may therefore be termed the *overall efficiency* of the engine.

The symbol to represent efficiency is η (the Greek letter eta).

Example. The following data were taken during a one-hour trial run on a single-cylinder, single-acting, four-stroke diesel engine of cylinder diameter 175 mm and stroke 225 mm, the speed being constant at 1000 rev/min:

$$\text{Indicated mean effective pressure} = 5\cdot5 \text{ bar}$$
$$\text{Effective diameter of rope brake} = 1066 \text{ mm}$$
$$\text{Load on brake} = 400 \text{ N}$$
$$\text{Reading of spring balance} = 27 \text{ N}$$
$$\text{Fuel consumed} = 5\cdot7 \text{ kg}$$
$$\text{Calorific value of fuel} = 44\cdot2 \text{ MJ/kg}$$

Calculate the indicated power, brake power, specific fuel consumption per indicated kWh and per brake kWh, mechanical efficiency, indicated thermal efficiency and brake thermal efficiency.

$$
\begin{aligned}
\text{ip} &= p_m A L n \\
&= 5\cdot5 \times 10^2 \times 0\cdot7854 \times 0\cdot175^2 \times 0\cdot225 \times \frac{1000}{60 \times 2} \\
&= 24\cdot8 \text{ kW} \quad \text{Ans. (i)} \\
\text{bp} &= T\omega \\
&= (400 - 27) \times \frac{1\cdot066}{2} \times \frac{1000 \times 2\pi}{60} \\
&= 2\cdot082 \times 10^4 \text{ W} = 20\cdot82 \text{ kW} \quad \text{Ans. (ii)}
\end{aligned}
$$

Spec. fuel cons. (indicated)
$$= \frac{5\cdot7}{24\cdot8} = 0\cdot2298 \text{ kg/ind. kWh} \quad \text{Ans. (iii)}$$

Spec. fuel cons. (brake)

$$= \frac{5 \cdot 7}{20 \cdot 82} = 0 \cdot 2738 \text{ kg/brake kW h} \quad \text{Ans. (iv)}$$

$$\text{Mechanical effic.} = \frac{\text{brake power}}{\text{indicated power}}$$

$$= \frac{20 \cdot 82}{24 \cdot 8}$$

$$= 0 \cdot 8395 \text{ or } 83 \cdot 95\% \quad \text{Ans. (v)}$$

$$\text{Indicated thermal effic.} = \frac{3 \cdot 6 \, [\text{MJ/kW h}]}{\text{kg fuel/ind. kW h} \times \text{cal. value} \, [\text{MJ/kg}]}$$

$$= \frac{3 \cdot 6}{0 \cdot 2298 \times 44 \cdot 2}$$

$$= 0 \cdot 3544 \text{ or } 35 \cdot 44\% \quad \text{Ans. (vi)}$$

$$\text{Brake thermal effic.} = \frac{3 \cdot 6 \, [\text{MJ/kW h}]}{\text{kg fuel/brake kW h} \times \text{cal. value} \, [\text{MJ/kg}]}$$

$$= \frac{3 \cdot 6}{0 \cdot 2738 \times 44 \cdot 2}$$

$$= 0 \cdot 2975 \text{ or } 29 \cdot 75\% \quad \text{Ans. (vii)}$$

Alternatively,

$$\text{Brake thermal effic.} = \text{indicated thermal effic.} \times \text{mech. effic.}$$
$$= 0 \cdot 3544 \times 0 \cdot 8395$$
$$= 0 \cdot 2975 \text{ (as above)}$$

HEAT BALANCE

Of the total heat energy supplied to an engine, only a small proportion is converted into useful work. The heaviest losses are those due to the heat energy transferred to and carried away by the cooling water, and the heat energy remaining in the gases which are released from the cylinders and exhausted up the flue. A clear picture of the distribution of heat is shown by constructing a heat balance chart, based on taking the heat supplied in the fuel as 100%.

Fig. 38 illustrates an example of a heat balance of a four-stroke diesel engine. In this particular case the exhaust gases were passed through the tubes of a waste-heat steam boiler before escaping up the funnel, and 40% of the heat energy in the exhaust gases was usefully recovered in generating steam. The radiation loss is usually very small and in most cases it is included with the other losses.

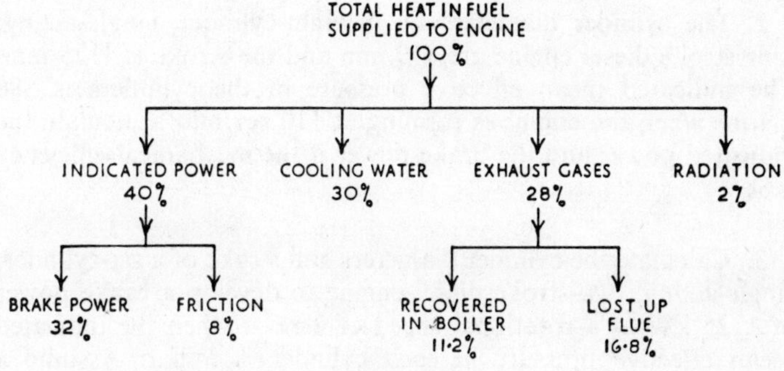

Fig. 38

TEST EXAMPLES 7

1. The area of an indicator diagram taken off a four-cylinder, single-acting, four-stroke, internal combustion engine when running at 5·5 rev/s is 390 mm², the length is 70 mm, and the scale of the indicator spring is 1 mm = 0·8 bar. The diameter of the cylinders is 150 mm and the stroke is 200 mm. Calculate the indicated power of the engine assuming all cylinders develop equal power.

2. The cylinder diameters of an eight-cylinder, single-acting, four-stroke diesel engine are 750 mm and the stroke is 1125 mm. The indicated mean effective pressure in the cylinders is 586 kN/m² when the engine is running at 110 rev/min. Calculate the indicated power and the brake power if the mechanical efficiency is 86%.

3. Calculate the cylinder diameters and stroke of a six-cylinder, single-acting, two-stroke diesel engine to develop a brake power of 1125 kW at a rotational speed of 2 rev/s when the indicated mean effective pressure in each cylinder is 5 bar. Assume a mechanical efficiency of 84% and make the length of the stroke 25% greater than the diameter of the cylinders.

4. In a single-acting, two-stroke, opposed-piston engine, the ratio of the masses of the top piston and its connected moving parts to the bottom piston and its moving parts, is 7·5 to 6, and the strokes of the pistons are in inverse ratio to their masses. The combined stroke of the top and bottom pistons is 2430 mm. The engine has six cylinders of 625 mm diameter and the indicated mean effective pressure is 6·5 bar when running at 1·75 rev/s. Assuming a mechanical efficiency of 90%, find the strokes of the top and bottom pistons, the indicated power and the brake power.

5. In an eight-cylinder, double-acting, two-stroke compression-ignition engine, the diameter of the cylinders is 700 mm, stroke 1350 mm, and piston rod diameter 250 mm. When running at 1·8 rev/s the indicated mean effective pressures above and below the pistons are 5·8 and 4·9 bar respectively. Calculate the indicated power developed above and below the pistons in each cylinder, the total engine indicated power and brake power, taking a mechanical efficiency of 80%.

6. The flywheel of a rope brake is 1·22 m diameter and the rope is 24 mm diameter. When the engine is running at 250 rev/min the load on the brake is 480 N on one end of the rope and 84 N on the other end. Calculate the brake power. If the rise in temperature of the brake cooling water is 18 K, calculate the quantity of water flowing through the brake in litres per hour assuming that the water carries away 90% of the heat generated at the brake. Take the specific heat of the water as 4·2 kJ/kg K.

7. A compression-ignition engine under test was coupled to a hydraulic dynamometer, the brake power formula in kW for this dynamometer is $W \times \text{rev/s}/300$ where W = brake load in newtons. When the engine was running at 50 rev/s the brake load was 180 N, mass flow of water through the brake 1218 kg/h with inlet temperature 16·4°C and outlet 36·9°C. Calculate what percentage of the heat generated at the brake is carried away by the cooling water.

8. During a Morse test on a four-cylinder petrol engine, the speed was kept constant at 24·5 rev/s by adjusting the brake and the following readings taken:

    With all cylinders firing, torque at brake = 193·8 N m
    „   no. 1 cyl. cut out   „ „ „ = 130·8 „
    „   no. 2 cyl. cut out   „ „ „ = 130·2 „
    „   no. 3 cyl. cut out   „ „ „ = 129·9 „
    „   no. 4 cyl. cut out   „ „ „ = 131·1 „
Calculate the bp, ip, and mechanical efficiency.

9. When developing a certain power, the specific fuel consumption of an internal combustion engine is 0·255 kg/kWh (brake) and the mechanical efficiency is 86%. Calculate (i) the indicated thermal efficiency, and (ii) the brake thermal efficiency, taking the calorific value of the fuel as 43·5 MJ/kg. If 35 kg of air are supplied per kg of fuel, the air inlet being at 26°C and exhaust at 393°C, find (iii) the heat energy carried away in the exhaust gases as a percentage of the heat supplied, taking the specific heat of the gases as 1·005 kJ/kg K.

10. A diesel engine uses 27 tonne of fuel per day when developing 4960 kW indicated power and 4060 kW brake power. Of the total heat supplied to the engine, 31·7% is carried away by the cooling water and radiation, and 30·8% in the exhaust gases. Calculate the indicated thermal efficiency, mechanical efficiency, overall efficiency, specific fuel consumption (indicated), and the calorific value of the fuel.

11. The specific fuel consumption (brake) of an engine is 0·243 kg/kWh when developing 2600 kW brake power, the calorific value of the fuel being 42 MJ/kg. 40 tonne of lubricating oil circulate through the engine per hour, the inlet temperature is 24°C and outlet 49°C. Taking the specific heat of the lubricating oil as 2·1 kJ/kgK, calculate the heat carried away by it as a percentage of the heat supplied to the engine. Calculate also the mass flow (tonne/h) of cooling water passing through the oil cooler if the inlet and outlet temperatures of the water are 21°C and 31°C, taking the specific heat of the water as 4·2 kJ/kgK.

12. The mean effective pressure measured from the indicator diagram taken off a single-cylinder four-stroke gas engine was 3·93 bar when running at 5 rev/s and developing a brake power of 4·33 kW. The number of explosions per minute was 123 and the gas consumption 3·1 cubic metres per hour. The diameter of the cylinder is 180 mm, stroke 300 mm, and calorific value of the gas 17·6 MJ per cubic metre. Calculate the indicated power, mechanical efficiency, indicated thermal efficiency and brake thermal efficiency.

CHAPTER 8

IDEAL CYCLES

Having seen the practical working cycles of the more common types of internal combustion engines, we shall now proceed to study the ideal or theoretical cycles of these engines.

In order to make comparisons of the efficiencies of the various cycles we make the assumption that they all work as closed-circuit hot-air engines. Thus it is imagined that the cylinder contains air only as the "working fluid" which never leaves the cylinder but is made to do work on the piston by, firstly, absorbing heat energy from an external hot supply, converting as much of this heat energy as possible into mechanical work, and finally rejecting the unused heat energy to an external cold "sink". Any compression or expansion of the air during the cycle is assumed to be purely adiabatic. The efficiency of an ideal cycle is, therefore, termed the *ideal thermal efficiency*, or, since air is assumed to be the working fluid, it is also termed the *air-standard efficiency*.

In any cycle:

Heat converted into work = heat supplied — heat rejected
and,

$$\text{Thermal efficiency} = \frac{\text{heat converted into work}}{\text{heat supplied}}$$

$$= \frac{\text{heat supplied} - \text{heat rejected}}{\text{heat supplied}}$$

$$= 1 - \frac{\text{heat rejected}}{\text{heat supplied}}$$

(Note that the heat converted into work, i.e. work done, is represented by the area of the pV diagram.)

CONSTANT VOLUME CYCLE

This is also known as the OTTO cycle and is the basis on which petrol, paraffin, and gas engines usually work.

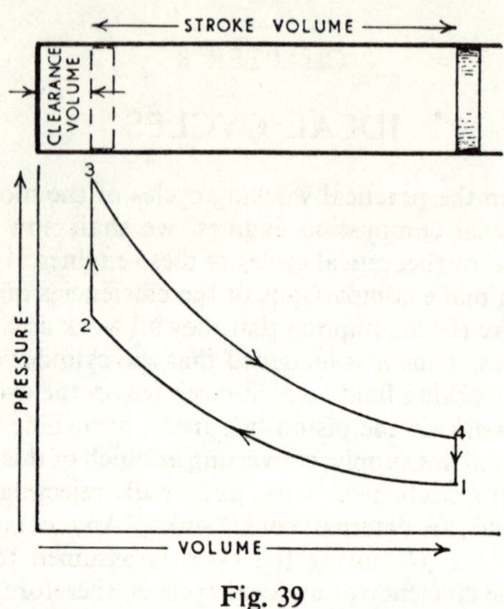

Fig. 39

Designating in sequence, the four cardinal state points of the cycle as 1, 2, 3 and 4 respectively (Fig. 39), the cycle of operations commence with a volume of air V_1 at pressure p_1 and temperature T_1. The piston moves inward and the air is compressed adiabatically to a volume V_2 and the pressure and temperature rise to p_2 and T_2. Heat energy is now given from some outside source and it is assumed that the air receives this heat instantaneously so that there is no time for any change of volume to occur. The pressure and temperature consequently rise to p_3 and T_3 while the volume remains unchanged and, therefore, V_3 is equal to V_2. Adiabatic expansion of the air now takes place while the piston is pushed outward on its power stroke, the volume increasing to V_4 which is the same as the initial volume V_1 and the pressure and temperature during expansion falling to p_4 and T_4. Finally, the cycle is completed by the air rejecting heat (theoretically instantaneously) at constant volume, to an outside source, which causes the pressure and temperature to fall to their initial values of p_1 and T_1.

It should be noted that, since the compression and expansion of the air is adiabatic, then there is no exchange of heat during these operations. This means that all the heat supplied takes place at constant volume between the state points 2 and 3, and all the heat rejected takes place at constant volume between the state points 4 and 1.

Heat supplied or rejected = mass × spec. ht. × temp. change
hence,

$$\text{heat supplied} = m \times c_V \times (T_3 - T_2)$$
$$\text{heat rejected} = \dot{m} \times c_V \times (T_4 - T_1)$$

therefore:

$$\text{Ideal thermal efficiency} = 1 - \frac{\text{heat rejected}}{\text{heat supplied}}$$

$$= 1 - \frac{m\, c_V(T_4 - T_1)}{m\, c_V(T_3 - T_2)}$$

$$= 1 - \frac{T_4 - T_1}{T_3 - T_2}$$

This is the general expression for the ideal thermal efficiency of an engine working on the constant volume cycle.

This efficiency can be expressed in terms of the ratio of compression, as follows,

$$\text{ratio of compression} = r = \frac{V_1}{V_2}$$

and, ratio of expansion $= \dfrac{V_4}{V_3}$

In this case the ratios of compression and expansion are the same because:

$$V_1 = V_4 \quad \text{and} \quad V_2 = V_3$$

Also, since $\dfrac{T_2}{T_1} = \left\{\dfrac{V_1}{V_2}\right\}^{\gamma-1} = r^{\gamma-1}$

and, $\dfrac{T_3}{T_4} = \left\{\dfrac{V_4}{V_3}\right\}^{\gamma-1} = r^{\gamma-1}$

then, $\dfrac{T_2}{T_1} = \dfrac{T_3}{T_4} = r^{\gamma-1}$

hence, $T_3 = T_4 r^{\gamma\,1} \quad \text{and} \quad T_2 = T_1 r^{\gamma\,1}$

therefore, $T_3 - T_2 = r^{\gamma-1}\,(T_4 - T_1)$

Substituting this value of $(T_3 - T_2)$ into the general expression for the thermal efficiency, $(T_4 - T_1)$ cancels, leaving:

$$\text{Ideal thermal efficiency} = 1 - \frac{1}{r^{\gamma-1}}$$

If γ is taken as 1·4 (for air) then this is also the Air Standard Efficiency.

$$\text{also, since } r^{\gamma-1} = \frac{T_2}{T_1} = \frac{T_3}{T_4}$$

$$\text{then, Ideal thermal efficiency} = 1 - \frac{T_1}{T_2}$$

$$= 1 - \frac{T_4}{T_3}$$

On examination of the above expression it will be seen that the greater the value of r, the greater will be the efficiency, hence the trend for higher ratios of compression in modern petrol engines. The majority of petrol engines, however, take in a mixture of petrol vapour and air during the induction stroke and this is compressed during the compression stroke. Being an explosive mixture it will burst into flame without the assistance of an electric spark or other

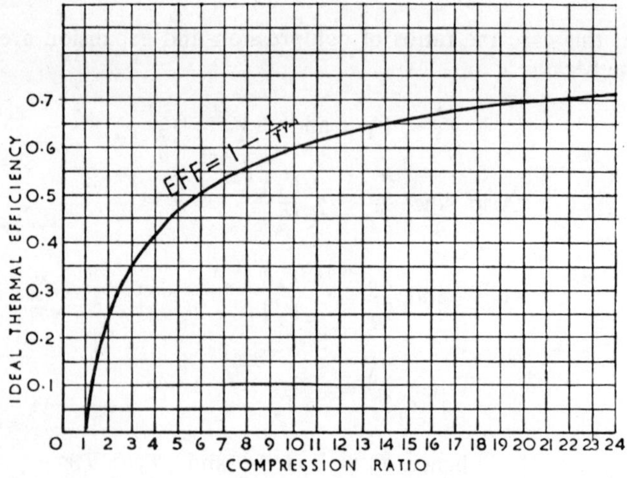

Fig. 40

means if it reaches its temperature of spontaneous ignition. Therefore, if the ratio of compression is too high for the grade of petrol used, preignition can take place.

Fig. 40 shows the relationship between the ideal efficiency and the ratio of compression in a constant volume cycle.

From the graph we see that, although the efficiency increases with higher compression ratios, the rate of increase in efficiency becomes less as the compression ratio is increased, and there is no appreciable gain by increasing the compression ratio above about 16 to 1.

Example. The compression ratio of an engine working on the constant volume cycle is 9·3 to 1. At the beginning of compression the temperature is 31°C and at the end of combustion the temperature is 1205°C. Taking compression and expansion to be adiabatic and the value of γ as 1·4, calculate (i) the temperature at the end of compression, (ii) the temperature at the end of expansion, (iii) the theoretical thermal efficiency.

Referring to Fig. 39:

$$V_1 = 9{\cdot}3 \quad \text{and} \quad V_4 = 9{\cdot}3$$
$$V_2 = 1 \quad \text{and} \quad V_3 = 1$$

$$T_1 = 31 + 273 = 304\,\text{K}$$
$$T_3 = 1205 + 273 = 1478\,\text{K}$$

COMPRESSION PERIOD:

$$\frac{T_2}{T_1} = \left\{\frac{V_1}{V_2}\right\}^{\gamma-1}$$

$$\therefore T_2 = 304 \times 9{\cdot}3^{0\cdot4} = 741{\cdot}8\,\text{K}$$

$\therefore$ temperature at end of compression

$$= 741{\cdot}8 - 273 = 468{\cdot}8°\text{C} \quad \text{Ans. (i)}$$

EXPANSION PERIOD:

$$\frac{T_4}{T_3} = \left\{\frac{V_3}{V_4}\right\}^{\gamma-1}$$

from which T_4 can be calculated, but, as ratios of compression and expansion are the same, and follow the same law, then an easier method is:

$$\frac{T_2}{T_1} = \frac{T_3}{T_4}$$

$$T_4 = \frac{304 \times 1478}{741 \cdot 8} = 605 \cdot 6 \, K$$

$\therefore$ temperature at end of expansion
$$= 605 \cdot 6 - 273 = 332 \cdot 6°C \quad \text{Ans. (ii)}$$

The theoretical efficiency can now be calculated from any of the expressions given above, thus,

$$\text{Efficiency} = 1 - \frac{T_4 - T_1}{T_3 - T_2} \quad \text{or} \quad 1 - \frac{1}{r^{\gamma - 1}}$$

$$\text{or} \quad 1 - \frac{T_1}{T_2} \quad \text{or} \quad 1 - \frac{T_4}{T_3}$$

Taking the last expression,
$$\text{Efficiency} = 1 - \frac{605 \cdot 6}{1478} = 1 - 0 \cdot 4099$$

$$= 0 \cdot 5901 \quad \text{or} \quad 59 \cdot 01\% \quad \text{Ans. (iii)}$$

DIESEL CYCLE

The term *constant pressure cycle* refers to one wherein the pressure remains constant during the two periods when heat energy is supplied and rejected. In the diesel cycle however, heat is supplied at constant pressure but rejection of heat takes place at constant volume. Thus, the diesel cycle, that upon which slow-speed diesel engines operate, is usually referred to as the *modified constant pressure cycle*.

Referring to Fig. 41, the cycle of operations commence with a volume of air V_1 at a pressure p_1 and temperature T_1. The air is compressed adiabatically to a volume V_2 and the pressure and

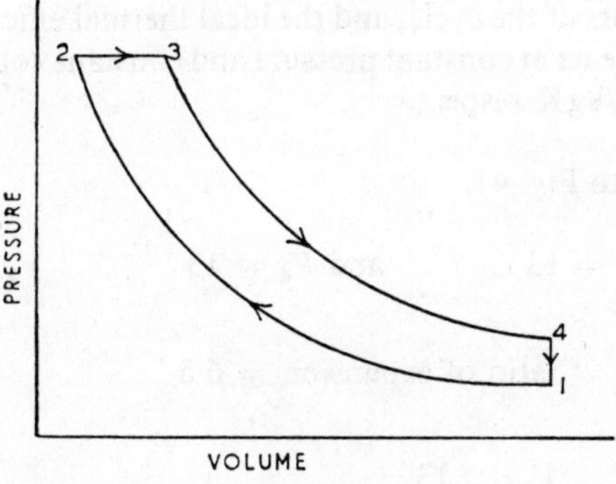

Fig. 41

temperature rise to p_2 and T_2. The piston is now at the top (inward end) of its stroke and heat is supplied at such a rate to maintain the pressure constant as the piston moves down the cylinder for a fraction of the power stroke. At the end of the heat supply period the volume is V_3, the temperature has been further increased to T_3, and the pressure represented by p_3 is the same as p_2. The air now expands adiabatically for the remainder of the power stroke until the final volume V_4 is the same as the initial volume V_1, the pressure and temperature falling during expansion to p_4 and T_4. Finally, the cycle is completed by the rejection of heat at constant volume to the initial conditions.

$$\text{Thermal efficiency} = 1 - \frac{\text{heat rejected}}{\text{heat supplied}}$$

$$= 1 - \frac{mc_V\,(T_4 - T_1)}{mc_P\,(T_3 - T_2)}$$

$$= 1 - \frac{1}{\gamma}\left\{\frac{T_4 - T_1}{T_3 - T_2}\right\}$$

This is the general expression for the ideal thermal efficiency of a diesel engine.

Example. The compression ratio in a diesel engine is 13 to 1 and the ratio of expansion is 6·5 to 1. At the beginning of compression the temperature is 32°C. Assuming adiabatic compression and expansion, calculate the temperatures at the three remaining cardinal points of the cycle, and the ideal thermal efficiency, taking the specific heats at constant pressure and constant volume as $1 \cdot 005$ and $0 \cdot 718$ kJ/kg K respectively.

Referring to Fig. 41,

$$V_1 = 13 \qquad \text{and } V_4 = 13 \qquad V_2 = 1$$

$$\frac{V_4}{V_3} = \text{ratio of expansion} = 6 \cdot 5$$

$$\therefore V_3 = \frac{V_4}{6 \cdot 5} = \frac{13}{6 \cdot 5} = 2$$

$$T_1 = 32 + 273 = 305 \, \text{K}$$

$$\gamma = \frac{c_P}{c_V} = \frac{1 \cdot 005}{0 \cdot 718} = 1 \cdot 4$$

FIRST STAGE, ADIABATIC COMPRESSION:

$$\frac{T_2}{T_1} = \left\{ \frac{V_1}{V_2} \right\}^{\gamma - 1}$$

$$\therefore T_2 = 305 \times 13^{0 \cdot 4} = 850 \cdot 9 \, \text{K}$$

$\therefore$ temperature at end of compression

$$= 850 \cdot 9 - 273 = 577 \cdot 9 °\text{C}. \quad \text{Ans. (i)}$$

SECOND STAGE, HEATING AT CONSTANT PRESSURE:

$$\frac{T_3}{T_2} = \frac{V_3}{V_2} \quad \text{(Charles' law)}$$

$$T_3 = 850 \cdot 9 \times 2 = 1701 \cdot 8 \, \text{K}$$

$\therefore$ temperature at end of combustion

$$= 1701 \cdot 8 - 273 = 1428 \cdot 8 °\text{C}. \quad \text{Ans. (ii)}$$

THIRD STAGE, ADIABATIC EXPANSION:

$$\frac{T_4}{T_3} = \left\{\frac{V_3}{V_4}\right\}^{\gamma-1}$$

$$T_4 = 1701 \cdot 8 \times \left\{\frac{2}{13}\right\}^{0 \cdot 4}$$

$$= \frac{1701 \cdot 8}{6 \cdot 5^{0 \cdot 4}} = 804 \cdot 8\,\mathrm{K}$$

$\therefore$ temperature at end of expansion

$$= 804 \cdot 8 - 273 = 531 \cdot 8°\mathrm{C}. \quad \text{Ans. (iii)}$$

Ideal thermal efficiency

$$= 1 - \frac{\text{heat rejected}}{\text{heat supplied}}$$

$$= 1 - \frac{1}{\gamma}\left\{\frac{T_4 - T_1}{T_3 - T_2}\right\}$$

$$= 1 - \frac{1}{1 \cdot 4}\left\{\frac{804 \cdot 8 - 305}{1701 \cdot 8 - 850 \cdot 9}\right\}$$

$$= 1 - \frac{1}{1 \cdot 4} \times \frac{499 \cdot 8}{850 \cdot 9}$$

$$= 1 - 0 \cdot 4196$$
$$= 0 \cdot 5804 \quad \text{or} \quad 58 \cdot 04\%. \quad \text{Ans. (iv)}$$

In the ideal diesel cycle, where the compression and expansion are both adiabatic, the thermal efficiency can be expressed in terms of the ratio of compression and a comparison can then be made with the efficiency of a constant volume cycle of the same ratio of compression.

Expressing all temperatures in terms of T_1, substituting and simplifying:

$$\frac{T_2}{T_1} = \left\{\frac{V_1}{V_2}\right\}^{\gamma-1} = r^{\gamma-1}$$

$$\therefore T_2 = T_1 r^{\gamma-1}$$

$$\frac{T_3}{T_2} = \frac{V_3}{V_2}$$ let this ratio of burning period volumes be represented by ρ (the Greek letter rho), then,

$$\frac{T_3}{T_2} = \rho$$

$$\therefore T_3 = T_2\rho = T_1 r^{\gamma-1}\rho$$

$$\frac{T_4}{T_3} = \left\{\frac{V_3}{V_4}\right\}^{\gamma-1}$$

Since $\dfrac{V_3}{V_2} = \rho$ and $\dfrac{V_4}{V_2} = r$, then $\dfrac{V_3}{V_4} = \dfrac{\rho}{r}$

$$\therefore \frac{T_4}{T_3} = \left\{\frac{\rho}{r}\right\}^{\gamma-1}$$

$$T_4 = T_3 \times \left\{\frac{\rho}{r}\right\}^{\gamma-1}$$

$$= T_1 r^{\gamma-1}\rho \times \left\{\frac{\rho}{r}\right\}^{\gamma-1}$$

$$= T_1 \rho^{\gamma}$$

Ideal thermal efficiency

$$= 1 - \frac{1}{\gamma}\left\{\frac{T_4 - T_1}{T_3 - T_2}\right\}$$

$$= 1 - \frac{1}{\gamma}\left\{\frac{T_1\rho^{\gamma} - T_1}{T_1 r^{\gamma-1}\rho - T_1 r^{\gamma-1}}\right\}$$

$$= 1 - \frac{1}{\gamma} \times \frac{1}{r^{\gamma-1}}\left\{\frac{\rho^{\gamma} - 1}{\rho - 1}\right\}$$

If $\gamma = 1\cdot4$, this is also the Air Standard Efficiency.

Comparing this expression with the ideal efficiency of the constant volume cycle in terms of r, it will be seen that, for the same ratio of compression, the constant volume cycle has the higher thermal efficiency. This does not mean, however, that a petrol engine working on the constant volume cycle is more efficient than a diesel engine working on the modified constant pressure cycle, because, in the former, an explosive mixture is compressed and there is a limit to the ratio of compression, whereas air only is compressed in a diesel engine and the ratio of compression can be as high as required.

DUAL COMBUSTION CYCLE

In most high-speed compression-ignition engines, combustion takes place partly at constant volume and partly at constant pressure, and therefore the cycle is referred to as *dual-combustion*.

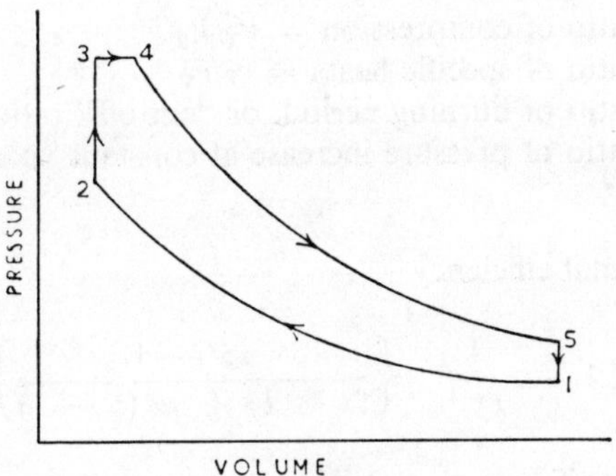

Fig. 42

Fig. 42 shows the ideal dual-combustion cycle. Commencing with a volume of air, V_1 at pressure p_1 and temperature T_1, the air is compressed adiabatically to a volume V_2 and the pressure and temperature rise to p_2 and T_2. Heat energy is now supplied at constant volume, the pressure and temperature is increased to p_3 and T_3 while the volume remains unchanged so that V_3 is equal to V_2. The supply of heat energy is continued at such a rate as to maintain the pressure constant while the piston moves outward until the volume is V_4, the temperature is further increased to T_4 and the pressure p_4 remains the same as p_3. Now adiabatic expansion takes place until the volume V_5 is the same as the initial volume V_1, the pressure and temperature falling due to expansion to p_5 and T_5. Finally, heat is rejected at constant volume and the pressure and temperature fall to the initial conditions of p_1 and T_1.

Ideal thermal efficiency

$$= 1 - \frac{\text{heat rejected}}{\text{heat supplied}}$$

$$= 1 - \frac{mc_V\,(T_5 - T_1)}{mc_V\,(T_3 - T_2) + mc_P\,(T_4 - T_3)}$$

$$= 1 - \frac{(T_5 - T_1)}{(T_3 - T_2) + \gamma\,(T_4 - T_3)}$$

The above expression can be converted into terms of the ratio of compression in a similar manner as the previous cycles, thus,

$$\text{If } r = \text{ratio of compression} = V_1/V_2$$
$$\gamma = \text{ratio of specific heats} = c_P/c_V$$
$$\rho = \text{ratio of burning period, or "cut-off" ratio} = V_4/V_3$$
$$\alpha = \text{ratio of pressure increase at constant volume} = p_3/p_2$$

Ideal thermal efficiency

$$= 1 - \frac{1}{r^{\gamma-1}}\left\{\frac{\alpha\rho^\gamma - 1}{(\alpha - 1) + \gamma\alpha\,(\rho - 1)}\right\}$$

Note (i) if $\alpha = 1$, the above becomes a pure diesel cycle.
     (ii) if $\rho = 1$, it becomes a pure constant-volume cycle.

Since compression-ignition oil engines depend upon the temperature of the air at the end of compression to ignite the fuel injected into the cylinder, the compression-ratio must be fairly high, usually not less than about 12 to give the necessary temperature rise during compression.

The higher compression pressures developed in this type of engine limit the use of constant volume combustion, since the maximum pressure in the cycle is limited by the consideration of strength.

As the maximum pressure is limited, increasing the compression ratio reduces the amount of fuel burned at constant volume, so that more must be burned at constant pressure and thus the gain due to increased compression ratio is partly nullified.

∫CARNOT CYCLE

This is a purely theoretical cycle devised by the French scientist Sadi Carnot. Although it is not possible from practical considerations for an engine to work on this cycle, it has a higher theoretical thermal efficiency than any other working between the same temperature limits and, therefore, provides a useful standard for comparing the performance of other heat engines.

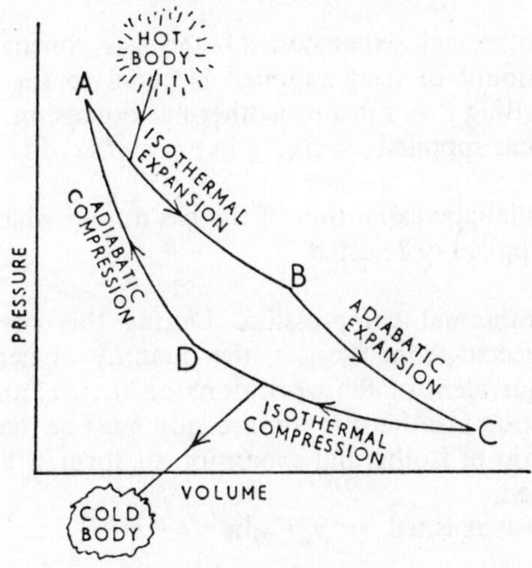

Fig. 43

Referring to the pV diagram, Fig. 43, it is usual to explain this cycle by commencing at state point A. Gas has been previously compressed in the cylinder by the piston moving inward and, at A, the piston is at the "top" of its stroke. The pressure and temperature are high, the value of the latter being represented by T_1. As the piston is pushed outward, doing work, heat is supplied to the gas from an external hot source at such a rate as to maintain its temperature constant, and during this period the gas therefore expands isothermally until point B is reached. At this point the heat supply is cut off and no heat is given to or rejected from the gas as the piston moves on to the end of the stroke at C. Hence during this period, the gas expands adiabatically as it does work and therefore the temperature falls. The temperature of the gas at point C is represented by T_2. The piston now moves inward to compress the gas from C to D and during this period it is assumed that any generated heat due to compression can flow out of the gas into a cold "sink". That is, the gas rejects heat energy to a cold external source at such a rate to maintain the temperature constant at T_2. This is isothermal compression. At point D, the flow of heat out of the gas is stopped and from D to A the gas is compressed adiabatically while the piston completes its stroke, the temperature of the gas rising to the initial temperature T_1.

Hence, the four stages of the Carnot cycle are briefly as follows:

A to B   Isothermal expansion of the gas during which the amount of heat supplied is equal to the work done. Letting r = ratio of isothermal expansion,
Heat supplied $= p_A V_A \ln r = mRT_1 \ln r$

B to C   Adiabatic expansion of the gas during which no heat is supplied or rejected.

C to D   Isothermal compression. During this period heat is rejected from the gas, the quantity of heat being the equivalent of the work done on the gas and, since the ratio of isothermal compression must be the same as the ratio of isothermal expansion to form a closed cycle, then,
Heat rejected $= p_C V_C \ln r = mRT_2 \ln r$

D to A   Adiabatic compression during which no heat is supplied or rejected.

Therefore,

$$\text{Thermal efficiency} = \frac{\text{heat supplied} - \text{heat rejected}}{\text{heat supplied}}$$

$$= 1 - \frac{\text{heat rejected}}{\text{heat supplied}}$$

$$= 1 - \frac{mRT_2 \ln r}{mRT_1 \ln r}$$

$$= 1 - \frac{T_2}{T_1}$$

$$= \frac{T_1 - T_2}{T_1}$$

This expression for the Carnot Efficiency shows that, to obtain the highest efficiency, heat should be taken in at the highest possible temperature (T_1) and rejected at the lowest possible temperature (T_2). This conclusion is applicable in the design of any heat engine.

f REVERSED CARNOT CYCLE

The Carnot cycle is theoretically reversible and if applied in reverse manner would act as a refrigerator by taking heat from a cold region and maintaining it at a low temperature as follows:

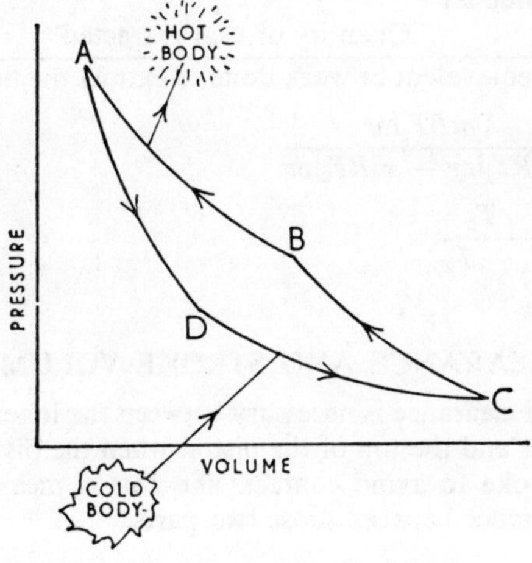

Fig. 44

Referring to Fig. 44 and commencing at state point A, the four stages of the reversed Carnot cycle consists of:

(i) Work done by the gas while it expands adiabatically from A to D and the temperature falls from T_1 to T_2. No heat is given to or taken from the gas during this process.

(ii) Further work done by the gas as it expands isothermally from D to C, a quantity of heat is taken in by the gas (from the cold body) equal to the work done, to maintain the temperature constant at T_2.

(iii)   Adiabatic compression of the gas from C to B, no heat being given to or taken from the gas, therefore the temperature increases from T_2 to T_1.

(iv)   Isothermal compression from B to A during which heat is rejected from the gas (to the hot body) to maintain the temperature constant at T_1.

Thus an engine working on the reversed Carnot cycle would require to be driven and, as heat would be continually taken from a cold region and sent out to a hotter region, it would therefore act as a refrigerating machine. The measure of the "efficiency" of refrigeration is known as the *coefficient of performance*, and its theoretical value is:

$$\frac{\text{Quantity of heat extracted}}{\text{Heat equivalent of work done to extract the heat}}$$

$$= \frac{mRT_2\ln r}{mRT_1\ln r - mRT_2\ln r}$$

$$= \frac{T_2}{T_1 - T_2}$$

CLEARANCE AND STROKE VOLUME

Mechanical clearance is necessary between the inner face of the cylinder cover and the top of the piston when the piston is at the top of its stroke to avoid contact, and this is measured as the minimum distance between those two parts.

The *clearance volume* is the volume of the enclosed space above the piston when at the top of its stroke, including all cavities up to the valve faces when the valves are closed. Thus, in a steam reciprocating engine, it includes the ports or passages as far as the slide valve face. In internal combustion engines the clearance volume is the combustion space to accommodate sufficient air for the complete combustion of the fuel and to limit the rise of temperature during burning. It is designed as near a spherical space as practicable, in many cases by concave piston tops and concave cylinder covers.

The *stroke volume*, sometimes termed the *swept volume*, is the volume swept out by the piston as it moves through one complete stroke. It is, therefore, equal to the product of the cross-sectional area of the cylinder and the length of the stroke.

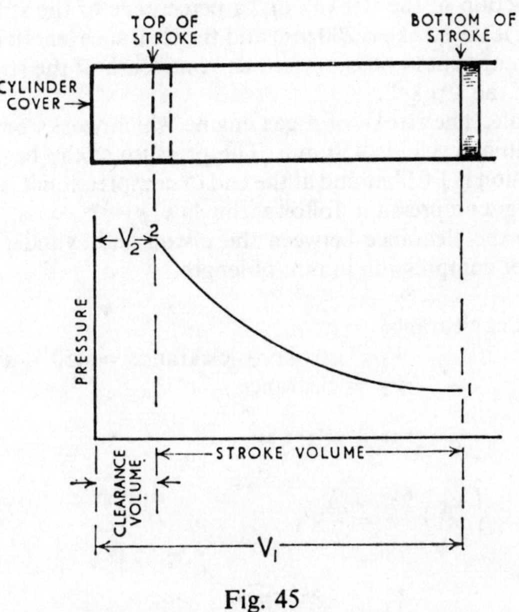

Fig. 45

Since the ratio of compression is the ratio of the volume at the beginning of compression to the volume at the end of compression, then, referring to Fig. 45,

$$r = \frac{\text{initial volume}}{\text{final volume}}$$

$$= \frac{V_1}{V_2} = \frac{\text{clearance vol.} + \text{stroke vol.}}{\text{clearance volume}}$$

Therefore, the magnitude of the clearance volume affects the ratio of compression (and expansion) and can be adjusted by shims under the foot of the connecting rod or plates in the clearance space.

Dividing stroke-volume by the cross-sectional area of the cylinder gives the length of the stroke. Similarly, dividing clearance-volume by the cross-sectional area of the cylinder gives the clearance in terms of length. Hence, for convenience, if the stroke is expressed in millimetres, the clearance may be expressed as a length in millimetres. Alternatively, the clearance may be expressed as "a fraction of the stroke" or "a percentage of the stroke", for instance, if the stroke is 200 mm and the clearance length is 20 mm, the clearance could be expressed as "one-tenth of the stroke", or, "10% of the stroke".

Example. The stroke of a gas engine which works on the constant volume cycle, is 450 mm. The pressure at the beginning of compression is 1·01 bar and at the end of compression it is 11·1 bar, Assuming compression follows the law $pV^{1\cdot36} = $ a constant, calculate the clearance between the piston and cylinder cover at the end of compression in mm of length.

$$\text{Let clearance} = c$$
$$V_1 = \text{stroke} + \text{clearance} = 450 + c$$
$$V_2 = \text{clearance} = c$$

$$p_1 V_1^{1\cdot36} = p_2 V_2^{1\cdot36}$$

$$\left\{\frac{V_1}{V_2}\right\}^{1\cdot36} = \frac{p_2}{p_1}$$

$$\frac{V_1}{V_2} = \sqrt[1\cdot36]{\frac{p_2}{p_1}}$$

$$\frac{450 + c}{c} = \sqrt[1\cdot36]{\frac{11\cdot1}{1\cdot01}}$$

$$\frac{450 + c}{c} = 5\cdot826$$

$$450 + c = 5\cdot826c$$
$$450 = 4\cdot826c$$
$$c = 93\cdot26 \text{ mm} \quad \text{Ans.}$$

TEST EXAMPLES 8

1. A petrol engine working on the constant volume cycle has a compression ratio of 9 to 1. If the pressure and temperature of the petrol-vapour-air mixture at the beginning of compression are 0·98 bar and 38°C respectively, calculate the pressure and temperature at the end of compression assuming it follows the law $pV^{1·34}$ = a constant.

2. The stroke of an internal combustion engine is 75 mm, the diameter of the cylinder is 70 mm, and the clearance volume at the end of compression is 36 cm³. Assuming compression follows the law $pV^{1·37} = C$, calculate the pressure at the end of compression if the initial pressure is 0·97 bar.

3. The pressure and temperature of the air at the beginning of compression in a diesel engine are 1·1 bar and 35°C respectively, and the clearance volume is equal to 7·5% of the piston swept volume. Calculate the pressure and temperature at the end of compression assuming the law of compression is $pV^{1·36} = C$.

4. The stroke of a petrol engine is 87·5 mm and the clearance is equal to 12·5 mm. A compression plate is now fitted which has the effect of reducing the clearance to 10 mm. Assuming the compression period to be the whole stroke, the pressure at the beginning of compression as 0·97 bar, and the law of compression $pV^{1·35} = C$, calculate the pressure at the end of compression before and after the compression plate is fitted.

5. The stroke of the piston in an internal combustion engine is 880 mm and the clearance is equal to 80 mm. The law of compression is $pV^{1·38} = C$ and the pressure at the end of compression is 32 bar. Find the increase of the final pressure caused by reducing the clearance by 5 mm.

6. The ratio of compression in a diesel engine is 16 to 1 and the temperature of the air at the beginning of compression is 49°C. Calculate the temperature at the end of compression assuming it follows the law $pV^{1·34}$ = a constant.

7. When the piston is moving up in a two-stroke diesel engine, the scavenge ports are closed when the piston is 675 mm from the top of its stroke, the pressure and temperature of the air in the

cylinder then being 13·8 kN/m² gauge and 43°C. The clearance is equal to 65 mm and the diameter of the cylinder is 650 mm. Calculate the mass of air compressed in the cylinder taking the atmospheric pressure as 101·3 kN/m² and R for air as 0·287 kJ/kgK.

8. Gas is compressed in an internal combustion engine according to the law $pV^{1·36} = C$. If the initial and final temperatures of the gas are 30°C and 382°C respectively, calculate the compression ratio.

9. The compression ratio of a petrol engine working on the constant volume cycle is 8·5. The pressure and temperature at the beginning of compression are 1 bar and 40°C and the maximum pressure of the cycle is 31 bar. Taking compression to follow the law $pV^{1·35} = C$, calculate (i) the pressure at the end of compression, (ii) temperature at end of compression, (iii) temperature at end of combustion.

10. In an internal combustion engine working on the constant volume cycle, the pressure, volume and temperature at the beginning of compression are 0·99 bar, 113 litres, and 48°C respectively, and the ratio of compression is 10. During combustion at constant volume the gas receives 95 kJ of heat. Taking compression according to the law $pV^{1·37} = C$, specific heat at constant volume 0·712 kJ/kgK and R (the characteristic gas constant) as 0·29 kJ/kgK, calculate: (i) mass of gas compressed, (ii) pressure and temperature at end of compression, (iii) pressure and temperature at end of combustion.

11. The compression ratio of a diesel engine is 15 to 1. Fuel is admitted for one-tenth of the power stroke and combustion takes place at constant pressure. Exhaust commences when the piston has travelled nine-tenths of the stroke. At the beginning of compression the temperature of the air is 41°C. Assuming compression and expansion to follow the law $pV^{n} = C$ where n is 1·34, calculate the temperatures at the end of compression, end of combustion, and beginning of exhaust.

12. An engine operates on the constant volume cycle and has a ratio of compression of 7 to 1. Another engine, working on the diesel cycle, has a compression ratio of 14 to 1, and fuel is admitted at constant pressure for 6% of the stroke. Compare the Air Standard Efficiencies of the two engines by use of the following formulae:

$$\text{A.S.E. of cons. vol. cycle} = 1 - \frac{1}{r^{\gamma-1}}$$

$$\text{A.S.E. of diesel cycle} = 1 - \frac{1}{\gamma} \times \frac{1}{r^{\gamma-1}} \left\{ \frac{\rho^{\gamma}-1}{\rho-1} \right\}$$

$$\text{where } r = \text{ratio of compression}$$
$$\gamma = 1\cdot4$$
$$\rho = \text{fuel cut-off ratio}$$

13. The compression ratio of a compression-ignition engine is 14 to 1, the diameter of the cylinder is 500 mm and the stroke/bore ratio is 1·2 to 1. At the beginning of compression the pressure and temperature of the air in the cylinder is 1·03 bar, and 51°C, and compression follows the law $pV^n = $ constant, where $n = 1\cdot35$. Assuming compression takes place over the whole stroke, calculate (i) the pressure at the end of compression, (ii) temperature at end of compression, (iii) mass of air compressed, (iv) work done on the air during compression, (v) change of internal energy, and (vi) heat exchange during compression. Take the values: R for air $= 0\cdot287$ kJ/kgK, and c_V for air $= 0\cdot718$ kJ/kgK.

14. The compression ratio of an engine working on the dual-combustion cycle is 10·7. The pressure and temperature of the air at the beginning of compression is 1 bar and 32°C. The maximum pressure and temperature during the cycle is 41 bar and 1593°C. Assuming adiabatic compression and expansion, calculate the pressures and temperatures at the remaining cardinal points of the cycle and the ideal thermal efficiency. Take the values, $c_V = 0\cdot718$, and $c_P = 1\cdot005$ kJ/kgK.

CHAPTER 9

RECIPROCATING AIR COMPRESSORS

Compressed air, at various pressures, is used on board ship for many purposes such as scavenging, supercharging, and starting diesel engines, and as the operating fluid for many automatic control systems. Air compressors to produce medium and high pressures are usually of the reciprocating type and may be single or multi-stage. Rotary types are common for large quantities of air at low pressures.

Fig. 46 shows diagrammatically a single-stage, single-acting reciprocating compressor, and its pV indicator diagram illustrating the cycle.

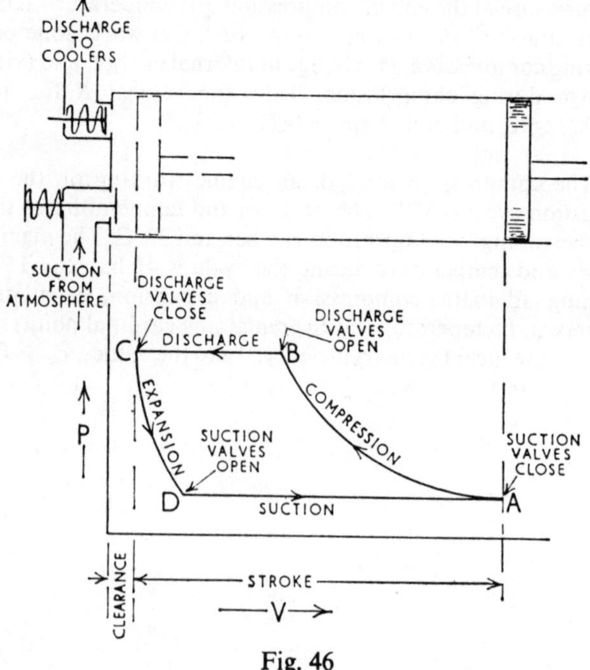

Fig. 46

Commencing at point A, the cycle of operations is as follows: A to B, compression period; with all valves closed the piston moves inward and the air which was previously drawn into the cylinder is compressed. Compression continues until the air pressure is sufficiently high to force the discharge valves open against their pre-set compression springs. Thus, at point B, the discharge valves open, and the compressed air is discharged at constant pressure for the remainder of the inward stroke, *i.e.*, from B to C. At point C the piston has completed its inward stroke and changes direction to move outward. Immediately the piston begins to move back there is a drop in pressure of the compressed air left in the clearance space, the discharge valves close, and from C to D this air expands. At point D the pressure has fallen to less than the atmospheric pressure and the lightly-sprung suction valves are opened by the greater pressure of the atmospheric air. Air is drawn into the cylinder for the remainder of the outward stroke, *i.e.*, from D to A.

Example. The stroke of the piston of an air compressor is 250 mm and the clearance volume is equal to 6% of the stroke volume. The pressure of the air at the beginning of compression is 0·98 bar and it is discharged at 3·8 bar. Assuming compression to follow the law $pV^n = $ constant, where $n = 1·25$, calculate the distance moved by the piston from the beginning of its pressure stroke before the discharge valves open and express this as a percentage of the stroke.

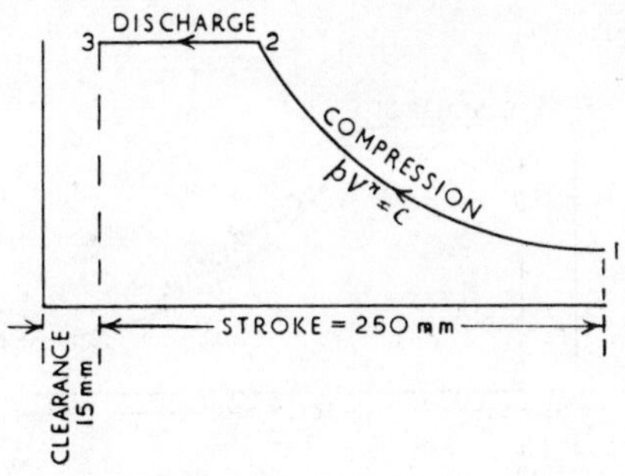

Fig. 47

$$\text{Clearance length} = 6\% \text{ of } 250 = 15$$
$$p_1 = 0·98 \qquad p_2 = 3·8 \qquad V_1 = 250 + 15 = 265$$
$$p_1 V_1^{1·25} = p_2 V_2^{1·25}$$
$$0·98 \times 265^{1·25} = 3·8 \times V_2^{1·25}$$

$$V_2^{1.25} = \frac{0.98 \times 265^{1.25}}{3.8}$$

$$V_2 = 265 \times \sqrt[1.25]{\frac{0.98}{3.8}} = 89.6$$

Distance moved by piston from beginning of stroke to point where discharge valves open is represented by $V_1 - V_2$

$$V_1 - V_2 = 265 - 89.6 = 175.4 \text{ mm} \quad \text{Ans. (i)}$$

Expressed as a percentage of the stroke of 250 mm

$$= \frac{175.4}{250} \times 100 = 70.16\% \quad \text{Ans. (ii)}$$

Example. The diameter of an air compressor cylinder is 140 mm, the stroke of the piston is 180 mm, and the clearance volume is 77 cm³. The pressure and temperature of the air in the cylinder at the end of the suction stroke and beginning of compression is 0·97 bar and 13°C. The delivery pressure is constant at 4·2 bar. Taking the law of compression as $pV^{1.3} = $ constant, calculate (i) for what length of the stroke air is delivered, (ii) the volume of air delivered per stroke, in litres, and (iii) the temperature of the compressed air.

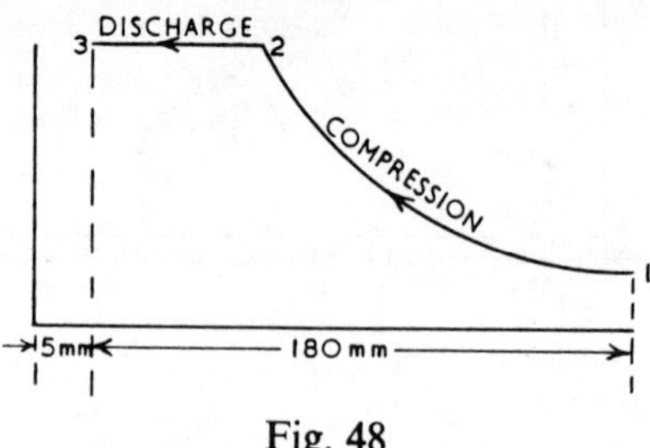

Fig. 48

$$\text{Clearance length [mm]} = \frac{\text{clearance volume [mm}^3\text{]}}{\text{area of cylinder [mm}^2\text{]}}$$

$$= \frac{77 \times 10^3}{0.7854 \times 140^2} = 5 \text{ mm}$$

$$p_1 V_1^{1 \cdot 3} = p_2 V_2^{1 \cdot 3}$$
$$0 \cdot 97 \times (180 + 5)^{1 \cdot 3} = 4 \cdot 2 \times V_2^{1 \cdot 3}$$

$$V_2^{1 \cdot 3} = \frac{0 \cdot 97 \times 185^{1 \cdot 3}}{4 \cdot 2}$$

$$V_2 = 185 \times \sqrt[1 \cdot 3]{\frac{0 \cdot 97}{4 \cdot 2}}$$

$$= 59 \cdot 94 \text{ mm}$$

$$\text{Delivery period} = V_2 - V_3 = 59 \cdot 94 - 5$$
$$= 54 \cdot 94 \text{ mm of stroke.} \quad \text{Ans. (i)}$$

$$\text{Volume delivered} = \text{area} \times \text{length}$$
$$= 0 \cdot 7854 \times 140^2 \times 54 \cdot 94$$
$$= 8 \cdot 457 \times 10^5 \text{ mm}^3$$

$$10^6 \text{ mm}^3 = 1 \text{ litre}$$
$$\text{Volume delivered} = 0 \cdot 8457 \text{ litre} \quad \text{Ans. (ii)}$$

$$\frac{p_1 V_1}{T_1} = \frac{p_2 V_2}{T_2}$$

$$\frac{0 \cdot 97 \times 185}{(13 + 273)} = \frac{4 \cdot 2 \times 59 \cdot 94}{T_2}$$

$$T_2 = \frac{286 \times 4 \cdot 2 \times 59 \cdot 94}{0 \cdot 97 \times 185} = 401 \cdot 2 \text{ K}$$

$$\therefore \text{ Temperature at end of compression}$$
$$= 401 \cdot 2 - 273 = 128 \cdot 2°\text{C} \quad \text{Ans. (iii)}$$

EFFECT OF CLEARANCE

Clearance is necessary between the piston face and the cylinder head and valves to avoid contact. This should be kept to a minimum because the volume of compressed air left in the clearance space at the end of the inward stroke must be expanded on the outward stroke to below atmospheric pressure before the suction valves can open, thus affecting the volume of air drawn into the cylinder during the suction stroke. This effect is illustrated below.

An ideal compressor would have, theoretically, no clearance as in Fig. 49. The suction valves would open immediately the piston began to move on its outward stroke and air would be drawn into the cylinder for the whole stroke.

Fig. 50 shows the effect of a small clearance. The small volume of compressed air left in the clearance space is quickly expanded to sub-atmospheric pressure to allow the suction valves to open, and air is drawn into the cylinder from D to A which is a large proportion of the full stroke E to A.

Fig. 51 shows the effect of excessive clearance. There is a comparatively large volume of air left in the clearance space and it requires a considerable movement of the piston on its outward stroke to expand this air to a pressure below atmospheric. The suction period D to A is an inefficient proportion of the stroke E to A.

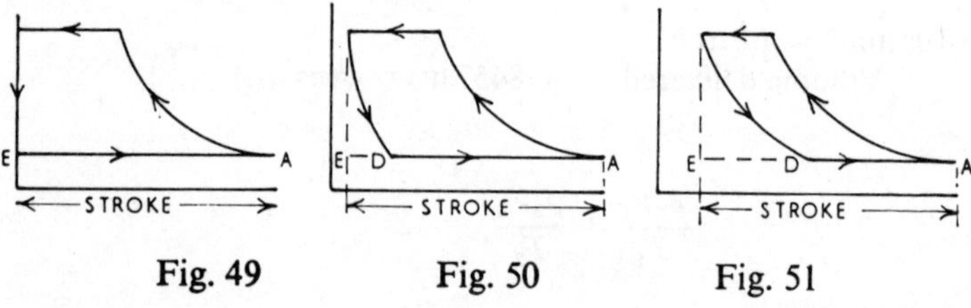

Fig. 49 Fig. 50 Fig. 51

The ratio between the volume of air drawn into the cylinder during the suction stroke and the full stroke volume swept out by the piston, is the *volumetric efficiency* of the compressor.

$$\text{Volumetric efficiency} = \frac{\text{volume of air drawn in per stroke}}{\text{stroke volume}}$$

and this is represented by the ratio $\dfrac{\text{DA}}{\text{EA}}$

WORK DONE PER CYCLE

NEGLECTING CLEARANCE

As previously shown, the area of a pV diagram represents work done, if the pressure is in kN/m^2 and the volume in cubic metres then the area of the pV diagram and the work done per cycle is in kJ.

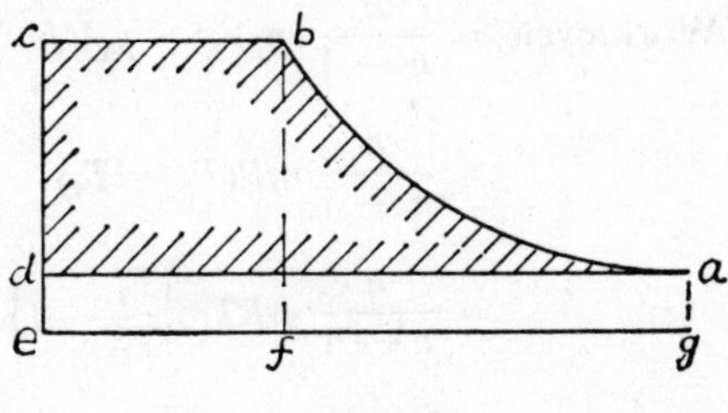

Fig. 52

Referring to Fig. 52 (neglecting clearance):

Net area = Net work done on the air per cycle
Area $abcd$ = Area $bcef$ + Area $abfg$ — Area $adeg$

Let p_2 = delivery pressure, V_2 = volume bc
p_1 = suction pressure, V_1 = volume da

Area under curve (see Chapter 6) $= \dfrac{p_2 V_2 - p_1 V_1}{n-1}$ therefore:

$$\text{Work/cycle} = p_2 V_2 + \frac{p_2 V_2 - p_1 V_1}{n-1} - p_1 V_1$$

$$= \frac{p_2 V_2(n-1) + p_2 V_2 - p_1 V_1 - p_1 V_1(n-1)}{n-1}$$

$$= \frac{n p_2 V_2 - p_2 V_2 + p_2 V_2 - p_1 V_1 - n p_1 V_1 + p_1 V_1}{n-1}$$

$$= \frac{n}{n-1}(p_2 V_2 - p_1 V_1) \quad \cdots \quad \cdots \quad \text{(i)}$$

$$p_1 V_1 = m R T_1 \qquad p_2 V_2 = m R T_2$$

$$\frac{T_2}{T_1} = \left\{\frac{p_2}{p_1}\right\}^{\frac{n-1}{n}} = \left\{\frac{V_1}{V_2}\right\}^{n-1}$$

By substitution, the work per cycle can be expressed in other terms to suit available data, such as:

$$\text{Work/cycle} = \frac{n}{n-1}(p_2V_2 - p_1V_1)$$

$$= \frac{n}{n-1}mR(T_2 - T_1)$$

$$= \frac{n}{n-1}mRT_1\left\{\frac{T_2}{T_1} - 1\right\} \quad \text{or}$$

$$\frac{n}{n-1}p_1V_1\left\{\frac{T_2}{T_1} - 1\right\}$$

$$= \frac{n}{n-1}mRT_1\left[\left\{\frac{p_2}{p_1}\right\}^{\frac{n-1}{n}} - 1\right] \quad \text{or}$$

$$\frac{n}{n-1}p_1V_1\left[\left\{\frac{p_2}{p_1}\right\}^{\frac{n-1}{n}} - 1\right]$$

$$= \frac{n}{n-1}mRT_1\left[\left\{\frac{V_1}{V_2}\right\}^{n-1} - 1\right] \quad \text{or}$$

$$\frac{n}{n-1}p_1V_1\left[\left\{\frac{V_1}{V_2}\right\}^{n-1} - 1\right]$$

Example. The cylinder of a single-acting compressor is 225 mm diameter and the stroke of the piston is 300 mm. It takes in air at 0·96 bar and delivers it at 4·8 bar and makes four delivery strokes per second. Assuming that compression follows the law $pV^n = $ constant, and neglecting clearance, calculate the theoretical power required to drive the compressor when the value of the index of the law of compression is, (i) 1·2, (ii) 1·35.

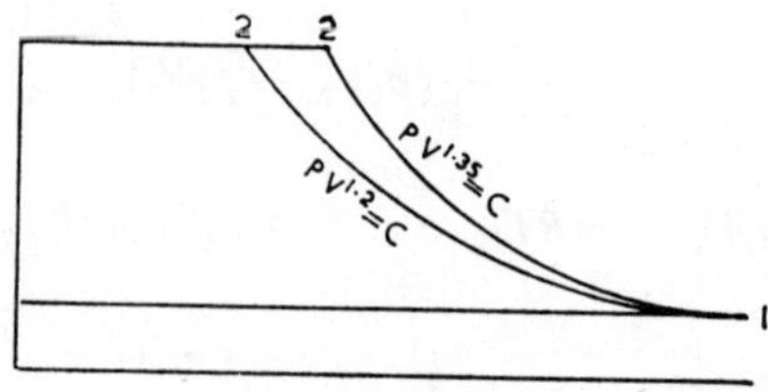

Fig. 53

$$V_1 = 0.7854 \times 0.225^2 \times 0.3 = 0.01193 \text{ m}^3$$

When $n = 1.2$:

$$p_1 V_1^{1.2} = p_2 V_2^{1.2}$$

$$0.96 \times 0.01193^{1.2} = 4.8 \times V_2^{1.2}$$

$$V_2^{1.2} = \frac{0.96 \times 0.01193^{1.2}}{4.8} = \frac{0.01193^{1.2}}{5}$$

$$V_2 = \frac{0.01193}{^{1.2}\sqrt{5}} = 0.00312 \text{ m}^3$$

Work done per cycle $= \dfrac{n}{n-1}(p_2 V_2 - p_1 V_1)$

$$= \frac{1.2}{0.2}(4.8 \times 10^2 \times 0.00312 - 0.96 \times 10^2 \times 0.01193)$$

$$= 2.114 \text{ kJ}$$

$$\begin{aligned}
\text{Power [kW]} &= \text{work done per second [kJ/s]} \\
&= \text{kJ/cycle} \times \text{cycle/s} \\
&= 2.114 \times 4 = 8.456 \text{ kW.} \quad \text{Ans. (i)}
\end{aligned}$$

Note that although any convenient units can be used for pressure and volume when they appear on both sides of an equation, such as pressure in bars, and volume represented by length on a pV diagram (as in the two previous examples), it is again emphasised that the pressure must be in kN/m² (1 bar $= 10^2$ kN/m²) and the volume in m³ when kJ of work is required by their product.

When $n = 1.35$:

$$p_1 V_1^{1.35} = p_2 V_2^{1.35}$$

$$0.96 \times 0.01193^{1.35} = 4.8 \times V_2^{1.35}$$

$$V_2 = \frac{0.01193}{^{1.35}\sqrt{5}} = 0.003621 \text{ m}^3$$

Work done per cycle $= \dfrac{n}{n-1}(p_2 V_2 - p_1 V_1)$

$$= \frac{1.35}{0.35}(4.8 \times 10^2 \times 0.003621 - 0.96 \times 10^2 \times 0.01193)$$

$$= 2.287 \text{ kJ}$$

$$\text{Power} = 2.287 \times 4 = 9.148 \text{ kW.} \quad \text{Ans. (ii)}$$

Note that the *mass* of air delivered per stroke is the same in each case. The greater volume indicated by the value of V_2 when n is 1·35 is due purely to the air being at a higher temperature at the end of compression than it is when $n = 1\cdot2$.

It can also be seen from the above that the nearer isothermal compression can be approached, the less work will be required to compress and deliver a given mass of air.

WORK DONE PER CYCLE

INCLUDING CLEARANCE

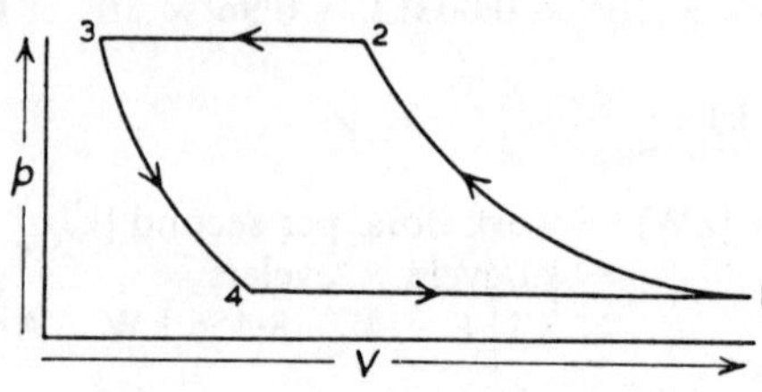

Fig. 54

Referring to Fig. 54, at the end of the compression and delivery stroke, the clearance space of volume V_3 is full of air at the delivery pressure of p_2 and temperature T_2. As the piston moves outward, the air does work on the piston as it expands to volume V_4 and the pressure falls to p_4, theoretically when p_4 is equal to p_1 the suction valves lift to begin the suction period.

The indicated work done by the piston on the air is therefore represented by the net area 1234.

Work/cycle

$$= \frac{n}{n-1}(p_2 V_2 - p_1 V_1) - \frac{n}{n-1}(p_3 V_3 - p_4 V_4)$$

$$= \frac{n}{n-1} p_1 V_1 \left[\left(\frac{p_2}{p_1}\right)^{\frac{n-1}{n}} - 1 \right] - \frac{n}{n-1} p_4 V_4 \left[\left(\frac{p_3}{p_4}\right)^{\frac{n-1}{n}} - 1 \right]$$

Taking the index of expansion to be equal to the index of compression and assuming $p_4 = p_1 \quad p_3 = p_2 \quad T_4 = T_1 \quad T_3 = T_2$ then,

Work/cycle

$$= \frac{n}{n-1} p_1 V_1 \left[\left(\frac{p_2}{p_1}\right)^{\frac{n-1}{n}} - 1 \right] - \frac{n}{n-1} p_1 V_4 \left[\left(\frac{p_2}{p_1}\right)^{\frac{n-1}{n}} - 1 \right]$$

$$= \frac{n}{n-1} p_1 (V_1 - V_4) \left[\left(\frac{p_2}{p_1}\right)^{\frac{n-1}{n}} - 1 \right]$$

$V_1 - V_4$ is the volume of air taken in per cycle at the pressure p_1 and temperature T_1, letting $m = $ mass of this air:

$$\text{Work/cycle} = \frac{n}{n-1} m R T_1 \left[\left(\frac{p_2}{p_1}\right)^{\frac{n-1}{n}} - 1 \right]$$

Comparing this with the work/cycle when there is no clearance, we see that for a given mass the work is the same in each case, thus, the work done per unit mass of air delivered is not affected by the size of the clearance space.

MULTI-STAGE COMPRESSION

By compressing the air in more than one stage and intercooling between stages, the practical compression curve can approach the isothermal more closely, hence reducing the work required per kg of air compressed. Figs. 55 and 56 show the pV diagrams neglecting clearance for two-stage and three-stage compression respectively, the shaded areas representing the work saved in each case compared with single-stage compression.

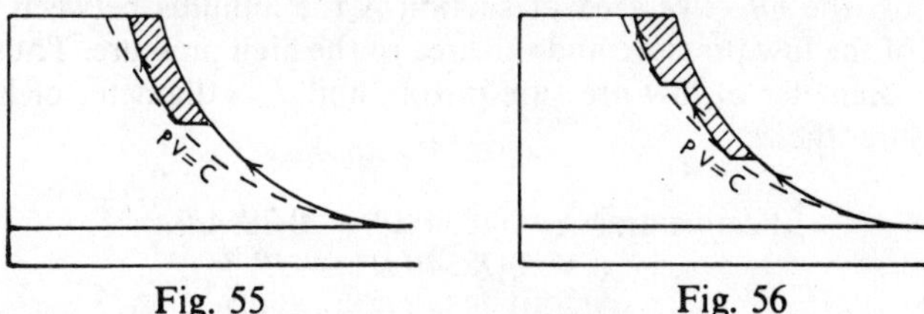

Fig. 55   Fig. 56

To obtain maximum efficiency from a multi-stage compressor, that is, to do the least work to compress and deliver a given mass of air, (i) the air should be intercooled to as near the initial temperature as possible, and (ii) the pressure ratio in each stage should be the same.

Multi-stage compressors may consist simply of separate compressor cylinders, or may be arranged in tandem. Fig. 57 shows a diagrammatic arrangement of a three-stage tandem air compressor.

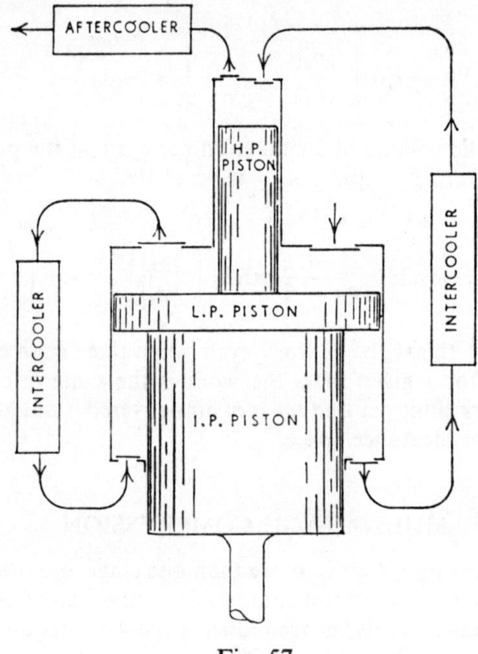

Fig. 57

It is important to note that, when calculating the volume of atmospheric air drawn into the low-pressure cylinder of a tandem compressor, the *effective* area of suction is the annulus between the area of the low pressure and the area of the high pressure. Thus, if D = diameter of low pressure piston, and d = diameter of high pressure, then:

$$\text{Effective area} = 0{\cdot}7854\,D^2 - 0{\cdot}7854d^2$$
$$= 0{\cdot}7854\,(D^2 - d^2)$$
$$= 0{\cdot}7854\,(D + d)\,(D - d)$$

Example. In a single-acting three-stage tandem air compressor, the piston diameters are 70, 335 and 375 mm diameter respectively, the stroke is 380 mm, and it is driven directly from a motor running at 250 rev/min. The suction pressure is atmospheric (1·013 bar) and the discharge is 45 bar gauge pressure. Assuming that the air delivered to the reservoirs is cooled down to the initial suction temperature and taking the volumetric efficiency as 90%, calculate the volume of compressed air delivered to the reservoirs per minute.

$$\text{Effective area of L.P.} = 0.7854\,(0.375^2 - 0.07^2)$$
$$= 0.7854 \times 0.445 \times 0.305 \text{ m}^2$$
$$\text{Stroke volume} = \text{area} \times \text{stroke}$$

$\therefore$ volume [m³] of air drawn in per stroke
$$= 0.7854 \times 0.445 \times 0.305 \times 0.38 \times 0.9$$
$$= 0.03646 \text{ m}^3$$

air drawn in per minute
$$= 0.03646 \times 250 = 9.115 \text{ m}^3$$

This is the volume of air taken into the compressor per minute at 1·013 bar. The volume delivered per minute at 46·013 bar is at the same temperature, therefore, since pressure × volume = constant, volume varies inversely as the absolute pressure, then:

$$\text{Volume delivered} = 9.115 \times \frac{1.013}{46.013}$$

$$= 0.2006 \text{ m}^3/\text{min.} \quad \text{Ans.}$$

Example. The cylinders of a single-acting two-stage tandem air compressor are 55 and 215 mm diameter respectively, and the stroke is 230 mm. It is connected to three air storage bottles of equal size of internal dimensions 300 mm. diameter and 1·5 m long overall with hemispherical ends. Taking atmospheric pressure as 1 bar and assuming a volumetric efficiency of 0·88, calculate the time required to pump up the bottles to a pressure of 31 bar gauge from empty, when running at 125 rev/min.

Length of cylindrical part of bottles
$$= \text{overall length} - \text{diameter}$$
$$= 1.5 - 0.3 = 1.2 \text{ m}$$

Volume of three bottles $= 3 \left(\frac{\pi}{6} \times 0.3^3 + \frac{\pi}{4} \times 0.3^2 + 1.2\right)$
$$= 3\pi \times 0.3^2 (0.05 + 0.3)$$
$$= 0.297 \text{ m}^3$$

To produce 0·297 m³ of air at an absolute pressure of $31 + 1 = 32$ bar, the volume of atmospheric pressure air (termed "free air") required, at the same temperature, is inversely proportional to the pressure:

Volume of atmospheric air

$$= 0.297 \times \frac{32}{1} = 9.504 \text{ m}^3$$

However, "empty" bottles contain their own volume of air at atmospheric pressure, therefore, volume of air to be taken into compressor from atmosphere

$$= 9.504 - 0.297 = 9.207 \text{ m}^3$$

Volume of air taken into compressor per minute
$$= 0.7854 \, (0.215^2 - 0.055^2) \times 0.23 \times 0.88 \times 125$$
$$= 0.8584 \text{ m}^3/\text{min}$$

$$\therefore \text{ Time required} = \frac{9.207}{0.8584} = 10.73 \text{ min. \quad Ans.}$$

TEST EXAMPLES 9

1. In a single-stage air compressor the diameter of the cylinder is 250 mm, the stroke of the piston is 350 mm, and the clearance volume is 900 cm³. Air is drawn in at a pressure of 0·986 bar and delivered at 4·1 bar. Taking the law of compression to be $pV^{1·25} = C$, calculate the distance travelled by the piston from the beginning of its compression stroke when the delivery valves open.

2. A compressor cylinder is 80 mm diameter and the stroke of the piston is 160 mm. The pressures at the beginning and end of compression are 1 bar and 5 bar respectively and the delivery valves open when the piston is 120 mm from the beginning of compression. If the law of compression is $pV^{1·3} = C$, find the clearance volume in cm³.

3. The internal volume of an air-storage vessel is 566 litres. Find the mass of air stored in it when the pressure is 40 atmospheres and the temperature is 26°C. Note: the mass of one cubic metre of air at atmospheric pressure and at 0°C is 1·293 kg.

4. A single-stage air compressor takes in 0·9 kg of air per minute at atmospheric pressure and 24°C and compresses it to 10 atmospheres of pressure, the law of compression being $pV^{1·2} = C$. Calculate (i) the volume of air drawn in per minute, (ii) the volume of air delivered per minute, and (iii) the temperature of the compressed air. Take the specific volume of air at atmospheric pressure and 0°C as 0·7734 m³/kg.

5. The volume of air in a single-stage compressor at the beginning of compression is 0·0424 m³ and its pressure is 1 bar. The clearance volume in the cylinder is 1230 cm³ and the discharge pressure is 8 bar. Calculate the volume of air delivered per stroke, in litres, if it is compressed (i) isothermally, (ii) adiabatically, taking $\gamma = 1·4$.

6. The diameter of the cylinder of an air compressor is 140 mm and the stroke of the piston is 200 mm. It takes in air at 0·98 bar and delivers it at 7 bar, the law of compression being $pV^{1·28} = C$. Calculate the volume of air delivered per stroke, in cubic centimetres, when the clearance volume is equal to (i) 10% of the stroke volume, (ii) 5% of the stroke volume.

7. The stroke of a compressor piston is 380 mm and the clearance volume is equal to 7% of the stroke volume. The pressure of the air at the beginning of the compression stroke is 0·99 bar. The delivery valves open when the piston has travelled 260 mm from the beginning of the compression stroke and the pressure of the air is then 4 bar. If compression follows the law $pV^n = C$, find the value of n.

8. A motor driven three-stage single-acting tandem air compressor runs at 170 rev/min. The high pressure and low pressure cylinder diameters are 75 mm and 350 mm diameter respectively, and the stroke is 300 mm. Find the time to pump up air reservoirs of 22 m³ total capacity from 19 bar gauge to 30 bar gauge, taking the volumetric efficiency as 0·92 and atmospheric pressure as 1·01 bar.

9. In a single-acting three-stage tandem air compressor, the diameters of the high pressure and low pressure pistons are 80 mm and 270 mm respectively, the stroke is 250 mm, and it runs at 110 rev/min. Taking atmospheric pressure as 1 bar and the volumetric efficiency as 0·9, find the time taken to pump up three air-bottles from atmospheric pressure to 40 bar gauge. The bottles have hemispherical ends and the internal dimensions of each are 450 mm diameter and 2·25 m long overall.

f 10. A single-stage single-acting air compressor, running at 5 rev/s, takes in 22·8 litres of atmospheric air per stroke at 1 bar and 25°C and compresses it to 5·7 litres according to the law $pV^{1·2} = C$. Calculate (i) the temperature of the air at the end of compression, (ii) the volume of air delivered per second to the reservoirs if it is cooled at constant pressure to its initial temperature through after-coolers, (iii) the mass of air delivered per second, (iv) the mass of sea water circulating through the cooler per second if the difference in temperature between inlet and outlet is 11°C. Take the values, R for air $= 0·287$ kJ/kg K, c_p for air $= 1·005$ kJ/kg K, spec. heat of sea water $= 4·12$ kJ/kg K.

f 11. In a two-stage compressor, 0·05 kg of air is taken in per stroke at 0·97 bar and 17°C and compressed in the first stage to 2·91 bar. It is then passed through the intercooler where it is cooled at constant pressure to its initial temperature. In the second stage, the air is further compressed to 7·3 bar and passed through the after-cooler where it is cooled at constant pressure to the initial temperature. Calculate (i) the percentage decrease in volume due to cooling at the end of each stage, (ii) the percentage decrease in volume if the air had been compressed in a single-stage from 0·97 to 7·3 bar and then finally cooled to its initial temperature of 17°C. Take R for air $= 0·287$ kJ/kg K and compression to follow the law $pV^n =$ constant where $n = 1·3$ in each stage.

f 12. The diameter of the cylinder of a single-acting air-compressor is 250 mm and the stroke of the piston is 400 mm. It runs at 2·5 rev/s taking in air at 1·01 bar and delivering it at 5·5 bar. Neglecting clearance, calculate the indicated power when the law of compression is $pV^{1·2} =$ a constant. Calculate also the percentage increase in power if the index of the law of compression increased to 1·3 without change of speed.

f 13. In a single-acting air compressor, the clearance volume is 364 cm^3, diameter of cylinder 200 mm, stroke 230 mm, and it runs at 2 rev/s. It receives air at 1 bar and delivers it at 5 bar, the index of compression and expansion being 1·28. Calculate the indicated power, the mean indicated pressure, and the volumetric efficiency.

f 14. The pressure and temperature of the air at the beginning of compression in a single-stage double-acting air compressor are 0·98 bar and 24°C, the pressure ratio is 4·55 and the index of compression and expansion is 1·25. The stroke is $1·2 \times$ cylinder diameter, clearance equal to 5% of the stroke, and the compressor runs at 8 rev/s. If it takes air from the atmosphere at 1·013 bar, 16°C, at the rate of 5 m^3/min, calculate (i) the compressor power, (ii) volumetric efficiency, (iii) the dimensions of the cylinder.

CHAPTER 10

STEAM

Under normal conditions of a steam engine plant, the engines consume steam at the same rate at which it is generated in the boilers, therefore the steam is generated at constant pressure.

The temperature at which water changes into steam, that is, the boiling point of the water, depends strictly upon the pressure exerted on it. A few examples are as follows:

PRESSURE [bar]	0·04	1·013	10	20	30
BOILING POINT [°C]	29	100	179·9	212·4	233·8

The temperature of the steam produced at any given pressure is the same as the temperature of the boiling point at the same pressure. Thus, if a boiler is working at a pressure of 30 bar the water begins to boil when its temperature reaches 233·8°C and the steam is generated at the same temperature.

DEFINITIONS. Steam which is in physical contact with the boiling water from which it has been generated is termed *saturated steam*, its temperature is the same as the boiling water and this is referred to as the *saturation temperature* (t_{sat}). The boiling water is referred to as *saturated liquid* because it is at the saturation temperature corresponding to that particular pressure (p_{sat}) under which evaporation takes place.

If the vapour produced is pure steam, it is called *dry saturated steam*. If the steam contains water (usually very fine particles held in suspension in the form of a mist) it is called *wet saturated steam* which is more often briefly referred to as *wet steam*. The quality of wet steam is expressed by its *dryness fraction* (x) which is the ratio of the mass of pure steam in a given mass of the steam-plus-water mixture.

In order to increase the temperature of steam above its saturation temperature, the steam must be taken away from direct contact with the water from which it was generated and heated externally, usually as it passes through superheater elements heated by high temperature flue gases. If the steam is still wet as it enters the superheaters, the particles of water must first be evaporated to produce dry steam before the temperature is increased. Steam whose temperature is higher than its saturation temperature (which depends upon its pressure) is termed *superheated steam*, and the difference in temperature between the superheated steam and its saturation temperature is referred to as the *degree of super-heat*.

FORMATION OF STEAM. Consider generating unit mass of steam in a boiler, at constant pressure p, from the initial stage of pumping in unit mass of the feed water.

Firstly, the water is forced into the boiler at pressure $p[\text{kN/m}^2]$ and, representing the volume of unit mass [1 kg] of the water as v_1 [m³] then the flow energy given to the water to enable it to enter the boiler is pv_1 [kJ].

Heat energy is then transferred to the water and its temperature is raised from feed temperature to evaporation temperature (boiling point). Most of the heat energy received by the water is to increase its temperature and therefore increase its stored up energy, called *internal energy* and represented by u, but some is used to do the work of increasing its volume from v_1 to v_f as it expands against the pressure p, the work done being $p\,(v_f - v_1)$. If u_1 represents the internal energy initially in the feed water, u_f the internal energy in the water at evaporation temperature, and h_f the heat energy supplied, then,

$$\begin{matrix} \text{heat energy} \\ \text{transferred} \end{matrix} = \begin{matrix} \text{increase in} \\ \text{internal energy} \end{matrix} + \begin{matrix} \text{external} \\ \text{work done} \end{matrix}$$

$$h_f = (u_f - u_1) + p\,(v_f - v_1)$$

On reaching the boiling point, the water evaporates into steam as it absorbs more heat energy, thus changing its physical state from a liquid into a vapour. During the evaporating process the pressure and temperature remain constant, and the steam produced is saturated steam at the same pressure and temperature as the boiling point of the water. During the change, considerable expansion takes place, the volume of the steam being many times greater than the volume occupied by the water from which it was generated. Some of the heat energy transferred during this stage is to evaporate the water into steam, freeing the molecules and increasing its internal energy from u_f to u_g. The remainder of the heat energy is used in expanding the volume from v_f to v_g against the pressure p, the work done being $p(v_g - v_f)$. Hence, if h_{fg} represents the heat energy transferred, then,

$$h_{fg} = (u_g - u_f) + p(v_g - v_f)$$

ENTHALPY. For a constant pressure process, the heat energy transferred, which is the sum of the internal energy and the work done, is termed *enthalpy*. Thus, enthalpy is an energy function of a constant pressure process and, in quantities per unit mass, is defined by the equation

$$h = u + pv$$

where h = specific enthalpy [kJ/kg]
　　　　u = specific internal energy [kJ/kg]
　　　　v = specific volume [m³/kg]
　　　　p = absolute pressure [kN/m²]

For a mass of m kg, the appropriate symbols are written in capitals:

$$H = U + pV$$

Suffixes distinguish between the properties of liquid and vapour, some of these have already been introduced in the previous chapter:

　　　　f refers to saturated liquid
　　　　g　　,,　　,,　　,,　　vapour
　　　　fg　　,,　　,, evaporation from liquid to vapour

Thus,

h_f = specific enthalpy of saturated liquid, that is, when the liquid is at its saturation temperature.

h_g = specific enthalpy of dry saturated vapour.

$h_{fg} = h_g - h_f$ = specific enthalpy of evaporation, that is, the change of specific enthalpy to evaporate unit mass of saturated liquid to dry saturated vapour.

h = specific enthalpy of either liquid or vapour at any other state.

Since there is no absolute value for internal energy, there is none for enthalpy, but changes of the property can be measured and the convenient datum of water at 0°C is chosen from which to measure enthalpy and other properties of water and steam. These values are set out in steam tables.

STEAM TABLES

Much experimental work has been done on the properties of steam and the results are published in various forms. Those used in this book are:

THERMODYNAMIC AND TRANSPORT PROPERTIES OF FLUIDS, SI UNITS arranged by Y. R. Mayhew and G. F. C. Rogers, published by Basil Blackwell, Oxford.

The student should have a copy by his side for reference while studying the following examples. Note that in these tables, p represents the pressure in bars. 1 bar = 10^5 N/m² = 10^2 kN/m².

The properties of saturated water and steam are tabulated on pages 2 to 5. The first table, on page 2, is for the convenience of reference to temperature up to 100°C. The remainder are set out with reference to pressure.

The properties of superheated steam at various pressures and temperatures are tabulated on pages 6 to 8.

WATER. The properties of liquids depend almost entirely on the temperature, therefore when looking up the properties of water, we look at its particular temperature and disregard its pressure.

Example. To read from the tables the specific enthalpy of water at,

$$
\begin{aligned}
&\text{(a) pressure of} \quad 1 \text{ bar and temperature} \quad 60°C \\
&\text{(b)} \quad \text{,,} \quad \text{,,} \quad 5 \text{ bar} \quad \text{,,} \quad \text{,,} \quad 88°C \\
&\text{(c)} \quad \text{,,} \quad \text{,,} \quad 10 \text{ bar} \quad \text{,,} \quad \text{,,} \quad 130°C \\
&\text{(d)} \quad \text{,,} \quad \text{,,} \quad 12 \text{ bar} \quad \text{,,} \quad \text{,,} \quad 188°C
\end{aligned}
$$

$$
\begin{aligned}
&\text{(a)} \quad 60°C: \text{see page 2,} \quad h = 251 \cdot 1 \text{ kJ/kg} \\
&\text{(b)} \quad 88°C: \quad \text{,,} \quad \text{,,} \quad 3, \quad h = 369 \quad \text{kJ/kg} \\
&\text{(c)} \quad 130°C: \quad \text{,,} \quad \text{,,} \quad 4, \quad h = 546 \quad \text{kJ/kg} \\
&\text{(d)} \quad 188°C: \quad \text{,,} \quad \text{,,} \quad 4, \quad h_f = 798 \quad \text{kJ/kg}
\end{aligned}
$$

Note that for case (d) we see that the temperature of 188°C is the saturation temperature corresponding to its pressure of 12 bar, therefore we use the suffix f to signify that the water is at its saturation temperature.

STEAM. Example. To find the temperature and specific enthalpy of saturated water and dry saturated steam at pressures of (a) 1·01325 bar, (b) 10 bar, (c) 20 bar, (d) 42 bar.

$$
\begin{aligned}
&\text{(a) } 1\cdot01325 \text{ bar, page 2,} \quad t_s = 100°C, \quad h_f = 419\cdot1, \quad h_g = 2675\cdot8 \\
&\text{(b) } 10 \text{ bar,} \quad \text{,,} \quad 4, \quad t_s = 179\cdot9°C, \quad h_f = 763, \quad h_g = 2778 \\
&\text{(c) } 20 \text{ bar,} \quad \text{,,} \quad 4, \quad t_s = 212\cdot4°C, \quad h_f = 909, \quad h_g = 2799 \\
&\text{(d) } 42 \text{ bar,} \quad \text{,,} \quad 5, \quad t_s = 253\cdot2°C, \quad h_f = 1102, \quad h_g = 2800
\end{aligned}
$$

WET STEAM. When steam is "wet" it means that it is a mixture of dry saturated steam and saturated water. Thus, if the dryness fraction is represented by x, then 1 kg of wet steam is composed of

x kg of dry saturated steam, and the remainder, which is $(1 - x)$ kg is water. Hence only x kg out of each 1 kg has been evaporated and the change of enthalpy to do this is $x \times$ specific enthalpy of evaporation, written xh_{fg}.

Therefore, specific enthalpy of wet steam of dryness fraction x is:
$$h = h_f + xh_{fg}$$

Example. Find the enthalpy and volume of 1 kg of wet saturated steam at a pressure of 0·2 bar and dryness fraction 0·85, and find also the additional heat energy required to completely dry the steam.

0·2 bar, page 3,
$$\begin{aligned}
h &= h_f + xh_{fg} \\
&= 251 + 0.85 \times 2358 \\
&= 251 + 2004.3 \\
&= 2255.3 \text{ kJ/kg} \quad \text{Ans. (i)}
\end{aligned}$$

v_g is the volume occupied by 1 kg of dry saturated steam, therefore the volume occupied by x kg of dry saturated steam is xv_g. The value of v_g is given on page 3 as 7·648 m³ for a pressure of 0·2 bar.

v_f is the volume occupied by 1 kg of saturated water, therefore the volume occupied by $(1 - x)$ kg of saturated water is $(1 - x)v_f$. The value of v_f is not listed on page 3 but a near figure can be found on page 10 which gives $v_f \times 10^2 = 0.1017$ for a pressure of 0·1992 bar.

Hence, volume of 1 kg of wet steam is
$$\begin{aligned}
v &= xv_g + (1 - x)v_f \\
&= 0.85 \times 7.648 + 0.15 \times 0.001\,017 \\
&= 6.5008 + 0.000\,152\,55
\end{aligned}$$

We see that the volume of the water is comparatively very small and therefore, for most practical cases can be neglected. Hence, the specific volume of wet steam is usually taken as,
$$v = xv_g$$
and for this example it is:
$$\begin{aligned}
v &= 0.85 \times 7.648 \\
&= 6.5008 \text{ m}^3/\text{kg} \quad \text{Ans. (ii)}
\end{aligned}$$

To dry the steam, $(1 - 0.85)$ kg of water is to be evaporated,
$$\begin{aligned}
\text{Increase of enthalpy} &= 0.15 \times h_{fg} \\
&= 0.15 \times 2358 \\
&= 353.7 \text{ kJ} \quad \text{Ans. (iii)}
\end{aligned}$$

Alternatively, it is the difference between the enthalpy of dry saturated steam and wet steam:

$$= h_g - h$$
$$= 2609 - 2255{\cdot}3$$
$$= 353{\cdot}7 \text{ kJ}$$

Example. If the specific enthalpy of wet saturated steam at a pressure of 11 bar is 2681 kJ/kg, find its dryness fraction.

The specific enthalpy of wet steam of dryness x is:

$$h = h_f + xh_{fg}$$

$$\text{hence,} \quad x = \frac{xh_{fg}}{h_{fg}} = \frac{h - h_f}{h_{fg}}$$

$$= \frac{2681 - 781}{2000} \text{ (page 4)}$$

$$= \frac{1900}{2000} = 0{\cdot}95 \quad \text{Ans.}$$

Example. Find the heat transfer required to convert 5 kg of water at a pressure of 20 bar and temperature 21°C into steam of dryness fraction 0·9, at the same pressure.

Water 21°C, page 2, $\quad h = 88 \text{ kJ/kg}$
Steam 20 bar, page 4,

$$h = h_f + xh_{fg}$$
$$= 909 + 0{\cdot}9 \times 1890 = 2610 \text{ kJ/kg}$$

Change of enthalpy, water to steam
$$= 2610 - 88 = 2522 \text{ kJ/kg}$$

Heat transfer $\quad =$ total change of enthalpy (for 5 kg)
$$= 5 \times 2522 = 12\,610 \text{ kJ} \quad \text{Ans.}$$

SUPERHEATED STEAM. As previously stated, steam is said to be *superheated* when its temperature is higher than the saturation temperature corresponding to its pressure. In practice, steam is superheated at constant pressure, the saturated steam being taken from the boiler steam space and passed through superheater tubes where it receives additional heat energy to dry the steam and raise its temperature. The properties of superheated steam are given on pages 6 to 8 of the tables.

Example. Find the specific volume and specific enthalpy of superheated steam (a) at a pressure of 10 bar and temperature 250°C, (b) at a pressure of 100 bar and temperature 400°C.

10 bar 250°C, page 7,

$$\left.\begin{array}{l} v = 0\cdot2328 \text{ m}^3/\text{kg} \\ h = 2944 \text{ kJ/kg} \end{array}\right\} \text{Ans. (a)}$$

100 bar 400°C, page 8,

$$\left.\begin{array}{l} v = 0\cdot02639 \text{ m}^3/\text{kg} \\ h = 3097 \text{ kJ/kg} \end{array}\right\} \text{Ans. (b)}$$

Note that at the higher pressures of 80 bar and upwards (page 8) the values of the specific volume are given as $v \times 10^2$, that is, the listed values are $10^2 \times$ actual value.

Example. Find the specific volume and specific enthalpy of steam at a pressure of 7 bar having 85°C of superheat, and determine the mean specific heat of the superheated steam over this range.

7 bar, page 7, $t_{\text{sat}} = 165°C, \quad h_g = 2764 \text{ kJ/kg}$
Temperature of superheated steam

$$= 165 + 85 = 250°C$$
$$v = 0\cdot3364 \text{ m}^3/\text{kg} \quad \text{Ans. (i)}$$
$$h = 2955 \text{ kJ/kg} \quad \text{Ans. (ii)}$$

Heat energy [kJ] added to superheat the steam

$$= \text{mass [kg]} \times c_\text{P}[\text{kJ/kg K}] \times \theta[\text{K}]$$
$$2955 - 2764 = 1 \times c_\text{P} \times 85$$
$$c_\text{P} = \frac{191}{85} = 2\cdot247 \text{ kJ/kg K} \quad \text{Ans. (iii)}$$

INTERPOLATION. If properties are required at an intermediate pressure or temperature to those given in the tables, an estimate can be made by taking values from the nearest listed pressure or temperature above and below that required and assume a linear variation between the two. The following demonstrate the usual methods.

Example. Find the enthalpy per kg of steam at a pressure of 8 bar and temperature 270°C.

Properties at 8 bar are given on page 7 but 270°C lies between the listed temperatures of 250 and 300°C.

$$\begin{aligned}
&\text{At 8 bar } 300°C, \quad h = 3057 \text{ kJ/kg} \\
&\text{,, ,, ,, } 250°C, \quad h = 2951 \\
\hline
&\text{Difference, increase } 50°C, \quad h = 106 \text{ increase}
\end{aligned}$$

Given temperature of 270°C is 20° higher than 250°C,

$$\text{difference for } 20° = \frac{20}{50} \times 106 = 42.4 \text{ kJ/kg}$$

$$\begin{aligned}
\therefore \text{ At 8 bar, } 270°C, \quad h &= 2951 + 42.4 \\
&= 2993.4 \text{ kJ/kg} \quad \text{Ans.}
\end{aligned}$$

Example. Find the specific enthalpy of steam at a pressure of 12 bar and temperature 350°C.

Values of pressures of 10 bar and 15 bar are given on page 7, temperature of 350°C is listed.

$$\begin{aligned}
&\text{At 10 bar } 350°C, \quad h = 3158 \text{ kJ/kg} \\
&\text{,, 15 bar } 350°C, \quad h = 3148 \\
\hline
&\text{Difference, increase } 5 \text{ bar} \quad\quad h = 10 \text{ kJ/kg } \textit{decrease}
\end{aligned}$$

Given pressure of 12 bar is 2 bar greater than 10 bar

$$\text{Difference for 2 bar increase} = \frac{2}{5} \times 10 = 4 \text{ kJ/kg less}$$

$$\begin{aligned}
\therefore \text{ at 12 bar } 350°C, h &= 3158 - 4 \\
&= 3154 \text{ kJ/kg.} \quad \text{Ans.}
\end{aligned}$$

Example. Find the specific enthalpy of steam at a pressure of 25 bar and temperature 380°C.

Nearest listed pressures and temperatures are 20 — 30 bar and 350 — 400°C, on page 7, this requires double interpolation.

$$\begin{aligned}
&\text{20 bar } 400°C, \quad h = 3248 \\
&\text{30 bar } 400°C, \quad h = 3231 \\
&\text{For 10 bar increase} \quad h = 17 \text{ kJ/kg decrease}
\end{aligned}$$

$$\text{,, } \quad 5 \text{ bar increase} \quad h = \frac{5}{10} \times 17 = 8.5 \text{ decrease}$$

$$\begin{aligned}
\therefore \text{ At 25 bar } 400°C, \quad h &= 3248 - 8.5 \\
&= 3239.5 \text{ kJ/kg} \quad \text{....} \quad\quad \text{....} \quad \text{(i)}
\end{aligned}$$

$$20 \text{ bar } 350°C, \quad h = 3138$$
$$30 \text{ bar } 350°C, \quad h = 3117$$
$$\text{For 10 bar increase} \quad h = 21 \text{ kJ/kg decrease}$$
$$\text{,, } 5 \text{ bar increase} \quad h = \frac{5}{10} \times 21 = 10.5 \text{ decrease}$$
$$\therefore \text{ At 25 bar } 350°C, \quad h = 3138 - 10.5$$
$$= 3127.5 \text{ kJ/kg} \quad .... \quad .... \quad \text{(ii)}$$
$$\text{From (i) 25 bar } 400°C, \quad h = 3239.5$$
$$\text{From (ii) 25 bar } 350°C, \quad h = 3127.5$$
$$\text{For increase of } 50°C, \quad h = 112 \text{ kJ/kg increase}$$
$$\text{,, \quad ,, \quad ,, } 30°C, \quad h = \frac{30}{50} \times 112 = 67.2 \text{ increase}$$
$$\therefore \text{ At 25 bar } 380°C, \quad h = 3127.5 + 67.2$$
$$= 3194.7 \text{ kJ/kg.} \quad \text{Ans.}$$

MIXING STEAM AND WATER

In Chapter 2, examples were given involving the mixing of ice and water, and metals in liquids at different temperature, and now further examples will be given to include steam. These have many practical applications as will be seen later.

The same principles apply, namely, that unless otherwise stated it is assumed there is no transfer of heat energy from or to an outside source during the mixing process. That is, the quantity of heat energy absorbed by the colder substance is equal to the quantity of heat energy lost by the hotter substance. In other words, when steam and water are mixed together, the total enthalpy of the water and steam before mixing is equal to the total enthalpy of the resultant mixture.

Example. 3 kg of wet steam at 14 bar and dryness fraction 0·95 are blown into 100 kg of water at 22°C. Find the resultant temperature of the water.

Water 22°C, page 2 of tables, $h = 92.2$ kJ/kg

Steam 14 bar, page 4,
$$h = h_f + x h_{fg}$$
$$= 830 + 0.95 \times 1960 = 2692 \text{ kJ/kg}$$

Before mixing:

$$\text{enthalpy of 100 kg water} = 100 \times 92.2 = \quad 9220 \text{ kJ}$$
$$\text{,, \quad ,, \quad 3 kg steam} = 3 \times 2692 = \quad 8076$$
$$\text{total enthalpy} = \overline{17296} \text{ kJ}$$

After mixing there are $100 + 3 = 103$ kg of water of total enthalpy 17 296 kJ, therefore specific enthalpy of resultant water is:

$$h = \frac{17296}{103} = 167 \cdot 9 \text{ kJ/kg}$$

From page 2 of tables, 167·5 kJ/kg (which is sufficiently close) corresponds to a water temperature of 40°C,

hence, resultant temperature = 40°C.   Ans.

An alternative method is to work on the principle of heat energy absorbed by the cold water is equal to the heat energy lost by the steam.

When heat is taken from the steam, it first condenses at its saturation temperature of 195°C (page 4 of tables, at 14 bar) into water at the same temperature, the heat energy lost being the change of enthalpy of condensation which is mass $\times xh_{fg}$. This water condensate then falls in temperature from 195°C to the final temperature of the mixture, the further heat energy lost being the product of the mass, specific heat, and fall in temperature.

The heat energy gained by the cold water is the product of its mass, specific heat, and rise in temperature.

Thus, let θ = final temperature of the water mixture, and assuming the mean specific heat of water to be 4·2 kJ/kg K, then:

Heat energy gained by water = Heat energy lost by steam

$$100 \times 4 \cdot 2\,(\theta - 22) = 3\{0 \cdot 95 \times 1960 + 4 \cdot 2(195 - \theta)\}$$
$$420\,\theta - 9240 = 3\{1862 + 819 - 4 \cdot 2\,\theta\}$$
$$420\,\theta - 9240 = 8043 - 12 \cdot 6\,\theta$$
$$432 \cdot 6\,\theta = 17283$$
$$\theta = 39 \cdot 95°C$$

Example. Steam is tapped from an intermediate stage of a steam turbine at a pressure of 2·5 bar, its dryness fraction being 0·95, and passed to a contact feed heater. The hotwell water at 48°C is pumped into the heater where it mixes with the heating steam. Estimate the temperature of the feed water leaving the heater when the amount of steam tapped off is 9% of the steam supplied to the engine.

Referring to Fig. 58, let the mass of steam supplied to the engine be 1 kg, then 0·09 kg is tapped off to the heater. This leaves $(1 - 0 \cdot 09) = 0 \cdot 91$ kg of steam to continue through the engine, into the condenser, as water into the hotwell, then

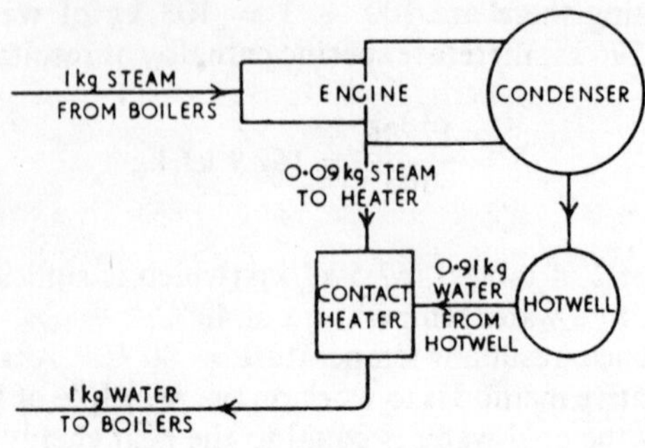

Fig. 58

pumped into the feed heater. Being a *contact* heater, the 0·09 kg of heating steam makes contact and mixes with the 0·91 kg of condensate, making 1 kg of feed water to be pumped back into the boilers.

Tables page 4, steam 2·5 bar, $h_f = 535$, $h_{fg} = 2182$
  „   „   2, water 48°C, $h = 200·9$
Enthalpy of steam entering heater:
$$= 0·09(h_f + x h_{fg})$$
$$= 0·09(535 + 0·95 \times 2182) = 234·7 \text{ kJ}$$
Enthalpy of water entering heater:
$$= 0·91 \times 200·9 = 182·8 \text{ kJ}$$
Total enthalpy of steam and water entering heater:
$$= 234·7 + 182·8 = 417·5 \text{ kJ}$$

Total enthalpy of water leaving heater is that of 1 kg of feed water, therefore the *specific* enthalpy of the feed water is 417·5 kJ/kg which we now look for in the tables and find the near figure of 417 kJ/kg on page 3 for a temperature of 99·6°C.

∴ Temperature of feed water $= 99·6°C$ Ans.

Alternatively for the latter part, we could assume a mean specific heat of 4·2 kJ/kg K for the feed water and take the enthalpy of water at θ°C as the heat energy to be transferred to it to raise its temperature from 0 to θ°C (mass $\times$ spec. heat $\times$ temp. rise), then:

$$1 \times 4·2 \times \theta = 417·5$$

$$\theta = \frac{417·5}{4·2} = 99·4°C$$

Example. Steam is bled from the main steam pipe line at a pressure of 17 bar and dryness fraction 0·98, to a surface feed heater, and the remainder passes through the engine. The condensate at 42°C from the engine condenser, and the drain from the feed heater, passes to the hotwell. Calculate the percentage of the main steam bled off to the heater if the temperature of the feed water to the boilers is 104·8°C.

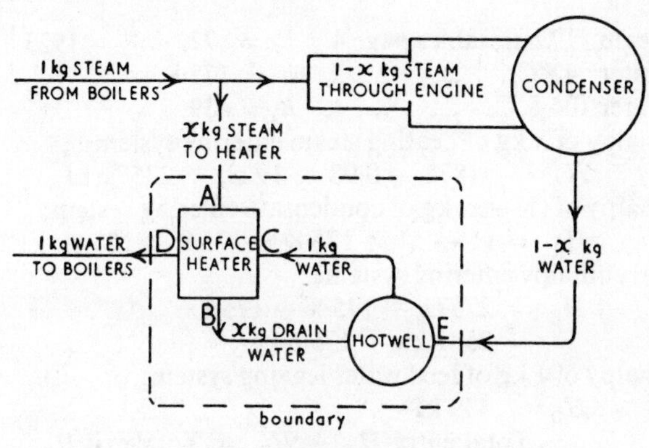

Fig. 59

The arrangement is illustrated in Fig. 59. Being a *surface* heater, there is no mixing of the heating steam and feed water. In this type, the water passes through nests of tubes which are heated on the outside by the steam.

Let H_A = enthalpy of steam entering heater.

H_B =   ,,   ,, drain water leaving heater.

H_C =   ,,   ,, water entering heater from hotwell.

H_D =   ,,   ,, water leaving heater (feed to boilers).

H_E =   ,,   ,, water entering hotwell from condenser.

Enthalpy of drain water and condensate entering hotwell
$$= \text{Enthalpy of water leaving hotwell}$$
$$H_B + H_E = H_C \quad \dots \quad \dots \quad \dots \quad \dots \quad \dots \quad \text{(i)}$$

Enthalpy of heating steam and hotwell water entering heater
$$= \text{Enthalpy of drain and feed water}$$
$$\text{leaving heater}$$
$$H_A + H_C = H_B + H_D \quad \dots \quad \dots \quad \dots \quad \dots \quad \text{(ii)}$$

Substituting value of H_C from (i) into (ii):
$$H_A + H_B + H_E = H_B + H_D$$
$$\therefore \ H_A + H_E = H_D$$

From this we can see that, since there is no heat energy transfer to or from an external source, the heater and hotwell with their connections can be considered as one common system as shown by the dotted boundary line on Fig. 59.

Let the mass of steam supplied from the boilers = 1 kg, then 1 kg of water is fed back to the boilers.

Let x kg of the supply steam be the amount bled off to the heater, then $(1 - x)$ kg is the mass of steam passed on to the engine.

Steam   17 bar, tables page 4. $h_f = 872$, $h_{fg} = 1923$
Water   42°C,   ,,   ,,   2, $h = 175 \cdot 8$
Water 104·8°C,   ,,   ,,   4, $h = 439$

Enthalpy of x kg of heating steam entering system:
$$H_A = x(872 + 0 \cdot 98 \times 1923) = 2757x \text{ kJ}$$

Enthalpy of $(1 - x)$ kg of condensate entering system:
$$H_E = (1 - x) \times 175 \cdot 8 = 175 \cdot 8 - 175 \cdot 8x \text{ kJ}$$

Total enthalpy entering system:
$$H_A + H_E = 2757x + 175 \cdot 8 - 175 \cdot 8x$$
$$= 2581 \cdot 2x + 175 \cdot 8 \text{ kJ} \qquad \text{.... ....} \quad \text{(i)}$$

Enthalpy of 1 kg of feed water leaving system:
$$H_D = 439 \text{ kJ} \qquad \text{.... ....} \quad \text{(ii)}$$

$$\text{Total entry } H_A + H_E = \text{Total exit } H_D$$
$$2581 \cdot 2x + 175 \cdot 8 = 439$$
$$2581 \cdot 2x = 263 \cdot 2$$
$$x = 0 \cdot 102$$

As a percentage of the main steam supply,
$$\text{Bled steam} = 10 \cdot 2\% \quad \text{Ans.}$$

THROTTLING OF STEAM

Throttling is the process of passing the steam through a restricting orifice or a partially opened valve which causes a "wire-drawing" effect and reduces the pressure. There is not sufficient time for any appreciable heat transfer to take place between the steam and its surrounds and, as there are no moving parts, expansion takes place freely and no work is done by the steam. The difference in the velocity of the steam before and after is sufficiently small to be negligible. Consequently, if there is no transfer or conversion of heat energy, then there is no change in the enthalpy of the steam, that is, enthalpy after throttling is equal to enthalpy before throttling.

One important application of throttling takes place in the

steam reducing valve. This is a specially designed valve connected to the high pressure steam range, the restricted valve opening is spring controlled and set to maintain a near steady reduced pressure at the outlet, the reduced pressure steam being more suitable for some pumps and other auxiliaries.

Example. Steam is passed through a reducing valve and reduced in pressure from 20 bar to 7 bar. Find the effect of throttling on the temperature and quality of the steam if the high pressure steam was (a) wet, having a dryness fraction of 0·9, (b) dry saturated, (c) superheated to a temperature of 300°C.

Tables pages 4 and 7,

7 bar, $t_s = 165°C$, $h_f = 697$, $h_{fg} = 2067$, $h_g = 2764$

20 bar, $t_s = 212·4°C$, $h_f = 909$, $h_{fg} = 1890$, $h_g = 2799$, $h_{sup} = 3025$

(a) 20 bar, dryness 0·9:

$$h = h_f + xh_{fg}$$
$$= 909 + 0·9 \times 1890 = 2610$$

2610 is less than h_g for 7 bar, therefore reduced steam is wet

$$\therefore \text{7 bar, dryness } x, \cdot h = h_f + xh_{fg}$$

$$\text{enthalpy before} = \text{enthalpy after}$$
$$2610 = 697 + x \times 2067$$
$$2067x = 1913$$
$$x = 0·9253$$

Effect of throttling:

Temperature is reduced from 212·4°C to 165°C
Dryness is increased from 0·9 to 0·9253 $\Big\}$ Ans. (a)

(b)   20 bar dry sat. $h_g = 2799$
7 bar dry sat. $h_g = 2764$
enthalpy before = enthalpy after

hence, enthalpy at 7 bar is 2799, which is 35 kJ/kg higher than its h_g and is therefore superheated.

From page 7 of tables, for 7 bar, $h = 2799$, lies between $h_g = 2764$ at 165°C and $h = 2846$ at 200°C by interpolation,

$$7 \text{ bar}, \quad h = 2846 \quad \text{for } 200°C$$
$$h = 2764 \quad \text{for } 165°C$$
$$\text{difference in } h = \quad 82 \text{ kJ for } 35°C$$
$$\text{difference for } 35 \text{ kJ} = \frac{35}{82} \times 35 = 14\cdot94° \text{ say } 15°$$

$\therefore$ reduced pressure steam has 15°C of superheat, its temperature being $165 + 15 = 180°C$.

Effect of throttling:

Temperature is reduced from 212·4°C to 180°C \
Steam has 15°C of superheat $\Big\}$ Ans. (b)

(c) 20 bar, temperature 300°C, $h = 3025$

(steam has $300 - 212\cdot4 = 87\cdot6°C$ of superheat)
$h = 3025$ is more than h_g for 7 bar, reduced pressure steam is therefore superheated.

$$\text{enthalpy before} = \text{enthalpy after}$$

From page 7 of tables, for 7 bar, $h = 3025$, lies between $h = 2955$ at 250°C and $h = 3060$ at 300°C.

$$h = 3060 \quad \text{for } 300°C$$
$$h = 2955 \quad \text{for } 250°C$$
$$\text{difference in } h = \quad 105 \text{ kJ for } 50°C$$

$\therefore$ difference in temperature for $(3025 - 2955) = 70$ kJ is

$$\frac{70}{105} \times 50 = 33\cdot3°C$$

$$\therefore \text{ temperature} = 250 + 33\cdot3 = 283\cdot3°C$$
$$\text{Degree of superheat} = 283\cdot3 - 165$$
$$= 118\cdot3°C.$$

Effect of throttling:

Temperature is reduced from 300°C to 283·3°C \
Degree of superheat is increased from 87·6°C to 118·3°C $\Big\}$ Ans. (c)

Note that in every case the effect of throttling is to (i) reduce the pressure, (ii) reduce the temperature, (iii) increase the quality, that is, either to produce drier steam or to increase the degree of superheat.

With regard to finding the degree of superheat at the lower pressure, if the mean specific heat of superheated steam over this range were given, the degree of superheat could be obtained from,

$$\text{Heat energy to raise temperature} = \text{mass} \times \text{spec. heat} \times \text{temp. rise}$$

therefore, for 1 kg,

$$\text{Increase of enthalpy above } h_g = c_P \times \text{degree of superheat}$$

or,

$$h - h_g = c_P(t - t_{sat})$$

where t = temperature of the steam

Taking the last case (c) as an example, if the mean specific heat was given as 2·21 kJ/kg K:

$$h - h_g = c_P \times \text{deg. of superheat}$$

$$\text{deg. of superheat} = \frac{3025 - 2764}{2·21} = 118·1°C$$

THROTTLING CALORIMETER

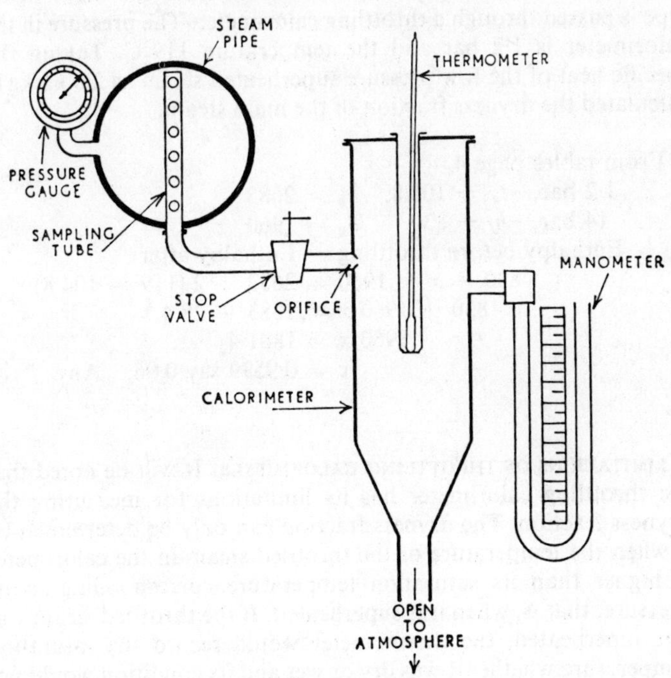

Fig. 60

The *throttling calorimeter* is an instrument for measuring the dryness fraction of steam. It consists of a calorimeter admitting steam through a very small orifice near the top and an exit open to atmosphere at the bottom. It is fitted with a thermometer pocket so that the temperature of the steam in the calorimeter can be observed, and a mercurial manometer to measure the pressure. The arrangement is illustrated diagrammatically in Fig. 60. In the steam pipe the tendency is for the drier steam to flow along the upper part of the pipe bore and the wetter (and heavier) steam along the lower part. An attempt is therefore made to obtain a reasonably true representative sample by fitting the sampling tube vertically in the pipe, the perforations facing the steam flow.

The principle of operation depends upon the fact that if steam is reduced in pressure by a throttling process, the enthalpy after throttling is equal to the enthalpy before throttling, as previously explained.

Example. Steam at a pressure of 14 bar from the main steam pipe is passed through a throttling calorimeter. The pressure in the calorimeter is 1·2 bar and the temperature 119°C. Taking the specific heat of the low pressure superheated steam as 2·0 kJ/kg K calculated the dryness fraction of the main steam.

From tables page 4,

$$1\cdot2 \text{ bar}, \quad t_s = 104\cdot8, \quad h_g = 2683$$
$$14 \text{ bar}, \quad h_f = 830, \quad h_{fg} = 1960$$
$$\text{Enthalpy before throttling} = \text{Enthalpy after}$$
$$830 + x \times 1960 = 2683 + 2\,(119 - 104\cdot8)$$
$$830 + 1960\,x = 2683 + 28\cdot4$$
$$1960\,x = 1881\cdot4$$
$$x = 0\cdot9599 \text{ say } 0\cdot96. \quad \text{Ans.}$$

LIMITATIONS OF THROTTLING CALORIMETER. It will be noted that the throttling calorimeter has its limitations for measuring the dryness fraction. The dryness fraction can only be determined by it when the temperature of the throttled steam in the calorimeter is higher than its saturation temperature corresponding to its pressure, that is, when it is superheated. If the throttled steam was not superheated, the thermometer would record the saturation temperature whether it was dry or wet and its condition would not be known.

For instance, referring to the previous example of main steam at 14 bar, if the throttled steam at 1·2 bar had no superheat, the temperature recorded by the thermometer would be the saturation temperature of 104·8°C and there is no knowing whether this throttled steam is dry or wet. If it was just dry and saturated, the equation would be:

$$\text{Enthalpy before} = \text{Enthalpy after}$$
$$830 + x \times 1960 = 2683$$
$$x = 0.9454$$

This is the limiting dryness fraction of the main stream. Above this, the steam in the calorimeter after throttling would be superheated and the dryness fraction of the main steam could be calculated as shown. Below this, the calorimeter steam would be wet of unknown dryness fraction, therefore the dryness of the main steam could not be obtained because of two unknowns in the one equation.

SEPARATING CALORIMETER

When very wet steam samples are to be tested, an approximate value for the dryness fraction may be determined by passing the steam through a *separating calorimeter*.

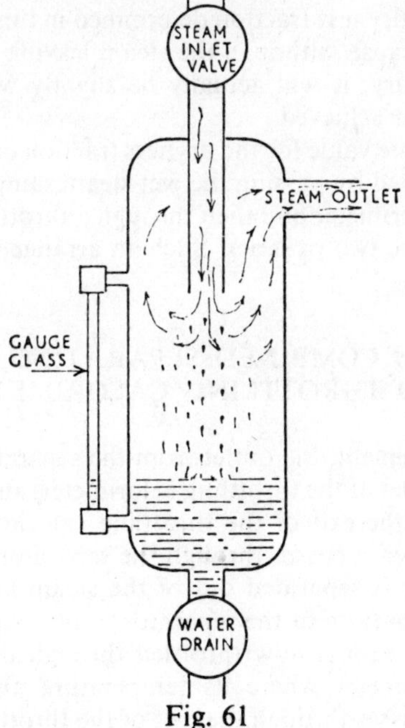

Fig. 61

This consists of a vessel with an inlet at the top leading into an open-ended internal tube, an outlet near the top, and a drain valve at the bottom. When steam is passed through it changes its direction of flow from the internal tube to the outlet which causes the heavier water particles to be thrown out of suspension by centrifugal action. The separated water collects in the bottom of the separator which is later drained off and measured, let this be m_1. The remaining steam passes out through the outlet and is led to a small condenser where it is condensed into water, collected and measured, let this be m_2:

From the sample of wet steam taken,

m_1 = mass of water separated from the sample

m_2 = mass of steam (assumed dry) remaining

Dryness fraction

$$= \frac{\text{mass of pure steam}}{\text{total mass of wet steam (steam + water mixture)}}$$

$$= \frac{m_2}{m_2 + m_1}$$

However, the dryness fraction determined in this manner is only approximate because, although the steam leaving the separator is assumed to be dry, it will actually be slightly wet since perfect separation is not achieved.

A more accurate value for the dryness fraction of very wet steam may be determined by passing the wet steam sample first through a separating calorimeter and then through a throttling calorimeter by connecting the two in series. Such an arrangement is called a:

ƒ COMBINED SEPARATING AND THROTTLING CALORIMETER

In this arrangement, the outlet from the separator is connected directly to the inlet of the throttling calorimeter, and the condenser is connected to the exit of the throttling calorimeter. When the sample of wet steam passes through the separator, some (but not all) of the water is separated out of the steam and this water is collected in the bottom of the separator to be measured (m_1). The partially dried steam is now throttled through the orifice in the throttling calorimeter where its temperature and pressure are noted. It then passes through the exit of the throttling calorimeter and into the condenser where it is condensed into water and collected for measuring (m_2).

Using the throttling calorimeter as previously explained, the dryness fraction of the steam leaving the separator and entering the throttling calorimeter is calculated, let this be x_2. Then each kg of steam passed into the throttling calorimeter contains x_2 kg of dry steam, and in m_2 kg there are $x_2 m_2$ kg of dry steam.

The total mass of the sample is $m_2 + m_1$. Since the dryness fraction of the sample is the mass of dry steam divided by the total mass of the wet steam sample, then,

$$\text{Dryness fraction of sample} = x = \frac{x_2 m_2}{m_2 + m_1}$$

Note that $\dfrac{m_2}{m_2 + m_1}$ would be the apparent dryness fraction if the separating calorimeter only was used, let this be represented by x_1 then we have:

$$x = x_1 \times x_2$$

In words this is:

$$\text{Dryness fraction} = \left\{ \begin{array}{l} \text{dryness} \\ \text{fraction as} \\ \text{measured} \\ \text{by separating} \\ \text{calorimeter} \end{array} \right\} \times \left\{ \begin{array}{l} \text{dryness} \\ \text{fraction as} \\ \text{measured} \\ \text{by throttling} \\ \text{calorimeter} \end{array} \right\}$$

f Example. Steam at a pressure of 8 bar was tested by passing a sample through a combined separating and throttling calorimeter. The mass of water collected in the separator was 0·25 kg and the mass of condensate collected from the condenser after throttling was 2·5 kg. The pressure of the steam in the throttling calorimeter was 1·1 bar and its temperature 106·8°C. Taking the specific heat of the superheated steam after throttling as 2·0 kJ/kg K, find the dryness fraction of the steam sample.

From tables page 4,
$$8 \text{ bar}, \quad h_f = 721, \qquad h_{fg} = 2048$$
$$1 \cdot 1 \text{ bar}, \quad t_s = 102 \cdot 3°C, \quad h_g = 2680$$

Dryness fraction by separating calorimeter:

$$x_1 = \frac{m_2}{m_2 + m_1} = \frac{2 \cdot 5}{2 \cdot 5 + 0 \cdot 25} = 0 \cdot 909$$

Dryness fraction by throttling calorimeter:
Enthalpy before throttling = Enthalpy after
$$721 + x_2 \times 2048 = 2680 + 2(106 \cdot 8 - 102 \cdot 3)$$
$$x_2 \times 2048 = 1968$$
$$x_2 = 0 \cdot 961$$
Dryness fraction of steam sample:
$$x = x_1 \times x_2 = 0 \cdot 909 \times 0 \cdot 961 = 0 \cdot 8736. \quad \text{Ans.}$$

AIR IN STEAM CONDENSERS

It was stated in Chapter 5 under Dalton's law of partial pressures that the pressure exerted in a vessel occupied by a mixture of gases, or a mixture of gases and vapours, is equal to the sum of the pressures that each would exert if it alone occupied the whole volume of the vessel. The pressure exerted by each gas is termed a *partial pressure*.

Applying this to steam condensers which contain a mixture of leakage air and steam:

$$\text{Total pressure of mixture} = \text{partial pressure due to air} + \text{partial pressure due to steam}$$

This enables an estimate to be made on the amount of air leakage into condensers.

Example. The pressure in a condenser is $0 \cdot 12$ bar and the temperature is $43 \cdot 8°C$. If the internal volume of the condenser is $8 \cdot 5 \text{ m}^3$, estimate the mass of air in the condenser, taking R for air $= 0 \cdot 287 \text{ kJ/kg K}$. If the dryness fraction of the steam is $0 \cdot 9$, calculate the mass of steam present.

Since there is always some water condensate in the condenser and the steam is in contact with this water, then the steam is at its saturation temperature and the pressure of the steam in the condenser depends upon this temperature. From the tables, page 3, we see that for a saturation temperature of $43 \cdot 8°C$ the pressure of the steam is $0 \cdot 09$ bar:

Partial pressure due to air = total pressure — steam pressure
$$= 0 \cdot 12 - 0 \cdot 09$$
$$= 0 \cdot 03 \text{ bar} = 3 \text{ kN/m}^2$$
From $pV = mRT$, mass of air present:

$$m = \frac{3 \times 8 \cdot 5}{0 \cdot 287 \times (43 \cdot 8 + 273)}$$

$$= 0 \cdot 2804 \text{ kg}. \quad \text{Ans. (i)}$$

Also from tables page 3, for saturation temperature of 43·8°C, specific volume of dry saturated steam $V_g = 16·2$ m³/kg.

For wet steam,

$$v = xv_g = 0·9 \times 16·2 = 14·58 \text{ m}^3/\text{kg}$$

$\therefore$ mass of steam in the condenser volume of 8·5 m³

$$= \frac{8·5}{14·58} = 0·583 \text{ kg.} \quad \text{Ans. (ii)}$$

TEST EXAMPLES 10

1. Steam at a pressure of 9 bar is generated in an exhaust gas boiler from feed water at 80°C. If the dryness fraction of the steam is 0·96, determine the heat transfer per kg of steam produced.

2. If wet saturated steam at 8 bar requires 82 kJ of heat per kg of steam to completely dry it, what is the dryness fraction of the steam?

3. Wet saturated steam at 17 bar, dryness 0·97, is produced from feed water at 85°C. Find (i) the heat energy supplied per kg, (ii) the percentage less heat energy required per kg of steam at the same pressure and quality if the feed temperature is increased to 110°C.

4. A turbo-generator is supplied with superheated steam at a pressure of 30 bar and temperature 350°C. The pressure of the exhaust steam from the turbine is 0·06 bar with a dryness fraction of 0·88. (i) Calculate the enthalpy drop per kg of steam through the turbine. (ii) If the turbine uses 0·25 kg of steam per second, calculate the power equivalent of the total enthalpy drop.

5. Steam enters the superheaters of a boiler at a pressure of 20 bar and dryness 0·98, and leaves at the same pressure at a temperature of 350°C. Find (i) the heat energy supplied per kg of steam in the superheaters, and (ii) the percentage increase in volume due to drying and superheating.

6. Wet steam at 17 bar, 0·95 dry, is produced in a boiler from feed water at 95°C. If the temperature of the feed remains constant, find the percentage extra heat energy required to produce (i) dry saturated steam, (ii) superheated steam at 350°C, at the same pressure.

7. One kg of wet steam at a pressure of 8 bar and dryness 0·94 is expanded until the pressure is 4 bar. If expansion follows the law $pV^n = $ a constant, where $n = 1·12$, find the dryness fraction of the steam at the lower pressure.

8. Two boilers of equal evaporative capacities generate steam at the same pressure of 15 bar to a common pipe line. One boiler produces superheated steam at 250°C and the other produces wet

steam. If the mixture is just dry and saturated, find the dryness fraction of the wet steam from the second boiler.

9. 0·2 m³ of steam at 7 bar and 0·95 dry is expanded to a pressure of 4·5 bar. Calculate the dryness fraction of the expanded steam if the expansion follows the law (i) pV = a constant, (ii) $pV^{1·14}$ = a constant.

10. A closed system consisting of one kg of superheated steam at 20 bar and 400°C is cooled at constant volume until the pressure is 12 bar. Find the condition of the steam at the lower pressure, and (ii) the transfer of heat energy from the steam.

11. 2·5 litres of superheated steam at 25 bar and 400°C is expanded in an engine to a pressure of 0·1 bar when its dryness fraction is 0·9. Find the final volume of the steam.

12. 1·5 kg of wet steam at a pressure of 5 bar and dryness 0·95 is blown into 70 litres of water at 12°C. Find the final temperature of the mixture.

13. Exhaust steam from an engine passes into a condenser at a pressure of 0·12 bar and dryness 0·88. The temperature of the condensate from the condenser is 40°C. The circulating water enters the condenser at 12°C and leaves at 29°C. Calculate (i) the mass of circulating water per kg of steam condensed. If 500 tonne of circulating water pass through the condenser per hour, find (ii) the power equivalent of the heat energy pumped overboard.

14. In an experiment to determine the dryness fraction of steam, a sample at a pressure of 1·1 bar was blown into a vessel containing 10 kg of water at 15°C. The final mass of water in the vessel was 10·75 kg and the final temperature 55°C. Find the dryness fraction of the steam, taking the water equivalent of the vessel as 0·45 kg.

15. Dry saturated steam at a pressure of 2·4 bar is tapped off the inlet branch of a low pressure turbine to supply heating steam in a contact feed heater. The temperature of the feed water inlet to the heater is 42°C and the outlet is 99·6°C. Find the percentage mass of steam tapped off.

16. 10% of the mass of steam supplied to an engine is bled off at an intermediate stage and led to a surface feed heater. The pressure of the steam to the heater is 2·9 bar and the drain from the heater passes as water to the hotwell. The temperature of the condensate from the condenser is 40°C and the temperature of the feed water to the boilers is 100°C. Calculate the dryness fraction of the steam supplied to the heater.

17. Wet saturated steam at 16 bar and dryness 0·98 enters a reducing valve and is throttled to a pressure of 8 bar. Find the dryness fraction of the reduced pressure steam.

18. A throttling calorimeter was fitted to a pipe carrying steam at 12 bar in order to measure the dryness fraction. The pressure in the calorimeter was 1·2 bar and its temperature was 116°C. Taking the specific heat of the superheated steam in the calorimeter as 2·0 kJ/kg K, find the dryness fraction of the main steam.

f19. A combined separating and throttling calorimeter was connected to a main steam pipe carrying steam at 15 bar and the following data recorded:

Mass of water collected in separator = 0·55 kg
Mass of condensate after throttling = 10 kg
Press. of steam in throttling calorimeter = 1·1 bar
Temp. ,, ,, ,, ,, ,, = 111°C

Taking the specific heat of the throttled superheated steam as 2·0 kJ/kg K, find the dryness fraction of the main steam.

f20. A vessel of volume 8·7 m³ contains air and dry saturated steam at a total pressure of 0·06 bar and temperature 29°C. Taking R for air as 287 J/kg K, calculate the mass of steam and the mass of air in the vessel.

f21. The pressure and temperature in a steam condenser is 0·08 bar and 39°C, and the dryness fraction of the steam is 0·86. Find the mass ratio of steam to air present in the condenser. R for air = 0·287 kJ/kg K.

f22. During the process of raising steam in a boiler, when the pressure was 1·9 bar gauge the temperature inside the boiler was 130°C, and when the pressure was 6·25 bar gauge the temperature was 165°C. If the volume of the steam space is constant at 4·25 m³ calculate the masses of steam and air present in each case. Take R for air = 0·287 kJ/kg K, atmospheric pressure = 1 bar, and assume the steam is dry in each case.

CHAPTER 11

f ENTROPY
(First Class Students Only)

We have seen that mechanical energy, being the product of two quantities, can be represented as an area on rectangular axes. Fig. 62 is a diagram in which the ordinates represent pressure in kN/m^2 and the abscissae represent volume in m^3, the area enclosed represents mechanical work, $kN/m^2 \times m^3 = kNm = kJ$. The availability to do work depends upon the magnitude of the pressure.

In a similar manner, heat energy can be represented as an area. The availability to transfer heat energy depends upon the temperature and therefore the ordinates of a heat energy diagram represent absolute temperature. The abscissae are termed *entropy*. The symbol for entropy is S, change of entropy may be written $\triangle S$ and the units are kilojoule per kelvin [kJ/K]. Specific entropy, which is entropy per unit mass, is represented by the symbol s and the units are kilojoule per kilogramme kelvin [kJ/kg K].

Thus, a diagram whose area represents heat energy, the ordinates represent absolute temperature (T) and the abscissae represent entropy (S), is referred to as a temperature-entropy (T-S) diagram, as shown in Fig. 63.

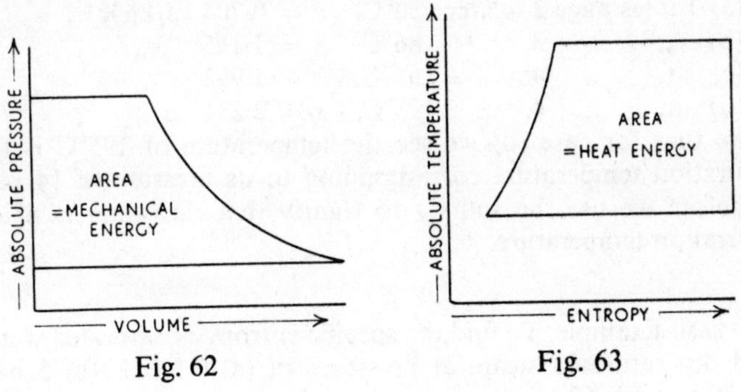

Fig. 62   Fig. 63

ENTROPY OF WATER AND STEAM

Subscripts to distinguish the specific entropy of saturated liquids and vapours are used as for other properties, explained in the previous chapter, thus:

$$s_f = \text{specific entropy of saturated liquid}$$
$$s_g = \quad ,, \quad ,, \quad ,, \quad ,, \quad \text{vapour}$$
$$s_{fg} = s_g - s_f = \text{specific entropy of evaporation from liquid to vapour}$$
$$s = \text{specific entropy of either liquid or vapour at any other state.}$$

As for internal energy and enthalpy, the values of specific entropy measured from the datum of water at $0°C$ are listed in the steam tables. The student should now refer to the booklet used in the previous chapter, Thermodynamic and Transport Properties of Fluids, SI units, by Y. R. Mayhew and G. F. C. Rogers, while studying the following exercises.

WATER. The properties of liquids depend almost entirely on temperature, therefore when looking up the properties of water we look for those against its particular temperature, disregarding its pressure unless it is saturated.

Example. To read from the tables the specific entropy of water at (a) pressure of 1 bar and temperature $50°C$

(b)	,,	,, 4 bar ,,	,,	$86°C$
(c)	,,	,, 8 bar ,,	,,	$165°C$
(d)	,,	,, 14 bar ,,	,,	$195°C$

(a) Tables page 2, water $50°C$, $s = 0.704$ kJ/kg K
(b) ,, ,, 3, ,, $86°C$, $s = 1.145$,,
(c) ,, ,, 4, ,, $165°C$, $s = 1.992$,,
(d) ,, ,, 4, ,, $195°C$, $s_f = 2.284$,,

Note that for case (d) we see the temperature of $195°C$ is the saturation temperature corresponding to its pressure of 14 bar, therefore we use the suffix f to signify that the water is at its saturation temperature.

STEAM. Example. To find the specific entropy of saturated water and dry saturated steam at pressures of (a) 0.1 bar, (b) 5 bar, (c) 20 bar, (d) 50 bar.

(a) Tables page 3, 0.1 bar, $s_f = 0.649$, $s_g = 8.149$
(b) ,, ,, 4, 5 bar, $s_f = 1.860$, $s_g = 6.822$
(c) ,, ,, 4, 20 bar, $s_f = 2.447$, $s_g = 6.340$
(d) ,, ,, 5, 50 bar, $s_f = 2.921$, $s_g = 5.973$

Example. To find the specific entropy of wet steam (a) pressure 0·2 bar, dryness 0·8, (b) pressure 6 bar, dryness 0·9, (c) pressure 44 bar, dryness 0·95.

Specific entropy of wet steam of dryness fraction x is obtained in a similar manner as for specific enthalpy, previously explained:

$$s = s_f + x s_{fg}$$

(a) Tables page 3, $\quad s = 0·832 + 0·8 \times 7·075 = 6·492$
(b)   „   „   4, $\quad s = 1·931 + 0·9 \times 4·830 = 6·278$
(c)   „   „   5, $\quad s = 2·849 + 0·95 \times 3·180 = 5·870$

Example. Using the following formula, calculate the entropy per kg of superheated steam at a pressure of 20 bar and temperature 300°C, taking the mean specific heat of water as 4·25 kJ/kg K and the mean specific heat of superheated steam as 2·58 kJ/kg K. Compare the result with the value of entropy given in the tables.

$$s = c_w \ln \frac{T_{sat}}{273} + \frac{h_{fg}}{T_{sat}} + c_{sup} \ln \frac{T}{T_{sat}}$$

where $\ln$ = abbreviation for natural logarithm
$\quad c_w$ = mean specific heat of water
$\quad c_{sup}$ = mean specific heat of superheated steam
$\quad T_{sat}$ = saturation temperature absolute
$\quad T$ = superheated steam temperature absolute

Superheated steam temp. $= 300 + 273 = 573$ K
Tables page 4, 20 bar, $\quad h_{fg} = 1890$
Saturation temp. $= 212·4 + 273 = 485·4$ K

$$s = c_w \ln \frac{T_{sat}}{273} + \frac{h_{fg}}{T_{sat}} + c_{sup} \ln \frac{T}{T_{sat}}$$

$$= 4·25 \ln \frac{485·4}{273} + \frac{1890}{485·4} + 2·58 \ln \frac{573}{485·4}$$

$$= 4·25 \ln 1·778 + 3·894 + 2·58 \ln 1·18$$
$$= 4·25 \times 0·5756 + 3·894 + 2·58 \times 0·1655$$
$$= 2·446 + 3·894 + 0·427$$
$$= 6·767 \text{ kJ/kg K.} \quad \text{Ans. (i)}$$

From tables, page 7, press. 20 bar, temp. 300°C:
$$s = 6·768 \text{ kJ/kg K.} \quad \text{Ans. (ii)}$$

A TEMPERATURE-ENTROPY DIAGRAM is a clear method of illustrating the process of generating steam.

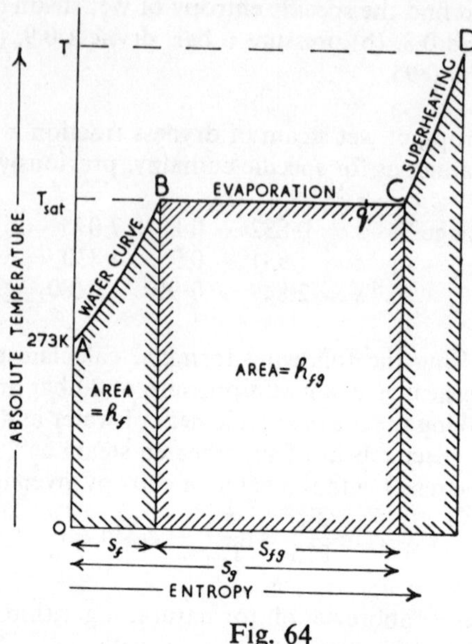

Fig. 64

Referring to Fig. 64, consider 1 kg of water commencing at a temperature of 0°C = 273 K, as heat energy is transferred to the water its temperature rises and its entropy is increased, the heating of the water up to its boiling point is represented by the water curve AB. At any point on this water curve, the ordinate is the absolute temperature of the water, and the abscissa is its specific entropy above that for water at 0°C. The heat energy transferred to the water to raise its temperature is the area of the diagram under the curve AB.

On reaching the saturation temperature corresponding to its pressure, point B, evaporation commences and this continues at constant temperature as the entropy is increased, therefore the evaporation process appears as a horizontal straight line. At point C evaporation is complete and dry saturated steam is produced. The latent heat energy of evaporation is represented by the area of the diagram under the evaporation line BC. Any intermediate point, such as q, along the evaporation line represents incomplete evaporation, that is, the steam is wet, the dryness fraction (x) is represented by the ratio Bq/BC and the heat energy of evaporation is the area under that part of the evaporation line, which is xh_{fg}.

When further heat energy is transferred to dry saturated steam at constant pressure, it becomes superheated and the temperature rises as its entropy increases. This is shown by the superheat curve CD, the area of the diagram under this curve is the heat energy added to superheat the steam.

TEMPERATURE-ENTROPY CHART

A temperature-entropy chart is a complete diagram drawn up within the range of temperatures normally required in steam engine practice. A simplified chart is illustrated in Fig. 65. For convenience, the temperature scale is in °C. Complete temperature-entropy charts to scale are obtainable at most students bookshops.

The WATER LINE gives the relation between the temperature of the water and its specific entropy. Curve AB on Fig. 64 is part of this line.

The DRY SATURATED STEAM LINE is produced by drawing a curve through points scaled off to the right from the water line, representing the increase in entropy for complete evaporation of one kg of water into steam at the various temperatures. Any point on this line represents dry saturated steam conditions at the given level of temperature. Point C on Fig. 64 lies on this line. Inside this line the steam is wet, outside this line the steam is superheated.

SUPERHEAT LINES slope steeply upwards from points on the dry saturated steam line. These give the relation between the temperature of the steam and its specific entropy when the steam is superheated at constant pressure. CD on Fig. 64 is one of these lines.

LINES OF CONSTANT DRYNESS FRACTIONS are drawn within the wet steam region, that is, between the two main boundary lines of the water and dry saturated steam. These are plotted by scaling off the values of xs_{fg} from the water line. For instance, if 0·9 of the values of specific entropy of complete evaporation for a series of temperatures were scaled off from the water line, the line drawn through these points is the dryness fraction line of 0·9. Lines for dryness fractions of 0·8, 0·7, etc., down to 0·1 are obtained in a similar manner.

CONSTANT VOLUME LINES are also included. Each constant

TEMPERATURE-ENTROPY CHART FOR STEAM
0·1 0·2 0·3 0·4 0·5 0·6 0·7 0·8 0·9
TEMPERATURE °C
300 280 260 240 220 200 180 160 140 120 100 80 60 40 20 0
PRESSURE IN BARS
80 70 60 50 40 30 20 10 5 1 0·1
WATER LINE
0·025
0·05
CONSTANT VOLUME LINES
0·1
0·2
0·4
0·8
DRYNESS FRACTION LINES
DRY SATURATED STEAM LINE
SUPERHEAT LINES
SPECIFIC ENTROPY kJ/kg K
0 1 2 3 4 5 6 7 8 9

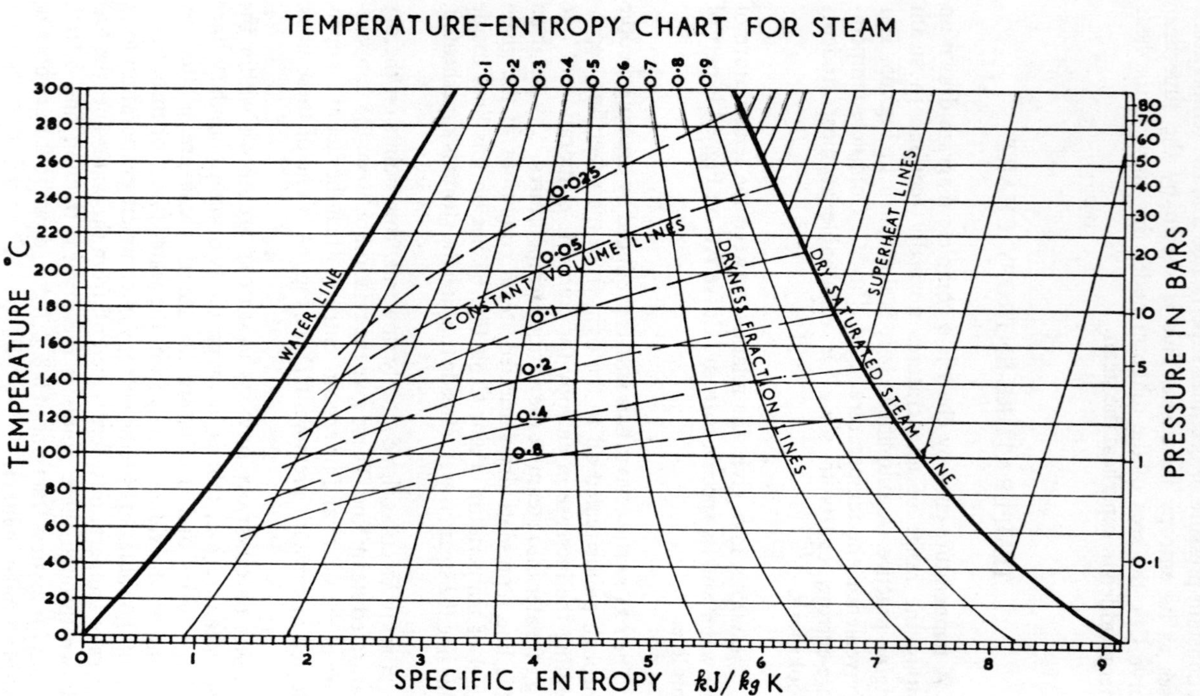

Fig. 65

volume line represents one particular volume occupied by one kg of saturated steam under varying conditions of quality from dry to very wet.

Taking a volume of 0·1 m³ as an example, we read from the tables that 0·1 m³ is the volume occupied by 1 kg of dry saturated steam when its temperature is 212·2°C. The curve for a volume of 0·1 m³ therefore begins at this point on the dry saturated steam line at the level of 212·2°C of temperature.

Now taking some other temperature, say 195°C, we note that the specific volume of dry saturated steam is given as 0·1408 m³/kg hence, for 1 kg of steam at 195°C to occupy a volume of 0·1 m³ it must be wet, and its dryness fraction will be 0·1 ÷ 0·1408 = 0·7103. At another temperature, say 165°C, v_g = 0·2728 m³/kg, for 1 kg of steam at 165°C to occupy a volume of 0·1 m³ the dryness fraction will be 0·1 ÷ 0·2728 = 0·3665. Plotting these three corresponding points of temperature and dryness fractions and drawing a curve through them, the constant volume line for 0·1 m³ is produced.

The above neglects the volume occupied by the water in the wet steam as being negligible, and obviously more than three plotted points are needed to obtain a true curve. Constant volume lines for various volumes are given covering such a range as likely to be met in practice.

ISOTHERMAL AND ISENTROPIC PROCESSES

During an isothermal process the temperature remains constant, therefore an isothermal operation is represented on a temperature-entropy diagram by a straight horizontal line.

During an adiabatic process, no heat transfer takes place to or from the surroundings, a reversible adiabatic process is *isentropic*, that is, it takes place without change of entropy. Therefore, the entropy after the process is equal to the entropy before, hence, an isentropic process appears on the temperature-entropy diagram as a straight vertical line.

As an example, Fig. 43 in Chapter 8 shows the pressure-volume diagram of the Carnot cycle for a gas, this on a temperature-entropy diagram appears as shown in Fig. 66, and, referring to this:

$$\text{Efficiency} = \frac{\text{heat supplied} - \text{heat rejected}}{\text{heat supplied}}$$

$$= 1 - \frac{\text{heat rejected}}{\text{heat supplied}}$$

$$= 1 - \frac{T_2(s_C - s_D)}{T_1(s_B - s_A)}$$

$$= 1 - \frac{T_2}{T_1} \text{ (as given in Chapter 8)}$$

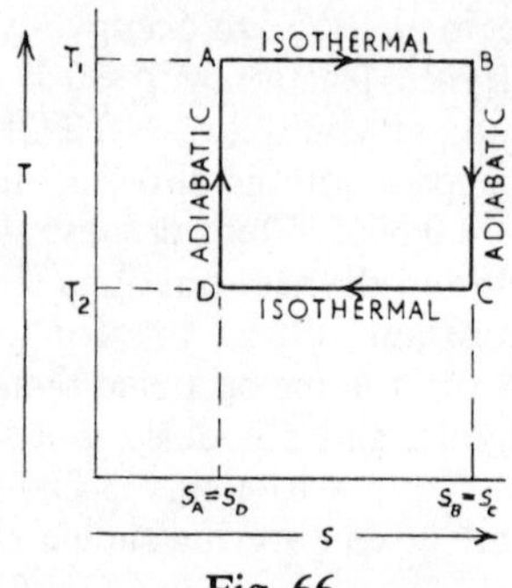

Fig. 66

EFFECT OF ISENTROPIC EXPANSION ON QUALITY. By drawing vertical lines down from the initial steam temperature on the temperature-entropy chart to represent isentropic expansion, it will be seen that dry saturated steam becomes wet during isentropic expansion, steam of normal wetness becomes wetter, and steam which is very wet initially becomes slightly dryer.

Example. Steam at a pressure of 14 bar expands isentropically to a pressure of 2·7 bar, find the dryness fraction at the end of expansion if the dryness of the steam at the beginning of expansion is (a) 1·0, (b) 0·8, (c) 0·6, (d) 0·4.

Tables page 4,

$$14 \text{ bar}, \quad s_f = 2\cdot284 \quad s_{fg} = 4\cdot185 \quad s_g = 6\cdot469$$
$$2\cdot7 \text{ bar}, \quad s_f = 1\cdot634 \quad s_{fg} = 5\cdot393$$

Entropy after expansion = Entropy before

(a)
$$1\cdot634 + x \times 5\cdot393 = 6\cdot469$$
$$x \times 5\cdot393 = 4\cdot835$$
$$x = 0\cdot8964. \quad \text{Ans. (a)}$$

(b)
$$1{\cdot}634 + x \times 5{\cdot}393 = 2{\cdot}284 + 0{\cdot}8 \times 4{\cdot}185$$
$$x \times 5{\cdot}393 = 3{\cdot}998$$
$$x = 0{\cdot}7415. \quad \text{Ans. (b)}$$

(c)
$$1{\cdot}634 + x \times 5{\cdot}393 = 2{\cdot}284 + 0{\cdot}6 \times 4{\cdot}185$$
$$x \times 5{\cdot}393 = 3{\cdot}161$$
$$x = 0{\cdot}5861. \quad \text{Ans. (c)}$$

(d)
$$1{\cdot}634 + x \times 5{\cdot}393 = 2{\cdot}284 + 0{\cdot}4 \times 4{\cdot}185$$
$$x \times 5{\cdot}393 = 2{\cdot}324$$
$$x = 0{\cdot}4309. \quad \text{Ans. (d).}$$

The above results can be checked graphically on the temperature-entropy chart, by drawing a vertical line from the temperature level corresponding to 14 bar, which is 195°C, and the appropriate dryness, down to the temperature level corresponding to 2·7 bar, which is 130°C, and reading off the dryness fraction at the lower temperature.

Example. Superheated steam at a pressure of 20 bar and temperature 300°C is expanded isentropically. (i) At what pressure will the steam be just dry and saturated? (ii) What will be the dryness fraction if it is expanded to a pressure of 0·04 bar?

Tables page 7, 20 bar 300°C, $s = 6{\cdot}768$
„   „   3, 0·04 bar, $s_f = 0{\cdot}422$ $s_{fg} = 8{\cdot}051$

(i) Since entropy at 20 bar 300°C is 6·768, dry saturated steam at a lower pressure is to have the same entropy.

Tables page 4 gives $s_g = 6{\cdot}761$ for 6 bar, therefore, pressure when steam is dry sat. = practically 6 bar. Ans. (i).

(ii)
$$\text{Final entropy} = \text{Initial entropy}$$
$$0{\cdot}422 + x \times 8{\cdot}051 = 6{\cdot}768$$
$$x \times 8{\cdot}051 = 6{\cdot}346$$
$$x = 0{\cdot}7881. \quad \text{Ans. (ii)}$$

ENTHALPY-ENTROPY CHART

A diagram drawn with specific entropy of steam as the base, and the vertical ordinates representing specific enthalpy, is a most useful chart for checking results of calculations. Fig. 65a is a simplified chart. Complete full-size enthalpy-entropy charts to scale are

ENTHALPY-ENTROPY CHART FOR STEAM
CONSTANT TEMPERATURE LINES t
CONSTANT PRESSURE LINES bar
SATURATION LINE
CONSTANT DRYNESS LINES
SPECIFIC ENTROPY kJ/kgK
SPECIFIC ENTHALPY kJ/kg

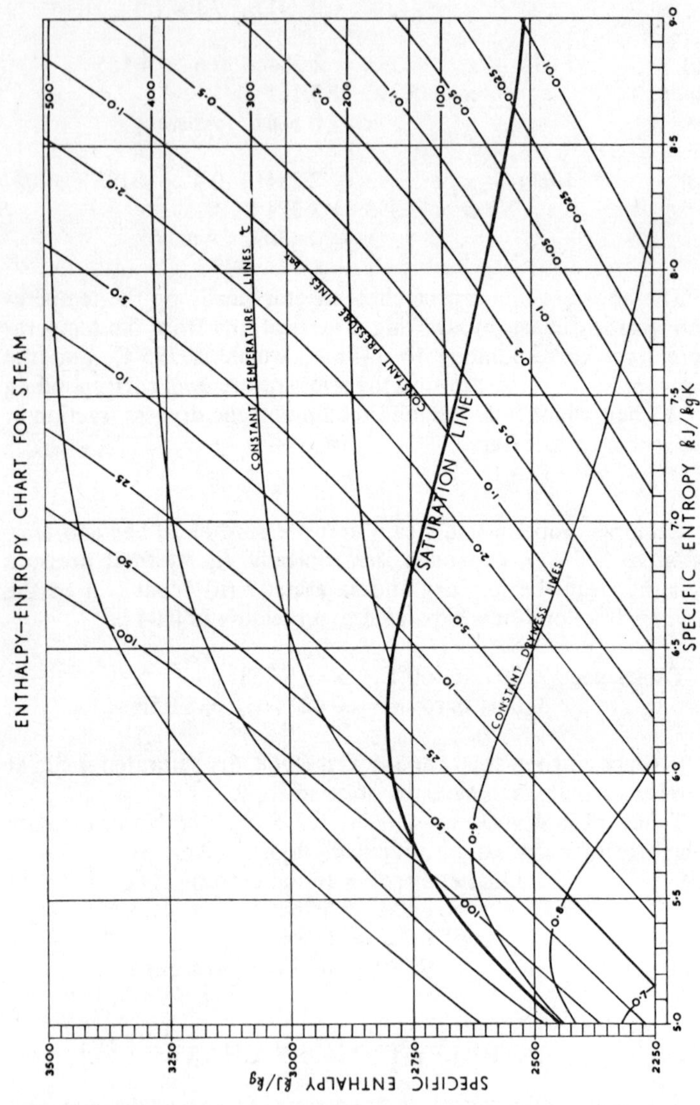

Fig. 65a

obtainable at most students bookshops and the student is advised to obtain a copy and use it whenever applicable.

That part under the saturation line is referred to as the wet steam region, and above it represents superheated conditions. Lines of constant dryness are plotted and drawn within the wet steam region, and constant temperature lines are drawn in the superheat region. Lines of constant pressure are drawn throughout the complete diagram.

Note the simplicity of reading off the enthalpy drop and final condition when steam is expanded isentropically, isentropic expansion being represented by drawing a straight vertical line from the initial pressure and condition of the steam to the final pressure line. Throttling, being a constant enthalpy process, is represented by a straight horizontal line from the initial pressure and condition to the final pressure line.

TEST EXAMPLES 11

f1. Find the entropy of one kilogramme of wet steam at a pressure of 17 bar if the dryness fraction is 0·95.

f2. Find the specific entropy of wet steam of temperature 195°C and dryness 0·9.

f3. Using the following formula, calculate the entropy per kg of superheated steam at 15 bar and 300°C, taking the mean specific heats of water and superheated steam as 4·24 kJ/kgK and 2·43 kJ/kgK respectively. Compare the result with the value given in the steam tables.

$$s = c_W \ln \frac{T_{sat}}{273} + \frac{h_{fg}}{T_{sat}} + c_{sup} \ln \frac{T}{T_{sat}}$$

where c_W = mean specific heat of water
c_{sup} = mean specific heat of superheated steam
T_{sat} = saturation temperature absolute
T = superheated steam temperature absolute

f4. Dry saturated steam at 5·5 bar is expanded isentropically to 0·2 bar, find the dryness fraction of the steam at the end of expansion.

f5. Superheated steam at 17 bar and 350°C is expanded in an engine and the final pressure is 1·7 bar. If the expansion is isentropic, find the dryness fraction of the expanded steam.

f6. One kilogramme of dry saturated steam at 22 bar is throttled to a pressure of 7 bar, then expanded at constant entropy to a pressure of 1·4 bar. Calculate (i) the degree of superheat at 7 bar, (ii) the increase in entropy, (iii) the dryness fraction at 1·4 bar.

CHAPTER 12

STEAM RECIPROCATING ENGINES

A steam engine is a machine which converts the heat energy in steam into mechanical energy. There are two distinct types of steam engine, namely the reciprocating steam engine and the steam turbine.

Because of their poor efficiency and low power/weight ratio in comparison with internal combustion engines and turbines, reciprocating steam engines are no longer being built. However, since there are still some in use, descriptive notes and calculations involved in the running of these engines are given here. Design work, such as the determination of cylinder diameters, valve diagrams, etc. in the previous edition of this volume are now excluded.

The general construction of a single cylinder steam reciprocating.engine is shown in Fig. 67. Steam from the boiler is admitted through the engine stop valve into the valve chest which houses the slide valve, which may be a flat type, or of circular section like a bobbin and termed a piston valve. Ports in the valve chest connect to passages leading to the top and bottom of the cylinder and a central port communicates with the exhaust range. The function of the slide valve is to open and close the ports to steam and exhaust so that the piston is pushed up and down in the cylinder. The reciprocating motion of the piston is converted into a rotary motion at the crank shaft by the usual connecting mechanism consisting of piston rod, connecting rod, and crank. The slide valve is actuated by an eccentric-sheave and strap, eccentric-rod and slide valve rod, this mechanism gives the slide valve a reciprocating motion from the rotary motion of the eccentric-sheave which is keyed to the crank shaft.

The piston rod passes through a neck bush, stuffing box and gland in the bottom of the cylinder to prevent leakage of steam as the rod reciprocates through the bottom of the cylinder. Likewise with the valve rod through the bottom of the valve chest. In high pressure engines, specially designed metallic packing is fitted in the stuffing boxes. In low pressure engines, soft packing is inserted which is squeezed up by the gland until steam tightness is obtained.

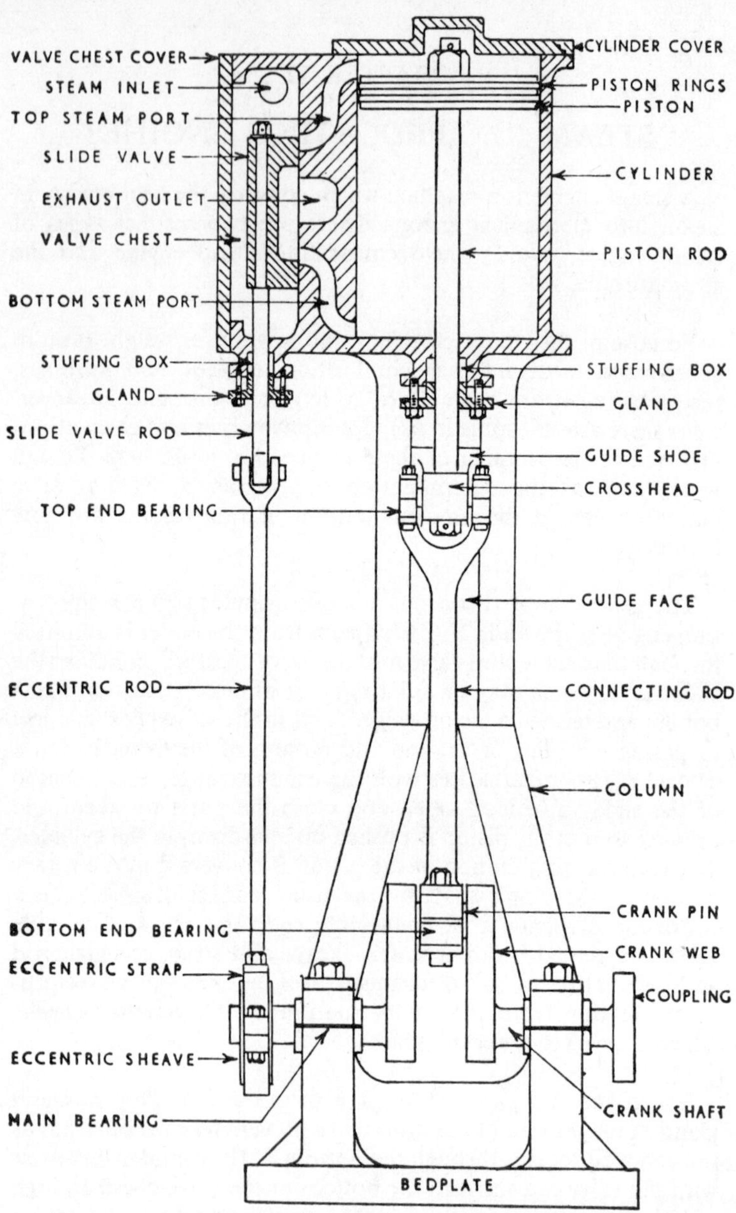

SINGLE CYLINDER STEAM ENGINE

Fig. 67

ACTION OF THE SLIDE VALVE

A simple slide valve is made of cast iron, rectangular in shape with a recess in its face and a hollow boss on the back through which is fitted the slide valve rod. A section through the axis of the valve resembles the letter D and is often referred to as the simple D slide valve.

Diagrammatic sketches of a simple slide valve showing its position relative to the piston and crank at various points in the cycle are given in Fig. 68 to which the following notes refer.

(i) Steam begins to be admitted to the top of the cylinder to start the piston on its outward stroke.

(ii) The port is at its maximum opening to steam at the top of the cylinder, piston moving outward.

(iii) The steam supply to the top of the cylinder is now cut off by the slide valve, the piston having travelled about half stroke, and the piston continues to be pushed forward by the expansion of the steam as it falls in pressure.

(iv) The steam is now at a low pressure and beginning to be released from the cylinder into the exhaust range. It will now exhaust for about three-quarters of the return stroke.

(v) The valve now closes the port to stop exhaust, the remainder of the steam trapped in the cylinder is compressed as the piston completes its return stroke and acts as a cushion to bring the reciprocating parts quietly to rest at the end of the stroke.

A similar cycle of events comprising steam admission, expansion, exhaust and compression takes place on the bottom side of the piston to make the engine "double-acting".

In high pressure steam engines a piston valve is fitted instead of a flat slide valve. Its function and operation is the same but, being cylindrical, has not the disadvantage of a flat valve which is subjected to a heavy steam load pressing it against the valve face and necessitating considerable force to move it against friction. Another advantage is that steam can be admitted "on the inside"

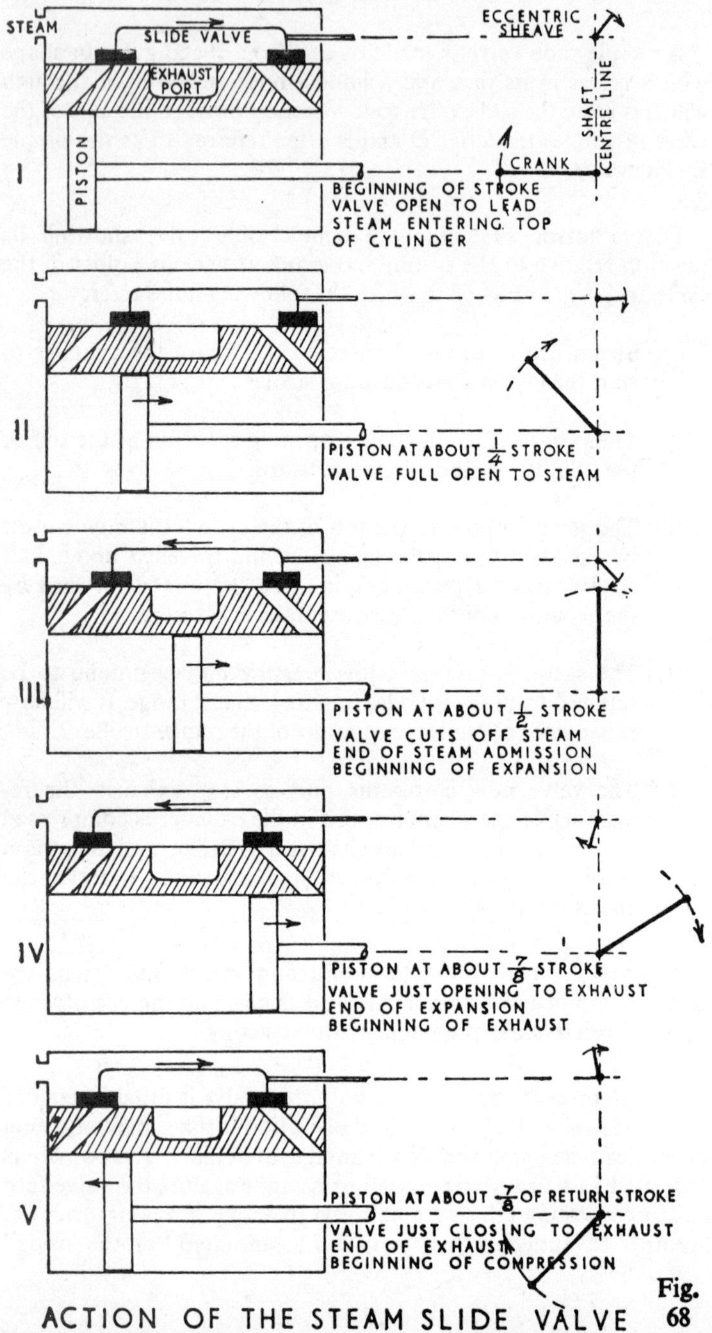

ACTION OF THE STEAM SLIDE VALVE

Fig. 68

of a piston valve, *i.e.* in the place of the exhaust of a slide valve, with exhaust over the outside edges; thus, by making the top of the valve slightly larger in diameter than the bottom, the extra upward steam thrust can compensate for the weight of the valve and gear to reduce wear on its bearings. Further, with steam admission on the inside, the valve rod gland is subjected only to the pressure of the exhaust steam and is therefore easier to keep steam tight.

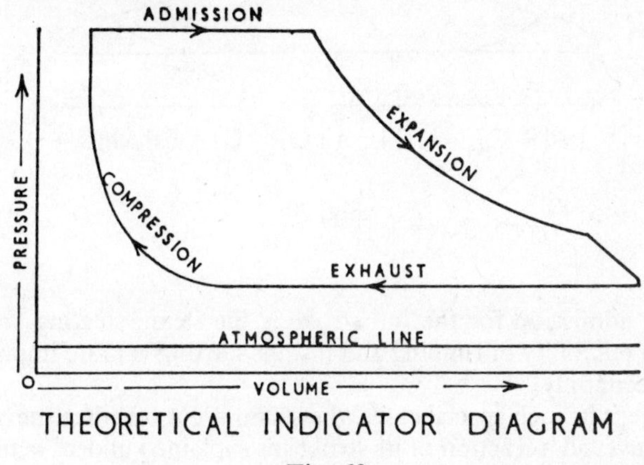

THEORETICAL INDICATOR DIAGRAM

Fig. 69

A pV diagram, showing the variation of steam pressure on one side of the piston as it moves the two strokes which constitute the working cycle, is shown in Fig. 69.

By means of indicator mechanism (described in Chapter 7), these diagrams are drawn to scale automatically on to cards when the engine is running. As all steam engines are double-acting, two diagrams, one from each end of the cylinder (top and bottom of the piston) are drawn on one card and appear as shown in Fig. 70.

The simplest type of slide valve is one in which the depths of the top and bottom bars of the valve are equal to the depths of the top and bottom steam ports in the valve chest face. In this case the valve is in mid-position (at half its travel) when the piston is at the beginning of its stroke. The eccentric sheave setting is therefore 90 degrees ahead of the crank in direction of rotation (90 degrees *behind* the crank for a piston valve with inside steam admission) so that the valve will open to steam and close again while the piston moves one complete stroke, and steam is admitted for the whole stroke of the piston. This is uneconomical as no use is made of the expansive properties of steam. One example of an engine with

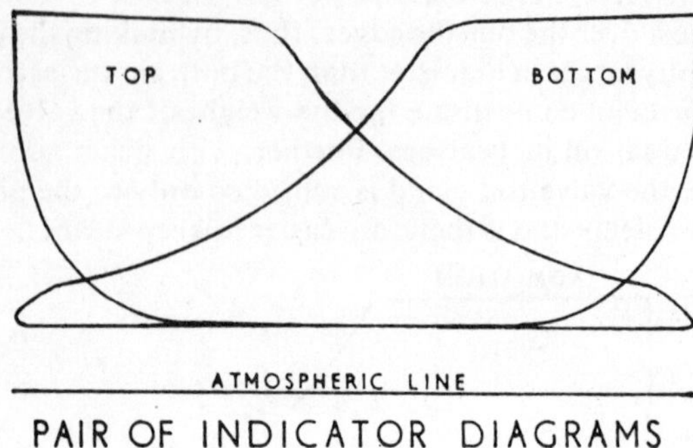

PAIR OF INDICATOR DIAGRAMS

Fig. 70

steam admission for the full stroke is the steam steering engine, where reliability of running and instant starting is more important than economy.

Except in such special cases, the steam is cut off when the piston has travelled a fraction of its stroke (as explained under "action of the slide valve"). Work continues to be done by the steam after cut off while it expands from its initial supply pressure to a pressure near exhaust, and thus, although the work done per stroke is less, more work is obtained per kg of steam used. To enable this to be done, the bars of the slide valve are made deeper than the steam ports. When the valve is in mid-position, the bars of the valve will therefore overlap the steam ports and the amounts of overlap on the steam edge is referred to as *steam lap*. Similarly, to cut off the exhaust from the cylinder before the end of the exhaust stroke, the valve is given *exhaust lap*.

With the crank on dead centre and the piston at the beginning of its stroke, the eccentric sheave setting must now be more than the normal 90 degrees ahead of the crank. It must be advanced sufficiently to move the valve through the distance equal to the steam lap, and a little further so that the port is actually open to steam to ensure a good head of steam in the cylinder to start the piston on its working stroke This latter quantity is referred to as the *lead* of the valve and it necessarily follows therefore, that for a valve with lead, the steam admission will actually commence before the piston reaches its dead centre.

The following usual brief definitions should now be readily understood.

STEAM LAP is the amount that the steam edge of the valve overlaps the steam port when the valve is in mid-position (See Fig 71).

EXHAUST LAP is the amount that the exhaust edge of the valve overlaps the steam port when the valve is in mid-position (See Fig. 71).

LEAD is the amount the port is open to steam when the piston is at the beginning of its stroke. (See Fig. 72).

ANGLE OF ADVANCE is the angle through which the eccentric sheave is advanced beyond its normal position of 90 degrees to the crank to allow for steam lap and lead.

THROW OF ECCENTRIC SHEAVE is the distance from the shaft centre to the geometrical centre of the sheave, and this is equal to half the travel of the slide valve. It will also be seen by reference to Fig. 74 that the full travel of the valve is equal to the difference between the thickest part and thinnest part of the sheave, *i.e.* *ab — cd*.

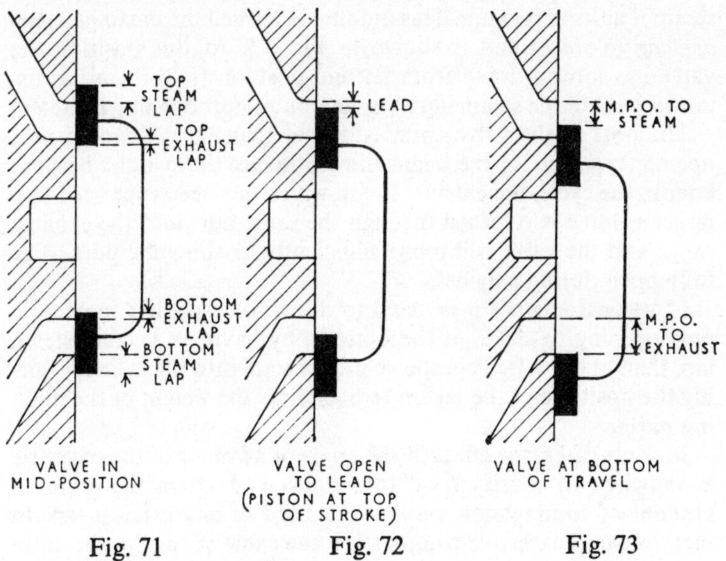

<table>
<tr><td>VALVE IN
MID-POSITION</td><td>VALVE OPEN
TO LEAD
(PISTON AT TOP
OF STROKE)</td><td>VALVE AT BOTTOM
OF TRAVEL</td></tr>
<tr><td>Fig. 71</td><td>Fig. 72</td><td>Fig. 73</td></tr>
</table>

It will be noted that when we speak of the full movement of the piston from one end of the cylinder to the other, it is referred to as the *stroke* of the piston, whereas the full movement of the valve

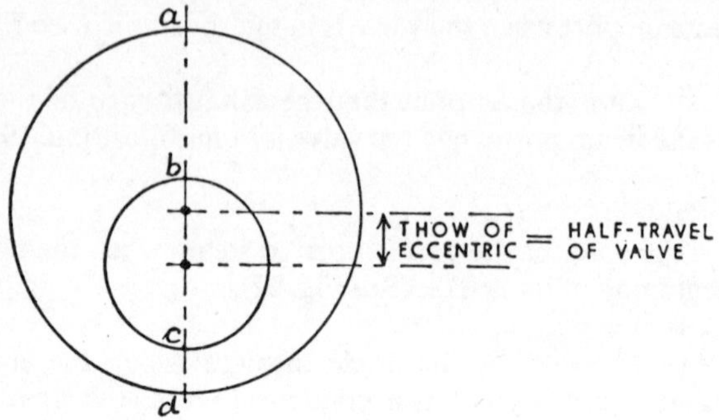

Fig. 74

from one extreme end to the other is referred to as the *travel* of the slide valve.

When the valve is at the bottom of its travel, the port opening to steam is at its maximum. This amount is termed the *maximum port opening to steam* and is shown in Fig. 73. At this position the valve has moved down from its mid-position (mid-travel) a distance equal to the steam lap plus maximum port opening to steam.

The port in the valve chest is deeper than the maximum port opening required for the steam admission into the cylinder because later in the cycle, the exhaust steam which has been expanded to a larger volume is returned through the same port into the exhaust range and the valve will move sufficiently to allow the port to be fully open during exhaust.

In vertical engines it is usual to allow for a greater maximum port opening to steam at the bottom, by having a smaller steam lap, than at the top. This allows more steam into the cylinder during the upstroke of the piston to overcome the weight of the moving parts.

In some cases, the effect of the angle of advance of the eccentric is sufficient to cause cut-off to exhaust and attain the required amount of compression without the need of any exhaust lap. In fact, to avoid excessive compression the exhaust edge of the valve may be cut away so that when the valve is in mid-position the port is actually open a little to exhaust. Such an amount of exhaust opening is referred to as "negative exhaust lap", or "minus exhaust lap" or "exhaust lead".

COMPOUNDING

Steam engines are usually constructed with two, three or four cylinders in series, their pistons being connected to a common crank shaft, so that the full expansion of the steam from boiler supply pressure to final exhaust pressure can be spread throughout those cylinders.

Taking as an example saturated steam supplied to the engine at a temperature of 195°C (14 bar) and exhaust at a temperature of 60°C (0·2 bar). This represents a temperature difference of 135°C. If only one cylinder was used for the expansion, the temperature of the cylinder walls and steam ports would be somewhere between 195 and 60°C but nearer the lower figure due to the longer period of exhaust, say about 120°C. When the supply steam at 195°C comes into contact with the colder metal, it causes a considerable proportion of the steam to condense before it can do work in the cylinder. This initial condensation results in a loss of efficiency.

In a two-cylinder engine, the total pressure drop can take place in two stages. The difference of 135°C between initial supply and final exhaust is divided into two separate temperature drops of 67·5°C. The high-pressure (H.P.) cylinder takes in the steam at 195°C and exhausts it at 127·5°C (2·5 bar), the working temperature of the cylinder and ports being about 155°C. The low-pressure (L.P.) cylinder takes in the steam exhausted from the H.P. at 127·5°C and exhausts it at 60°C, the working temperature of this cylinder being about 80°C. Thus there is less difference between the temperatures of the supply steam and the cylinder with consequent less initial condensation and higher efficiency.

A three-cylinder engine, in which the steam expansion and pressure drop occurs in three stages, is termed a *triple expansion engine*, and the temperature drop of the steam through each cylinder would be one-third of 135 = 45°C.

A four-cylinder engine in which the steam expansion takes place in four stages, is termed a *quadruple expansion engine*, and the steam temperature drop per cylinder would be one-quarter of 135°C.

Thus, the greater number of cylinders used to expand the steam from initial supply to final exhaust, the less the cooling effect and initial condensation of the steam. However, more cylinders mean a longer engine, more heavy parts to move and more frictional resistances to overcome, and therefore compounding the steam is usually limited to three cylinders in average sized marine engines, and four cylinders in the larger powered engines.

It must be understood that the terms "triple" and "quadruple" expansion refer to the number of expansion stages and not necessarily to the number of cylinders. For instance, a triple expansion engine could have four cylinders, the arrangement being, one H.P., one I.P., and two L.P. cylinders; the two L.P.'s take the steam simultaneously from the exhaust of the I.P., they can be made of moderate size and may be preferred over one large cumbersome L.P. cylinder. Such an engine is termed a "four-cylinder-triple", or "four-crank-triple".

In addition to the advantage of reducing cylinder condensation, as explained above, an engine with more than one cylinder gives a better balance and a more even turning moment at the crank shaft. Reheating of the steam can be arranged between the exhaust from one cylinder and the supply to the next, this improves the quality of the steam for expansion in succeeding cylinders The design of the engine is more satisfactory, the H.P. cylinder is made small and strong to take the high pressure steam, whereas the L.P. cylinder is made sufficiently large to accommodate the large volume of the steam at the end of expansion, but need not be so strong because the steam pressures at this stage are very low. Leakage steam past the pistons and valves is not entirely lost (except in the L.P.) since it will be available for work in the succeeding cylinders.

MEAN EFFECTIVE PRESSURE
AND POWER

The method of calculating the indicated power of reciprocating engines was dealt with in Chapter 7, the formula derived being:

$$\text{Indicated power} = p_m A L n$$

This is the power per cylinder in kilowatts when

p_m = indicated mean effective pressure [kN/m²]

A = area of piston [m²]

L = length of stroke [m]

n = number of power strokes per second.

It takes one revolution of the crank to complete one cycle of events, being double-acting, one cycle takes place on the top side of the piston and one on the bottom side, in every revolution. Hence the number of power strokes per second is twice the revolutions per second $n = 2 \times \text{rev/s}$.

The usual method of finding the mean effective pressure from the indicator diagram was also explained in Chapter 7. The area is obtained by means of a planimeter, the area divided by the length of the diagram gives the mean height, and the mean effective pressure is obtained by multiplying the mean height by the scale of the indicator spring.

In steam reciprocating engines, which are double-acting, it is usual to take the area of the diagram as the mean of the pair taken from the top and bottom sides of the piston.

Example. The area of an indicator card taken off a single-cylinder steam engine running at 3·5 rev/s is 1632 mm² and the length is 80 mm. The scale of the indicator spring is 0·25 bar per millimetre height of diagram. The bore of the cylinder is 200 mm and the piston stroke is 300 mm calculate the indicated power.

$$\text{Mean height of diagram} = \frac{\text{area}}{\text{length}}$$

$$= \frac{1632}{80} = 20\text{·}4 \text{ mm}$$

$$
\begin{aligned}
\text{Indicated } p_m &= \text{mean height} \times \text{spring scale} \\
&= 20\text{·}4 \times 0\text{·}25 \\
&= 5\text{·}1 \text{ bar} = 510 \text{ kN/m}^2 \\
\text{Indicated power} &= p_m A L n \\
&= 510 \times 0\text{·}7854 \times 0\text{·}2^2 \times 0\text{·}3 \times 3\text{·}5 \times 2 \\
&= 33\text{·}65 \text{ kW.   Ans.}
\end{aligned}
$$

HYPOTHETICAL MEAN EFFECTIVE PRESSURE

The hypothetical cycle which forms a basis for the comparison of actual cycles assumes that the steam is admitted into the cylinder at constant pressure, cut-off to take place instantaneously, the steam to expand from cut-off point to the end of the power stroke and to follow the law $pV = $ a constant, exhaust of the steam to begin at the end of the power stroke when there is an instantaneous drop of pressure to the exhaust (back) pressure, and exhaust to take place at constant pressure for the full return stroke.

The hypothetical diagram representing this is shown in Fig. 75 wherein p_1 is the supply pressure, p_2 the pressure at the end of expansion, p_b the back pressure of the exhaust, V_s is the stroke volume and V_c the clearance volume.

The hypothetical mean effective pressure is represented by the mean height of the diagram, which is the area divided by its length.

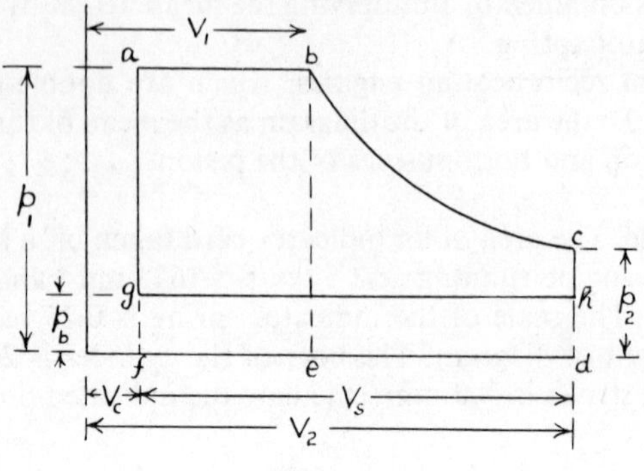

Fig. 75

We have previously seen (Chapter 6) that the area under a curve representing expansion following the law $pV = $ constant is given by $pV \ln r$, where r is the ratio of expansion, hence:

$$\text{Admission area } abef = p_1(V_1 - V_c)$$
$$\text{Expansion area } bcde = p_1 V_1 \ln r$$
$$\text{Back pressure area } ghdf = p_b(V_2 - V_c)$$

Net area representing work done $= abchg$
$$= p_1(V_1 - V_c) + p_1 V_1 \ln r - p_b(V_2 - V_c)$$

ratio of expansion $= r = V_2/V_1$

length of diagram $= V_s = V_2 - V_c$

$$\therefore \text{ hypothetical } p_m = \frac{p_1(V_1 - V_c) + p_1 V_1 \ln r - p_b(V_2 - V_c)}{V_2 - V_c}$$

$$= \frac{p_1(V_1 - V_c) + p_1 V_1 \ln r}{V_2 - V_c} - p_b \qquad \dots \text{ (i)}$$

If clearance is neglected, $V_c = 0$, then:

$$\text{hypothetical } p_m = \frac{p_1 V_1 + p_1 V_1 \ln r}{V_2} - p_b$$

$$= \frac{p_1}{r} + \frac{p_1}{r} \ln r - p_b$$

$$= \frac{p_1}{r}(1 + \ln r) - p_b \qquad \dots \qquad \dots \qquad \dots \text{ (ii)}$$

The above shows how the hypothetical mean effective pressure is obtained, using symbols and thereby deriving formulae. However, if the area under the curve, $pV \ln r$ is known, there is no necessity to memorise any complicated formulae as shown in examples to follow.

It will be seen that volumes can be represented by fractions because the units of volume cancel, it is usually most convenient to represent the stroke volume by unity and to express other volumes as fractions of the stroke volume.

The units of the calculated mean effective pressure will of course be the same as those in which the initial pressure and back pressure are expressed.

Example. Steam at 13·8 bar is admitted to an engine cylinder and cut off when the piston has travelled 0·4 of its stroke. The back pressure is 5·9 bar. The clearance volume is equal to 10% of the stroke volume. Calculate the hypothetical mean effective pressure.

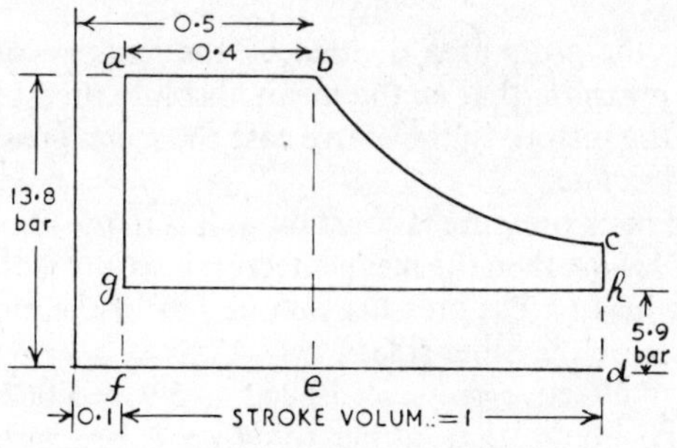

Fig. 76

Let stroke volume $= 1$

clearance volume $= 10\%$ of $1 = 0.1$

Volume of steam at beginning of expansion

$$= \text{volume up to cut-off point}$$
$$= 0.1 + 0.4 = 0.5$$

Volume of steam at end of expansion

$$= 0.1 + 1 = 1.1$$

$$\text{Ratio of expansion} = \frac{\text{volume at end of expansion}}{\text{volume at beginning of expansion}}$$

$$= \frac{1 \cdot 1}{0 \cdot 5} = 2 \cdot 2$$

$$
\begin{aligned}
\text{Admission area } abef &= 13 \cdot 8 \times 0 \cdot 4 = 5 \cdot 52 \\
\text{Expansion area } bcde &= p_1 V_1 \ln r \\
&= 13 \cdot 8 \times 0 \cdot 5 \times \ln 2 \cdot 2 \\
&= 13 \cdot 8 \times 0 \cdot 5 \times 0 \cdot 7885 \\
&= 5 \cdot 442 \\
\text{Gross area } abcdf &= 5 \cdot 52 + 5 \cdot 442 = 10 \cdot 962 \\
\text{Back press. area } ghdf &= 5 \cdot 9 \times 1 = 5 \cdot 9 \\
\text{Net area } abchg &= \text{gross area} - \text{back press. area} \\
&= 10 \cdot 962 - 5 \cdot 9 = 5 \cdot 062
\end{aligned}
$$

$$\text{Hypothetical } p_m = \frac{\text{area}}{\text{length}} = \frac{5 \cdot 062}{1}$$

$$= 5 \cdot 062 \text{ bar} \quad \text{Ans.}$$

Note that the *gross* area divided by the length would give the *mean gross pressure*, that is, the mean absolute pressure acting on that side of the piston. In the above case the mean gross pressure is 10·962 bar *absolute*.

When the back pressure is constant as it is in the above example, at 5·9 bar *absolute*, then the mean effective pressure is the difference between the mean gross pressure on one side of the piston and the back pressure on the other side:

$$\text{Mean effective press.} = 10 \cdot 962 - 5 \cdot 9 = 5 \cdot 062 \text{ bar}$$

Note particularly that a mean *effective* pressure is neither absolute nor gauge, because it is the difference between two pressures.

The difference between the actual indicator diagram and the hypothetical diagram is shown in Fig. 77. Wire-drawing of the steam through the ports during the admission period accounts for the sloping admission line. Near cut-off, the valve closes the port gradually and causes a further pressure drop before complete cut-off takes place. The volume at cut-off and during expansion is less than the hypothetical saturation curve due to initial condensation of the steam, however, the actual expansion curve

DIAGRAM FACTOR

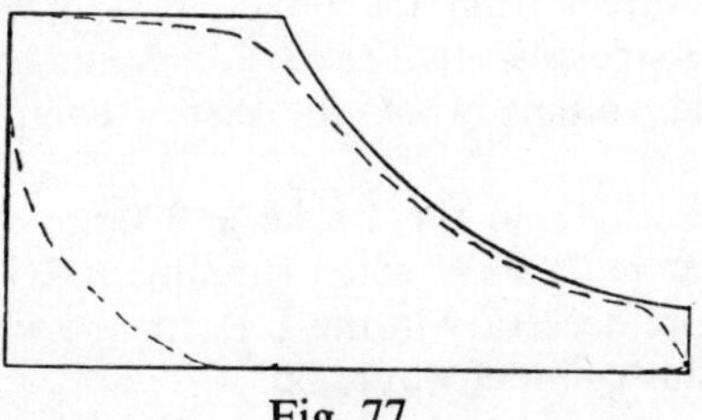

Fig. 77

usually approaches more closely to the theoretical curve near the end of expansion due to some re-evaporation. In practice, exhaust commences before the end of the power stroke and ceases before the end of the return stroke, as previously explained. Hence the result of these differences is to reduce the area of the diagram. Since the length remains the same, then the mean height, which represents the mean effective pressure, is reduced in the same ratio.

The ratio of the area of the actual diagram to the area of the hypothetical diagram, or the ratio of the actual mean effective pressure to the hypothetical mean effective pressure, is known as the *diagram factor*.

This value is usually between 0·6 and 0·75. Assuming a diagram factor of 0·7, the actual mean effective pressure in the last example would be 5·062 × 0·7 = 3·543 bar.

REFERRED MEAN PRESSURE

If we assume that the whole power of a compound, triple or quadruple expansion steam engine could be developed in one of its cylinders, the only cylinder which could accommodate the volume of steam after full expansion is the largest cylinder in the engine, namely the L.P.

Taking as an example a triple expansion engine of cylinder diameters 500, 870 and 1500 mm, and cut-off in the H.P. at 0·6 stroke, neglecting clearances, the overall ratio of expansion of the steam throughout the engine is:

$$\frac{\text{Full volume of L.P. cylinder}}{\text{Volume of H.P. cyl. up to cut-off}}$$

$$= \frac{0{\cdot}7854 \times 1{\cdot}5^2 \times \text{stroke}}{0{\cdot}6 \times 0{\cdot}7854 \times 0{\cdot}5^2 \times \text{stroke}} = 15$$

Thus, if we imagine the L.P cylinder to be of sufficient strength to take steam direct from the boiler, cut-off would be at one-fifteenth of the stroke so that the steam would be expanded to 15 times its initial volume when the piston completed its power-stroke.

Suppose the length of the stroke is 900mm and the required indicated power is 1850kW when running at 2·5 rev/s, the mean effective pressure necessary in the L.P. for the whole power to be developed in this cylinder would be:

$$p_m = \frac{\text{power}}{ALn}$$

$$= \frac{1850}{0·7854 \times 1·5^2 \times 0·9 \times 2·5 \times 2}$$

$$= 232·7 \text{ kN/m}^2 \quad \text{or} \quad 2·327 \text{ bar.}$$

This is termed the 'mean effective pressure all referred to the L.P. cylinder', or usually more briefly, the *referred mean pressure*, and is defined thus: The referred mean pressure is the mean effective pressure that would be required in the L.P. cylinder if the whole power of the engine be assumed to be developed in that cylinder.

If equal powers are developed in each cylinder, then, for a triple expansion engine, one-third of the total engine power is developed in each cylinder. The actual mean effective pressure in the L.P. will then be one-third of the referred mean pressure. In this example:

$$p_m \text{ in L.P.} = \tfrac{1}{3} \times 2·327 = 0·776 \text{ bar}$$

Since the power in the other two cylinders is to be the same as that in the L.P., then,

$$\text{H.P. power} = \text{I.P. power} = \text{L.P. power}$$

$$p_m \, ALn \text{ (H.P.)} = p_m ALn \text{ (I.P.)} = p_m ALn \text{ (L.P.)}$$

Cancelling quantities common to all terms:

$$p_m d^2 \text{(H.P.)} = p_m d^2 \text{(I.P.)} = p_m d^2 \text{(L.P.)}$$

Hence, to develop power in the H.P. and I.P. equal to the L.P. power:

$$p_m \text{ in H.P.} = \frac{p_m \text{(L.P.)} \times d^2 \text{(L.P.)}}{d^2 \text{(H.P.)}}$$

$$= \frac{0·776 \times 1·5^2}{0·5^2} = 6·984 \text{ bar}$$

$$p_m \text{ in I.P.} = \frac{0·776 \times 1·5^2}{0·87^2} = 2·308 \text{ bar}$$

Example. The cylinder diameters of a triple expansion engine are 650, 1100 and 1850 mm and the mean effective pressures are 5·86, 2·07 and 0·724 bar respectively when the mean piston speed is 3·86 m/s. Calculate the mean effective pressure referred to the L.P. cylinder, the power developed in each cylinder, and the total power of the engine.

p_m of H.P. referred to L.P.

$$= \frac{5·86 \times 0·65^2}{1·85^2} = 0·7233 \text{ bar}$$

p_m of I.P. referred to L.P.

$$= \frac{2·07 \times 1·1^2}{1·85^2} = 0·7318 \text{ bar}$$

Total mean effective pressure referred to L.P.
$$= 0·7233 + 0·7318 + 0·724 = 2·179 \text{ bar.} \quad \text{Ans. (i)}$$

Note that the mean piston speed in metres per second is the metres of distance travelled by the piston during one second, and is equal to, stroke length [m] $\times$ 2 $\times$ rev/s, which is $L \times n$ in the formula for indicated power of a steam reciprocating engine.

$$\text{Power} = p_m A L n$$
$$\text{H.P. power} = 586 \times 0·7854 \times 0·65^2 \times 3·86$$
$$= 750·6 \text{ kW.} \quad \text{Ans. (ii}\,a)$$
$$\text{I.P. power} = 207 \times 0·7854 \times 1·1^2 \times 3·86$$
$$= 759·5 \text{ kW.} \quad \text{Ans. (ii}\,b)$$
$$\text{L.P. power} = 72·4 \times 0·7854 \times 1·85^2 \times 3·86$$
$$= 751·3 \text{ kW.} \quad \text{Ans. (ii}\,c)$$
$$\text{Total power} = 750·6 + 759·5 + 751·3$$
$$= 2261·4 \text{ kW.} \quad \text{Ans. (iii).}$$

Check by assuming the total power is developed in the L.P. when the mean effective pressure in the L.P. is the referred mean pressure:
$$\text{Total power} = 217·9 \times 0·7854 \times 1·85^2 \times 3·86$$
$$= 2261 \text{ kW (as above).}$$

HYPOTHETICAL REFERRED MEAN PRESSURE. The expression for the hypothetical mean effective pressure derived on page 208 can also be applied to obtain the hypothetical mean effective pressure referred to the L.P., thus, neglecting clearances:

Hypothetical referred mean pressure

$$= \frac{p_1}{r}(1 + \ln r) - p_\mathrm{b}$$

where p_1 = initial pressure of steam supplied to H.P. cylinder

p_b = final back pressure (in the L.P. cylinder)

r = overall ratio of expansion throughout the engine which is,

$$\frac{\text{Volume of steam after expansion}}{\text{Volume of steam before expansion}}$$

$$= \frac{\text{Full volume of L.P. cylinder}}{\text{Volume of H.P. cylinder up to cut-off}}$$

The actual referred mean pressure is the hypothetical value multiplied by the overall diagram factor.

WILLANS' LAW

The usual methods of varying the power of a steam reciprocating engine are (i) by throttling the steam at the engine stop valve, (ii) by altering the cut-off point of the H.P. piston valve, or (iii) by a combination of (i) and (ii).

When the power is varied by throttling only, the mass of steam used by the engine varies proportionally as the indicated power. This is known as Willans' law and can be expressed by a linear equation:

$$m = a + bP$$

where m = mass of steam consumed (usually kg/h)

P = power developed

a and b = constants

The specific steam consumption is usually expressed in kilogrammes per kilowatt-hour [kg/kW h] and is a practical measure of the efficiency of the engine.

Example. The following data were taken during a trial run on an experimental quadruple expansion steam engine. Plot graphs of steam consumption in kg per hour, and specific steam consump-

tion in kg/kWh on a common base of indicated power, and find
Willans' law for this engine.

ind. power [kW]	20	40	60	80	100	120
consumption [kg/h]	310	450	590	730	870	1010

Dividing kg of steam per hour by ip to obtain specific steam
consumption kg/kWh = 15·5   11·25   9·83   9·13   8·7   8·42
The plotted graphs appear as in Fig. 78.

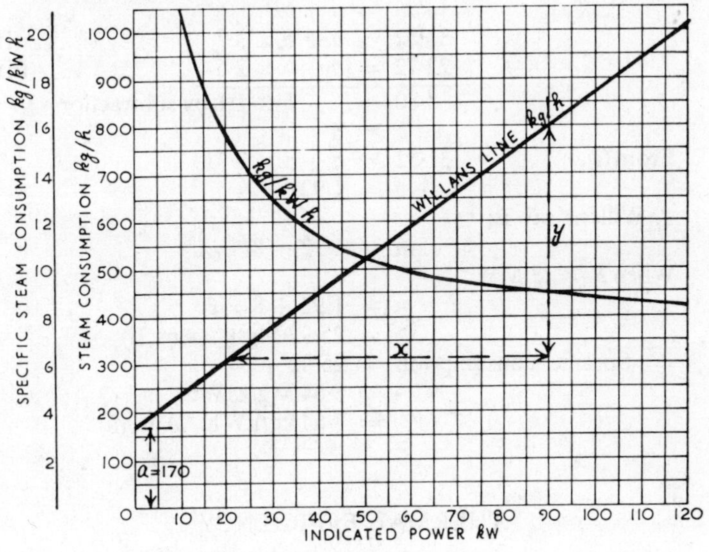

Fig. 78

Reading from the graph of Willans' line:

a = steam consumption in kg/h when the ip is theoretically
zero, this reads 170.

b = increase in steam consumption in kg/h for an increase of
1 kW, this is the slope of the line and obtained from $y \div x$
from any two points on the line.

From the points chosen as shown,

$$y \div x = 490 \div 70 = 7$$

$$\therefore \text{ Willans' law is, } m = 170 + 7P \quad \text{Ans.}$$

Example. The mass of steam used by an engine is 31·52 Mg
per hour when developing an indicated power of 3·7 MW, and

22·72 Mg per hour when the ip is 2·6 MW. Assuming the consumption of steam follows Willans' law, find (i) the mass of steam used per hour when the ip is 3 MW, and (ii) the specific consumption in kg/kWh at this power.

Putting in the pair of given quantities in the form of a simultaneous equation, solving the constants a and b, and expressing Willans' law for this engine:

$$m = a + bP$$
$$31\!\cdot\!52 = a + b \times 3\!\cdot\!7 \qquad \dots \quad \dots \quad \text{(i)}$$
$$22\!\cdot\!72 = a + b \times 2\!\cdot\!6 \qquad \dots \quad \dots \quad \text{(ii)}$$
$$\overline{8\!\cdot\!80 = \qquad b \times 1\!\cdot\!1} \text{ by subtraction}$$
$$\therefore b = 8$$

From (i),
$$31\!\cdot\!52 = a + 8 \times 3\!\cdot\!7$$
$$a = 1\!\cdot\!92$$

$\therefore$ Willans' law is,
$$m = 1\!\cdot\!92 + 8P$$

When $P = 3$ MW:

$$m = 1\!\cdot\!92 + 8 \times 3$$
$$= 25\!\cdot\!92 \text{ Mg/h} \quad \text{Ans. (i)}$$
$$\text{Specific consumption} = 25\!\cdot\!92 \div 3$$
$$= 8\!\cdot\!64 \text{ Mg/MWh}$$
$$= 8\!\cdot\!64 \text{ kg/kWh} \quad \text{Ans. (ii)}$$

THERMAL EFFICIENCY

The indicated thermal efficiency of a steam engine is the ratio of the heat energy converted into work in the cylinders to the heat energy supplied to the engine by the steam. The heat energy supplied by the steam is the heat energy given to it in the boilers. It is usual to assume that the boilers are fed with water at the temperature at which the steam is condensed in the condenser, any under-cooling of the condensate to a temperature below the exhaust steam temperature being disregarded since this is a condenser loss and not an engine loss of heat energy.

Thus, expressing power in kW and based upon a running time of one hour ($= 3600$ seconds):

Indicated thermal efficiency

$$= \frac{\text{indicated power [kW]} \times 3600}{\text{steam consumption [kg/h]} \times (h_1 - h_{f2})}$$

Where h_1 = specific enthalpy [kJ/kg] of supply steam.

h_{f2} = specific enthalpy [kJ/kg] of saturated water at
condenser pressure.

Dividing numerator and denominator by ip:

Indicated thermal efficiency

$$= \frac{3600}{\text{spec. steam cons. [kg/kWh]} \times (h_1 - h_{f2})}$$

Example. A steam reciprocating engine is supplied with dry saturated steam at a pressure of 14 bar and the pressure of the exhaust to the condenser is 0·2 bar. When developing an indicated power of 400 kW the mass of steam used by the engine is 4200 kg/h. Calculate the specific steam consumption in kg/kWh and the indicated thermal efficiency.

$$\text{Spec. steam cons.} = \frac{4200}{400} = 10\cdot5 \text{ kg/kWh} \quad \text{Ans. (i)}$$

Tables page 4, 14 bar, $h_g = 2790$
" 3, 0·2 bar, $h_f = 251$

Indicated thermal efficiency

$$= \frac{3600}{\text{spec. steam cons. [kg/kWh]} \times (h_1 - h_{f2})}$$

$$= \frac{3600}{10\cdot5 \times (2790 - 251)}$$

$$= 0\cdot1351 \text{ or } 13\cdot51\% \quad \text{Ans. (ii)}$$

f RANKINE EFFICIENCY

The ideal cycle adopted as a basis of comparison for the actual performance of steam engines (turbines as well as reciprocators) is known as the *Rankine cycle*, and the thermal efficiency of an engine working on the Rankine cycle is the *Rankine efficiency*. The ratio of the actual indicated thermal efficiency of an engine to the Rankine efficiency is termed the *relative efficiency* or *efficiency ratio*.

The Rankine cycle is illustrated in Fig. 79 on a pV diagram, and in Fig. 80 on a T-s diagram, and consists of:

(i) a to b, feed water pumped into the boiler and receiving heat energy as it is increased in pressure from p_2 to p_1 and in temperature from T_2 to T_1.

(ii) b to c. The water is completely evaporated into steam in the boiler at constant pressure p_1 and constant temperature T_1 and the steam is supplied to the engine as it is generated.

(iii) c to d. The steam supply to the engine is cut off from the boiler and expands isentropically in the engine from the highest to the lowest limits of pressure and temperature.

(iv) d to a. The steam is exhausted from the engine into the condenser and condensed into water at constant pressure p_2 and constant temperature T_2.

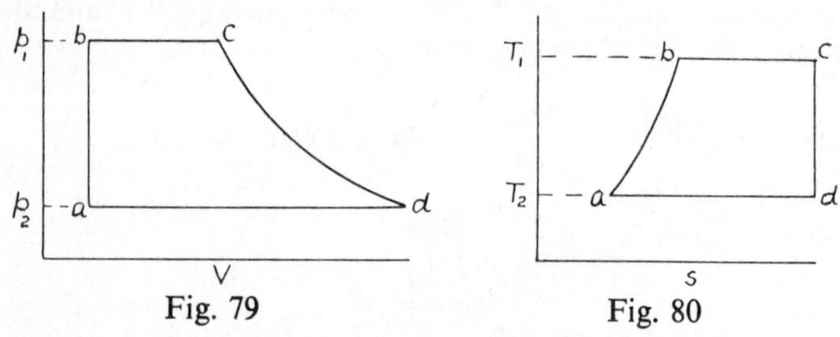

Fig. 79　　　　　Fig. 80

In the ideal engine there would be no energy losses due to friction or heat transfer, no leakage of steam, and no undercooling of the condensate in the condenser, *ie* the condensate would be at the saturation temperature corresponding to the condenser steam pressure.

Therefore the work which would be done in the ideal engine is equal to the heat energy given up by the steam on its passage through the engine. This is the enthalpy drop, *ie* the difference between the enthalpy of supply steam (h_1) and the enthalpy of the exhaust steam (h_2).

The heat energy supplied to the feed water in the boiler to produce the steam is the difference between the enthalpy of the supply steam (h_1) and the enthalpy of the feed water (h_{f2}), the feed water temperature being equal to the condensate temperature.

Thus, neglecting the work done by the feed pump in compressing the feed water as being comparatively small:

$$\text{Rankine efficiency} = \frac{h_1 - h_2}{h_1 - h_{f2}}$$

The value of h_1 depends upon whether the supply steam is wet, dry, or superheated. Since the exhaust steam is always wet, the value of h_2 will be obtained from $h_{f2} + xh_{fg2}$

Taking the engine in the previous example wherein dry saturated steam is supplied at 14 bar, and exhausted at 0·2 bar:

$$\text{Tables page 4, 14 bar,} \quad h_g = 2790 \quad s_g = 6{\cdot}469$$
$$\text{page 3, 0·2 bar,} \quad h_f = 251 \quad s_f = 0{\cdot}832$$
$$h_{fg} = 2358 \quad s_{fg} = 7{\cdot}075$$

Heat energy supplied to steam in boiler
$$= h_1 - h_{f2} = 2790 - 251 = 2539 \text{ kJ/kg}$$

To obtain the enthalpy of the steam after expansion (h_2) we must first find the dryness fraction of this steam,

entropy after expansion = entropy before
$$0{\cdot}832 + x \times 7{\cdot}075 = 6{\cdot}469$$
$$x \times 7{\cdot}075 = 5{\cdot}637$$
$$x = 0{\cdot}7968$$
$$\therefore h_2 = h_{f2} + xh_{fg2}$$
$$= 251 + 0{\cdot}7968 \times 2358$$
$$= 2130$$

Ideal work done
$$= h_1 - h_2 = 2790 - 2130 = 660 \text{ kJ/kg}$$

$$\therefore \text{Rankine efficiency} = \frac{660}{2539}$$

$$= 0{\cdot}2599 \quad \text{or} \quad 25{\cdot}99\%$$

$$\text{Efficiency ratio} = \frac{\text{indicated thermal efficiency}}{\text{Rankine efficiency}}$$

$$= \frac{0{\cdot}1351}{0{\cdot}2599} = 0{\cdot}5196$$

TEST EXAMPLES 12

1. The area of an indicator card taken off the high pressure cylinder of a quadruple expansion engine is 1445 mm², the length is 80 mm, and one millimetre of height represents 0·27 bar. The bore of the cylinder is 600 mm, piston stroke 1050 mm, and the engine runs at 1·65 rev/s. Calculate (i) the indicated power developed in this cylinder, (ii) the total indicated power of the engine assuming equal powers are developed in all cylinders.

2. During a test on a single-cylinder steam engine, the mean effective pressure was 4·05 bar when running at 3 rev/s. The brake load was 1·9 kN acting at an effective radius of 1·2 m. If the cylinder diameter is 0·275 m and the stroke 0·4 m, calculate (i) the indicated power, (ii) the brake power, (iii) the mechanical efficiency.

3. Steam at 7·5 bar is admitted to an engine cylinder and cut off at 0·34 of the stroke. The back pressure is 1·7 bar. Neglecting clearance, calculate the hypothetical mean gross pressure and mean effective pressure.

4. Steam is supplied to the cylinder of an engine at 8·3 bar and cut-off takes place when the piston has travelled 0·45 of its stroke. The clearance volume is equal to 5 % of the stroke volume. If the back pressure is 2·4 bar, calculate (i) the hypothetical mean effective pressure, (ii) the actual mean effective pressure taking a diagram factor of 0·72.

ƒ5. Steam at 15 bar is supplied to the high pressure cylinder of a steam engine and cut off at 0·45 stroke. The back pressure is 6·5 bar, clearance equal to 10 % of the stroke volume, and compression begins at 0·88 of the exhaust stroke. Assume expansion and compression to follow the law $pV = $ a constant and to continue to the ends of their respective strokes. Sketch the indicator diagram representing the above data and calculate the mean effective pressure.

6. The cylinders of a triple expansion engine are 600, 950 and 1500 mm diameter and the stroke is 1050 mm. When running at 1·8 rev/s the mean effective pressures in the cylinders are 5·17, 2·07 and 0·83 bar respectively. Calculate (i) the ip developed in each cylinder, (ii) the mean effective pressure referred to the L.P. and (iii) the total ip of the engine.

7. The diameters of the cylinders of a quadruple expansion engine are 680, 960, 1340 and 1900 mm respectively and the mean effective pressure referred to the L.P. is 3·1 bar when the mean piston speed is 4 m/s. Assuming equal powers are developed in all cylinders, calculate the mean effective pressure in each cylinder and the total indicated power of the engine.

f8. A steam engine consumes 13·3 Mg of steam per hour when developing an indicated power of 1·5 MW, and 18·7 Mg/h when the ip is 2·25 MW. Estimate Willans' law for this engine and find the steam consumption in Mg per hour and the specific consumption in kg/kW h when developing 2000 kW.

9. Steam at 15 bar 300°C is supplied to an engine and exhausted at 0·14 bar. The consumption of steam is 28 500 kg/h when the indicated power is 3000 kW. Calculate the specific steam consumption in kg/kW h and the indicated thermal efficiency.

f10. An engine is supplied with steam at a pressure of 15 bar and temperature 250°C, and the pressure of the exhaust is 0·16 bar. Assuming isentropic expansion, find (i) the dryness fraction of the steam after expansion, (ii) the Rankine efficiency.

CHAPTER 13

TURBINES

A steam turbine, like the steam reciprocating engine, is a machine for converting the heat energy in the steam into mechanical energy at the shaft, but the principles upon which these two engines work are entirely different. In the reciprocating engine the steam pressure acts as a static load on the piston to cause it to move up and down and this motion is converted from a reciprocating one into a rotary motion at the shaft by connecting rod and crank mechanism. In the turbine the rotor coupled to the shaft receives its rotary motion direct from the action of high velocity steam impinging on blades fitted into grooves around the periphery of the rotor, thus the action of the steam in a turbine is 'dynamic' instead of static as it is in the reciprocating engine.

There are two types of turbine, the impulse and the reaction. In both cases the steam is allowed to expand from a high pressure to a lower pressure so that the steam acquires a high velocity at the expense of pressure, and this high velocity steam is directed on to curved section blades which absorb some of its velocity; the difference is in the methods of expanding the steam.

In impulse turbines the steam is expanded in nozzles in which the high velocity of the steam is attained before it enters the blades on the turbine rotor, the pressure drop and consequent increase in velocity therefore takes place in these nozzles. As the steam passes over the rotor blades it loses velocity but there is no fall in pressure.

In reaction turbines, expansion of the steam takes place as it passes through the moving blades on the rotor as well as through the guide blades fixed to the casing.

THE IMPULSE TURBINE

As stated above, in the impulse turbine the high pressure steam passes into nozzles wherein it expands from a high pressure to a lower pressure and thus the heat energy in the steam is converted into velocity energy (kinetic energy). The high velocity steam is directed on to blades fitted around the turbine wheel, the blades being of curved section so that the direction of the steam is changed thereby imparting a force to the blades to push the wheel around. The simplest form of impulse turbine is the single-stage De

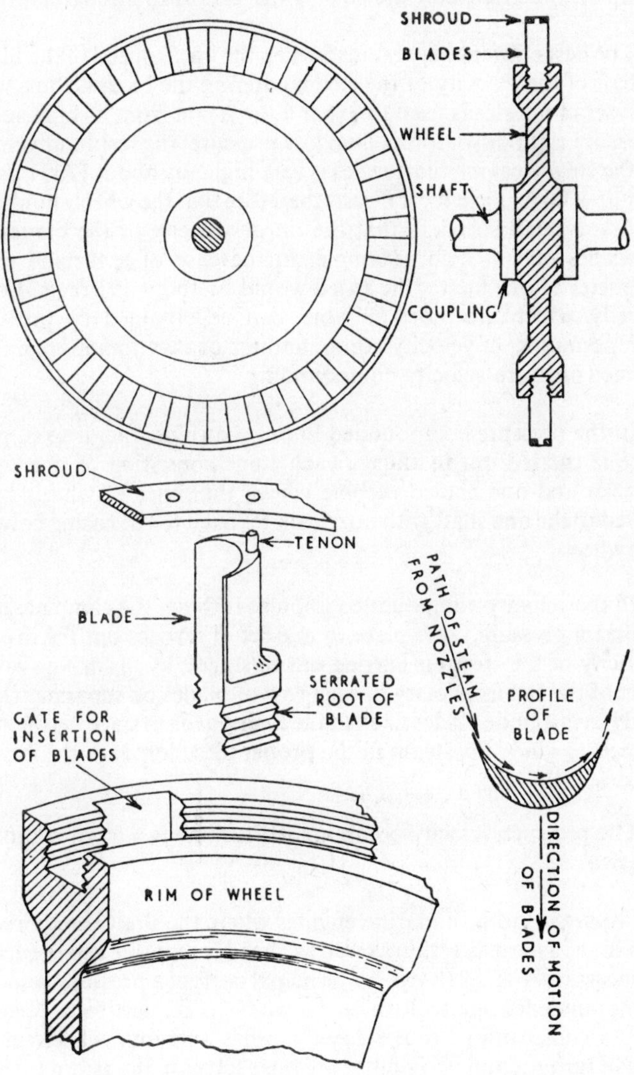

SINGLE STAGE IMPULSE TURBINE WHEEL

Fig. 81

Laval shown in Fig. 81. This consists of a solid wheel bolted by flanges to a shaft, blades of bronze or nickel steel are fixed into a groove around the rim of the wheel, caulked in, and a shroud or strap wrapped around the tips of the blades to strengthen them.

The best efficiency is obtained when the linear speed of the blades is half of the velocity of the steam entering the blades, thus, when one set of nozzles is used to expand the steam from its high supply pressure right down to the final low pressure, the resultant velocity of the steam leaving the nozzles is very high, say about 1200 m/s. To obtain a high efficiency it means therefore that the wheel should run at a very high speed so that the linear velocity of the blades approaches 600 m/s, for example, in the case of a turbine wheel diameter of one metre, the speed would be about 191 rev/s. Lower speeds, which are more suitable, can be obtained by pressure-compounding, or velocity-compounding, or a combination of these termed pressure-velocity-compounding.

In the pressure-compounded impulse turbine, the drop in pressure is carried out in stages, each stage consisting of one set of nozzles and one bladed turbine wheel, the series of wheels being keyed to the one shaft with nozzle plates fixed to the casing between the wheels.

In the velocity-compounded impulse turbine, the complete drop in steam pressure takes place in one set of nozzles but the drop in velocity of the steam is carried out in stages, by absorbing only a part of the steam velocity in each row of blades on separate wheels and having guide blades fixed to the casing at each stage between the wheels to guide the steam in the proper direction onto the moving blades.

The pressure-velocity-compounded turbine is a combination of the two.

CONSTRUCTION. In marine engines where the shaft is required to run in the astern direction as well as ahead, a separate astern turbine is necessary. Fig. 82 shows the principal parts of a pressure-velocity-compounded impulse turbine in which there are four pressure stages consisting of four sets of nozzles and four wheels in the ahead turbine, and two similar pressure stages in the astern turbine. Each wheel carries two rows of blades and there is one row of guide blades fixed to the casing protruding radially inwards between each row of moving blades to drop the velocity in two steps from each set

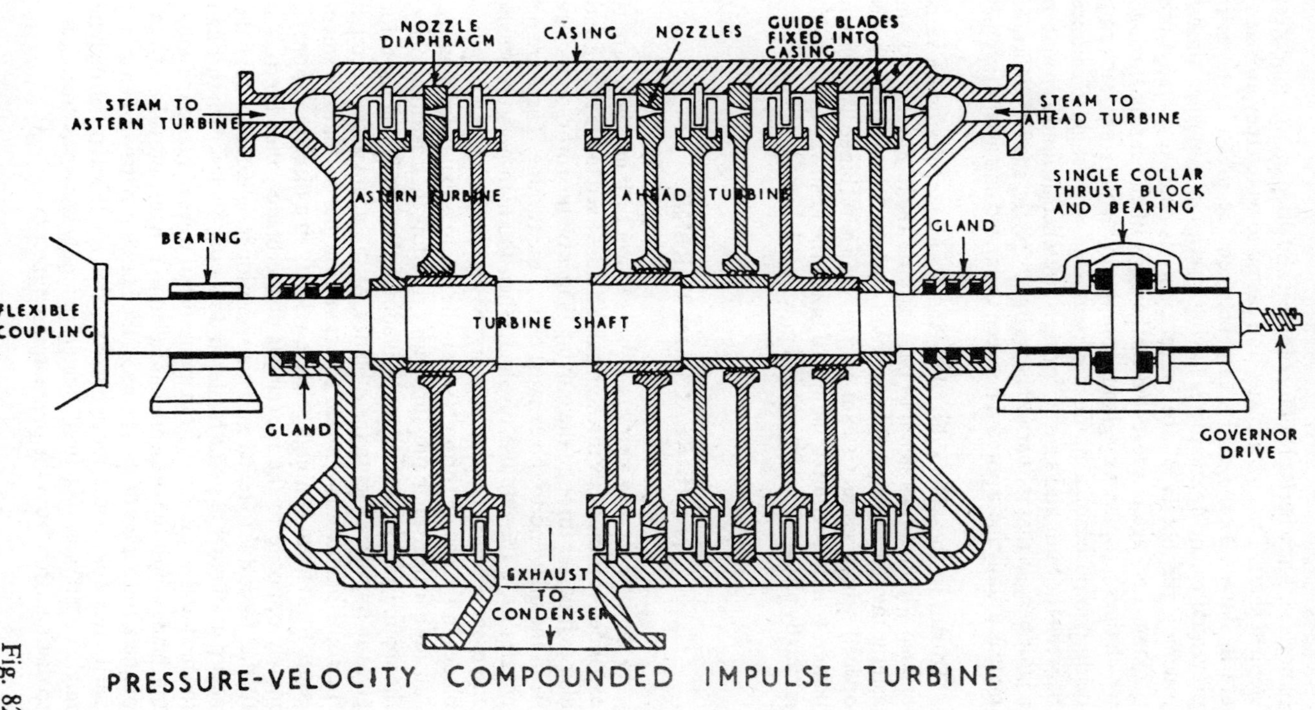

PRESSURE-VELOCITY COMPOUNDED IMPULSE TURBINE

Fig. 82

of nozzles. The wheels are of forged steel and fitted on to a mild steel stepped shaft. The nozzle plates and casing to which they are fixed, are in halves. Leakage of steam between the pressure stages is prevented by the nozzle plates having a series of thin rings almost touching the bosses of the wheels, this is known as labyrinth packing. Leakage is prevented at the ends of the casing through which the shaft passes by glands containing carbon rings, each ring being composed of a number of segments with slight clearance between the butts and a garter spring is stretched around a groove in a periphery so that the carbon bears lightly on the shaft. A deflector plate is usually incorporated between the ahead and astern turbine.

To run in the ahead direction, the ahead steam stop valve is opened which allows steam to pass into the ahead turbine. To run astern, the ahead stop valve is closed and the astern stop valve is opened which admits steam to the astern turbine. The astern power, which is required mainly to brake the headway of the ship, is usually about 70 % of the ahead power.

THE REACTION TURBINE

In the type usually known as the reaction turbine, the steam is expanded continuously through guide blades fixed to the casing and also as it passes through the moving blades on the rotor, on its way from the inlet end to the exhaust end of the turbine. There are no nozzles as in the impulse turbine. When the high pressure steam enters the reaction turbine, it is first passed through a row of guide blades in the casing through which the steam is expanded slightly, causing a little drop in pressure with a resulting increase in velocity, the steam being guided on to the blades in the first row of the rotor gives an impulse effect to these blades. As the steam passes through the rotor blades it is allowed to expand further so that the steam issues from them at a high relative velocity in a direction approximately opposite to the movement of the blades, thus exerting a further force due to reaction. This operation is repeated through the next pair of rows of guide blades and moving blades, then through the next and so on throughout a number of rows of guide and moving blades until the pressure has fallen to exhaust pressure. As explained, the action of the steam on the blades is partly impulse and partly reaction, and a more correct name for this type of turbine might be "impulse-reaction" but it is generally known as "reaction" to distinguish it from the pure impulse type.

TURBINE CASING
ALTERNATE ROWS OF (GUIDE) BLADES FIXED IN CASING
INLET STEAM
SINGLE COLLAR THRUST BLOCK AND BEARING
GLAND
BEARING
GLAND
FLEXIBLE COUPLING
TURBINE ROTOR
DUMMY PISTON
DOTTED LINES REPRESENT ROWS OF (MOVING) BLADES IN ROTOR
GOVERNOR DRIVE
EXHAUST TO CONDENSER
EQUALISING STEAM PIPE
REACTION TURBINE

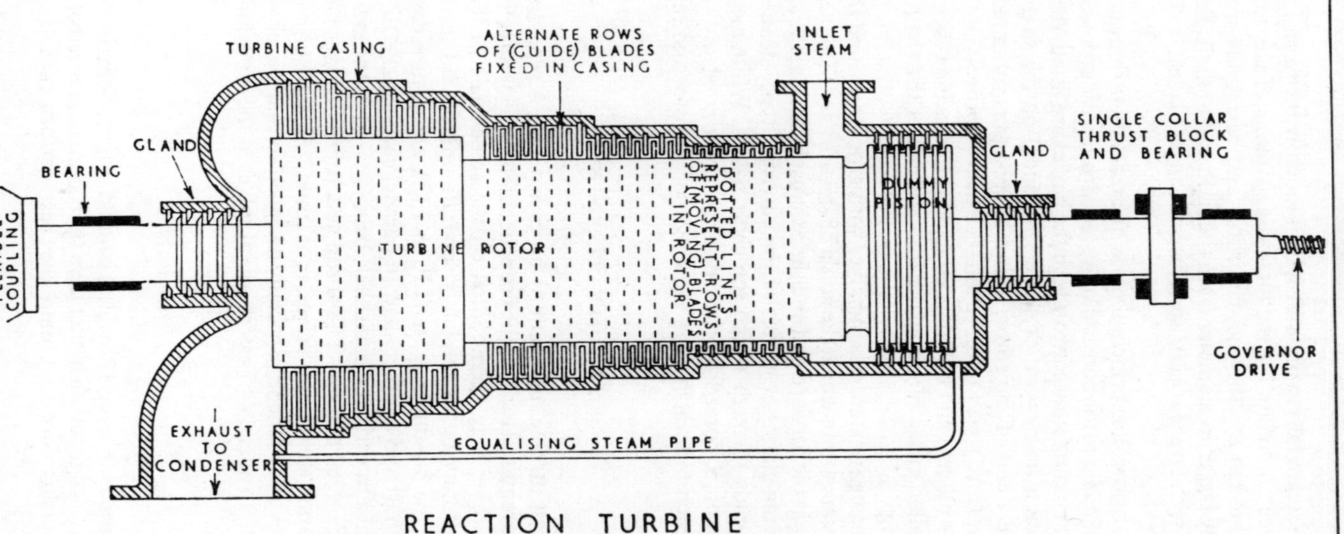

Fig. 83

As the steam falls in pressure it consequently increases in volume and to accommodate for the increasing volume of the steam as it passes through the turbine, the rotor and casing are made progressively larger in diameter from the H.P. end to the L.P., usually in steps, with larger area of annulus and longer blades.

CONSTRUCTION. The principal parts of a reaction turbine are shown in Fig. 83. The rotor consists of a steel drum with grooves around the outer circumference into which the blades are fitted, the drum is carried on a framework or spiders mounted on the shaft. At the steam inlet end of the rotor, a dummy piston is mounted for the steam pressure to act upon to balance the opposite axial thrust on the rotor blades; any unbalanced axial thrust being taken up by a single-collar thrust block at the forward end of the shaft, outside the casing.

The casing (or stator) is in halves with grooves around the inside circumference, into which the guide blades are fixed. The steam inlet branch is at the forward end and the exhaust branch at the after end. Carbon ring glands or glands of the labyrinth type, steam packed, seal the ends through which the shaft passes. White-metal lined brass bearings, forced lubricated, are incorporated in the casing outside each end which carry the weight of the rotor and shaft.

ASTERN RUNNING

As in all turbine installations on board ship, a separate astern turbine is required to drive the propeller in the reverse direction. This astern turbine may be mounted on the same shaft as the ahead turbine, it may be a completely separate unit geared to the main shaft, or the lay-out may be H.P. and I.P. ahead turbines on one turbine shaft and another shaft in parallel carrying the L.P. ahead and the astern turbines, each shaft being geared to the main shafting (see Figs. 82 and 84). The ahead and astern turbines have their own separate steam stop valves, or it could be one steam chest with two valves and branches, specially designed so that it is not possible to open both ahead and astern steam valves at the same time.

GEARING

The speed at which turbines should run to obtain the best efficiency, is high, much higher than the economical speed of a ship's propeller, therefore to obtain the best performance from both turbine and propeller, reduction gearing is interposed between the turbine shaft and the propeller shaft, to enable both to run at their best speeds.

Single reduction gearing consists of one stage of speed reduction in which a pinion on the turbine shaft meshes with a gear wheel on the main shaft.

Double reduction gearing consists of two stages of speed reduction by means of a pinion on the turbine shaft meshing with an intermediate wheel, on the intermediate shaft a secondary pinion meshes with the main gear wheel on the propeller shafting.

The teeth in the wheels are cut at an angle to the axis of the shaft (these are known as helical teeth) for smooth running, and all gear wheels are arranged in pairs to balance any axial thrust caused by

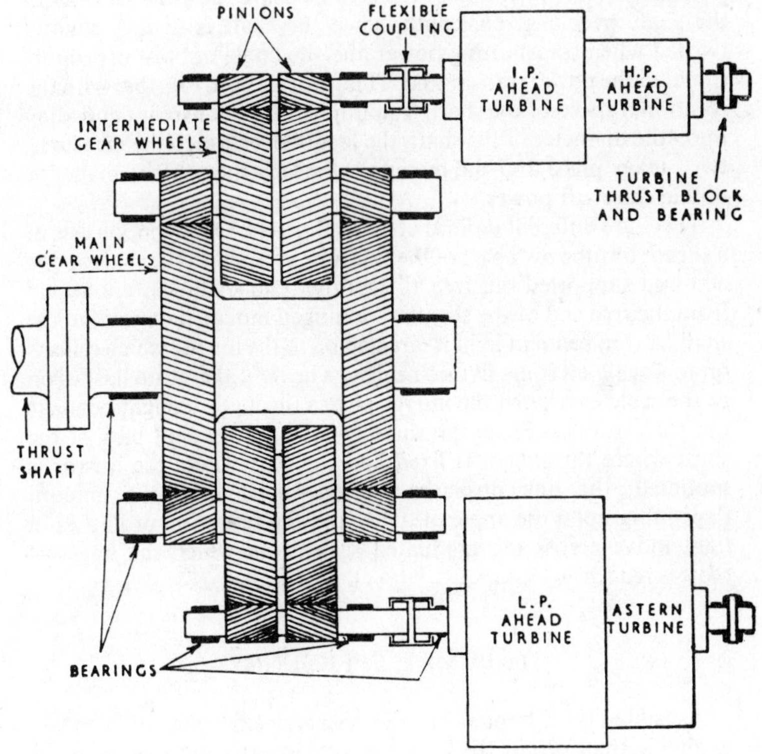

DOUBLE-REDUCTION GEARING

Fig. 84

the teeth being helical. Fig. 84 shows diagrammatically a typical turbine lay-out including astern turbine and double-reduction gearing. Note the flexible couplings between turbine shafts and pinion shafts, these are to prevent thermal expansion of the turbine or misalignment of its shaft affecting the perfect meshing of the gear wheel teeth.

POWER

Indicator cards cannot be taken off a turbine to obtain the indicated power as done in reciprocating engines therefore the power must be measured at the shaft, and this is done by fitting a torsionmeter on the shaft which measures the brake power. The principle upon which this works is to measure the angle of twist in the shaft over a given length, since the shaft is always slightly twisted when transmitting power and the angle of twist is proportional to the power transmitted. This information, together with the rotational speed of the shaft, is multiplied by a constant depending upon the diameter of the shaft, the length over which the measured twist takes place and the material of which the shaft is made, to obtain the shaft power.

There are different designs of torsionmeters. One type consists of a sheath or tube over part of the shafting, secured to the shaft at one end and supported but free at the other end. A link is connected from the free end of the sheath to a hinged mirror mounted on the shaft. A thin beam of light is directed on to the mirror which reflects on to a scale on some fixture nearby. The light therefore flashes on to the scale everytime the mirror passes through the light beam as the shaft rotates. Since the shaft twists between that part of the shaft where the sheath is fixed and the part where the mirror is mounted, the link displaces the mirror slightly, the amount depending upon the angle of twist, and the reflection of flashes of light move across the graduated scale from which the angle of twist is read.

THERMAL EFFICIENCY

As stated in Chapter 12, the thermal efficiency of a steam engine is the ratio of the heat energy converted into work in the engine to the heat energy supplied.

The heat energy is supplied by the steam and we take this as that transferred to the steam in the boilers from feed water at condenser condensate temperature, thus,

energy [kJ] supplied = steam consumption [kg] $\times$ $(h_1 - h_{f2})$ [kJ/kg]

where, h_1 = enthalpy per kg of supply steam.

h_{f2} = enthalpy per kg of water at saturation temperature corresponding to condenser pressure.

hence, on a time basis of one second:

$$\text{Thermal efficiency} = \frac{\text{power } [\text{kW} = \text{kJ/s}]}{\text{steam consumption } [\text{kg/s}] \times (h_1 - h_{f2})}$$

or, on a time basis of one hour (3600 seconds):

$$\text{Thermal efficiency} = \frac{\text{power } [\text{kW}] \times 3600}{\text{steam consumption } [\text{kg/h}] \times (h_1 - h_{f2})}$$

or, on a basis of one kW h:

$$\text{Thermal efficiency} = \frac{3600}{\text{specific steam cons. } [\text{kg/kW h}] \times (h_1 - h_{f2})}$$

*f*RANKINE EFFICIENCY. The thermal efficiency of a steam engine working on the ideal Rankine cycle, referred to as the Rankine Efficiency was also dealt with in Chapter 12 and expressed as:

$$\text{Rankine efficiency} = \frac{h_1 - h_2}{h_1 - h_{f2}}$$

where h_1 is the enthalpy per kg of steam supplied to engine.

h_2 is the enthalpy per kg of exhaust steam from engine.

h_{f2} is the enthalpy per kg of the saturated water at condenser pressure.

However, it is sometimes the practice in steam turbines to re-heat the steam between stages in order to avoid excessive wetness of the steam in the low pressure units. It also increases the power output without increasing the size of the turbine. Fig. 85 is a line diagram illustrating a simple reheat cycle wherein it shows the steam after passing through the high pressure stage being withdrawn and re-heated before passing on to the low pressure stage.

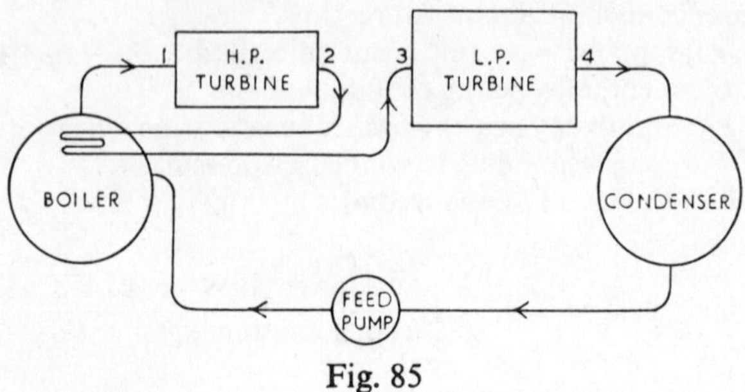

Fig. 85

Rankine efficiency

$$= \frac{\text{Heat energy given up by steam through engine}}{\text{Heat energy supplied by boiler and re-heater}}$$

$$= \frac{(h_1 - h_2) + (h_3 - h_4)}{(h_1 - h_{f4}) + (h_3 - h_2)}$$

ƒExample. In an ideal steam reheat cycle the steam is expanded in the first stage of a turbine from 40 bar 400°C to 3·0 bar. At this pressure the steam is passed through a reheater and its temperature is raised to 300°C at constant pressure. It then passes through the remainder of the turbine and expanded to 0·075 bar. Calculate the Rankine efficiency.

h_1 at 40 bar 400°C $= 3214$

Isentropic expansion from 40 bar 400°C to 3 bar:

$$\text{Entropy after expansion} = \text{Entropy before}$$
$$1·672 + x \times 5·321 = 6·769$$
$$x \times 5·321 = 5·097$$
$$x = 0·9579$$

h_2 at 3 bar, 0·9579 dry

$$= 561 + 0·9579 \times 2164 = 2634$$
$$h_3 \text{ at 3 bar 300°C} = 3070$$

Isentropic expansion from 3 bar 300°C to 0·075 bar:

$$\text{Entropy after expansion} = \text{Entropy before}$$
$$0·576 + x \times 7·674 = 7·702$$
$$x \times 7·674 = 7·126$$
$$x = 0·9287$$

h_4 at 0·075 bar, 0·9287 dry

$$= 169 + 0.9287 \times 2405 = 2403$$

h_{f_4} at 0·075 bar $= 169$

$$\text{Rankine efficiency} = \frac{(h_1 - h_2) + (h_3 - h_4)}{(h_1 - h_{f_4}) + (h_3 - h_2)}$$

$$= \frac{(3214 - 2634) + (3070 - 2403)}{(3214 - 169) + (3070 - 2634)}$$

$$= \frac{580 + 667}{3045 + 436} = \frac{1247}{3481}$$

$$= 0.3583 \text{ or } 35.83\% \quad \text{Ans.}$$

NOZZLES

The function of a steam nozzle is to produce a jet of high velocity steam which can be directed on to the blades of a turbine. The velocity is produced by allowing the steam to expand from a high pressure at the inlet to a lower pressure at the open exit, as no external work is done the decrease in enthalpy of the steam is converted into an increase in kinetic energy. The expansion of the steam is adiabatic because practically no heat energy is transferred in the very short time it takes the steam to pass along the nozzle. The ideal expansion would produce the maximum velocity at exit, this would be obtained if there were no friction and the expansion was isentropic.

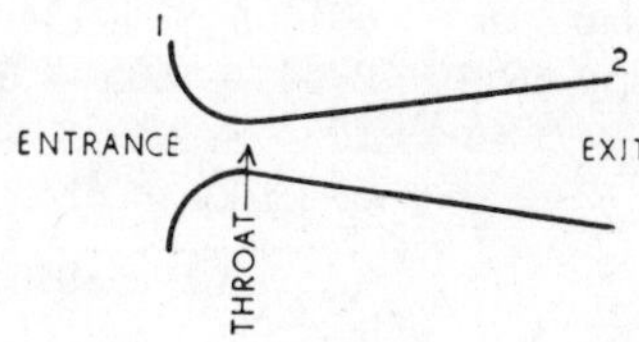

Fig. 86

Let conditions at entrance be:

Kinetic energy of steam $= \frac{1}{2} m v_1^2$

Enthalpy of steam $= H_1$

and conditions at exit:

Kinetic energy of steam $= \frac{1}{2} m v_2^2$

Enthalpy of steam $= H_2$

then, working in joules of energy,

$$\text{Gain in kinetic energy} = \text{Enthalpy drop}$$
$$\tfrac{1}{2}m(v_2{}^2 - v_1{}^2) = H_1 - H_2$$

for unit mass of steam,

$$\tfrac{1}{2}(v_2{}^2 - v_1{}^2) = h_1 - h_2$$

where h = specific enthalpy.

The inlet velocity of the steam is usually small compared with the exit velocity, therefore v_1 is negligible,

$$\tfrac{1}{2}v_2{}^2 = h_1 - h_2$$
$$v_2 = \sqrt{2(h_1 - h_2)}$$
$$\therefore \ v[\text{m/s}] = \sqrt{2 \times \text{spec. enthalpy drop [J/kg]}}$$

Note the units. Kinetic energy is in joules when m is the mass in kilogrammes and v is the velocity in m/s whereas specific enthalpy in the steam tables is given in kilojoules per kg. Hence, if h is to represent the specific enthalpy in kJ/kg as given in the tables instead of J/kg as above, then this must be multiplied by 10^3, thus

$$v[\text{m/s}] = \sqrt{2 \times 10^3 \times \text{spec. enthalpy drop [kJ/kg]}}$$

Since $\sqrt{2 \times 10^3} = 44 \cdot 72$, the above may be written

$$v = 44 \cdot 72 \sqrt{\text{spec. enthalpy drop}}$$

Example. Dry saturated steam enters a nozzle at 7 bar and leaves at 4 bar 0·98 dry. Find the velocity of the steam at exit.

$$\text{Tables page 4,} \quad 7 \text{ bar,} \quad h_g = 2764$$
$$4 \text{ bar,} \quad h_f = 605 \quad h_{fg} = 2134$$
$$\text{Spec. enthalpy drop} = 2764 - (605 + 0 \cdot 98 \times 2134)$$
$$= 2764 - 2696$$
$$= 68 \text{ kJ/kg} = 68 \times 10^3 \text{J/kg}$$
$$v = \sqrt{2 \times 68 \times 10^3} = 368 \cdot 7 \text{ m/s} \quad \text{Ans.}$$

MASS AND VOLUME FLOW. At any point along the nozzle:

$$\text{Volume flow [m}^3\text{/s]} = \text{area [m}^2\text{]} \times \text{velocity [m/s]}$$

$$\text{mass flow [kg/s]} = \frac{\text{volume flow [m}^3\text{/s]}}{\text{specific volume [m}^3\text{/kg]}}$$

$$\therefore \ \text{area [m}^2\text{]} \times \text{velocity [m/s]}$$
$$= \text{mass flow [kg/s]} \times \text{spec. vol. [m}^3\text{/kg]}.$$

Care must be taken to avoid confusion with the symbols. v usually represents velocity, and the same symbol is used to represent specific volume. In such cases where both velocity and specific volume would appear in the one expression or equation it is therefore advisable to write out the words fully or use an understandable abbreviation.

Taking the previous example further, if the exit area of the nozzle is 300 mm², the mass flow in kilogrammes per second can be calculated:

$$\text{Volume flow [m}^3\text{/s]} = \text{area [m}^2\text{]} \times \text{velocity [m/s]}$$
$$= 300 \times 10^{-6} \times 368 \cdot 7$$
$$= 0 \cdot 1106 \text{ m}^3/\text{s}$$

Tables page 4, 4 bar, $v_g = 0 \cdot 4623$ m³/kg

Spec. volume at 0·98 dry
$$= 0 \cdot 98 \times 0 \cdot 4623 = 0 \cdot 453 \text{ m}^3/\text{kg}$$

$$\text{mass flow} = \frac{0 \cdot 1106}{0 \cdot 453} = 0 \cdot 2442 \text{ kg/s.} \quad \text{Ans.}$$

f Example. Dry saturated steam enters a convergent-divergent nozzle at a pressure of 3·5 bar. The specific enthalpy drop between entrance and throat is 97 kJ/kg and the pressure there is 2·0 bar. The pressure at exit from the nozzle is 0·1 bar, the total possible specific enthalpy drop from entrance to exit is 534 kJ/kg and this is reduced by 12% due to the effect of friction in the divergent part of the nozzle. Calculate the nozzle area at the throat and the mouth to pass 0·113 kg of steam per second.

Tables page 4, 3·5 bar, $h_g = 2732$ $v_g = 0 \cdot 5241$

   2 bar, $h_f = 505$ $h_{fg} = 2202$ $v_g = 0 \cdot 8856$

   page 3, 0·1 bar, $h_f = 192$ $h_{fg} = 2392$ $v_g = 14 \cdot 67$

   h of steam at 2 bar $= h_g$ at 3·5 bar $- 97$

$$505 + x \times 2202 = 2732 - 97$$
$$x \times 2202 = 2130$$

dryness at throat $x = 0 \cdot 9674$

Spec. vol. of steam at throat
$$= 0 \cdot 9674 \times 0 \cdot 8856 = 0 \cdot 8567 \text{ m}^3/\text{kg}$$

Velocity through throat [m/s]

$$= \sqrt{2 \times \text{spec. enthalpy drop [J/kg]}}$$
$$= \sqrt{2 \times 10^3 \times 97} = 440 \cdot 4 \text{ m/s}$$

area [m²] × velocity [m/s] = mass flow [kg/s] × spec. vol. [m³/kg]

$$\therefore \text{Area [mm}^2] = \frac{0.113 \times 0.8567}{440.4} \times 10^6$$

$$= 219.9 \text{ mm}^2. \quad \text{Ans. (i)}$$

Spec. enthalpy drop between entrance and exit
$$= 0.88 \times 534 = 469.9 \text{ kJ/kg}$$
h of steam at 0·1 bar $= h$ at 3·5 bar $- 469.9$
$$192 + x \times 2392 = 2732 - 469.9$$
$$x \times 2392 = 2070$$
dryness at exit $x = 0.8654$
Spec. vol. of steam at exit
$$= 0.8654 \times 14.67 = 12.7 \text{ m}^3/\text{kg}$$
$$\text{Velocity at exit} = \sqrt{2 \times 10^3 \times 469.9}$$
$$= 969.4 \text{ m/s}$$
area [m²] × velocity [m/s] = mass flow [kg/s] × spec. vol. [m³/kg]

$$\therefore \text{Area [mm}^2] = \frac{0.113 \times 12.7}{969.4} \times 10^6$$

$$= 1480 \text{ mm}^2. \quad \text{Ans. (ii).}$$

VELOCITY DIAGRAMS FOR IMPULSE TURBINES

Fig. 87 illustrates the vector diagram of velocities at the entrance side of the rotor blades ("moving" blades). v_1 represents the

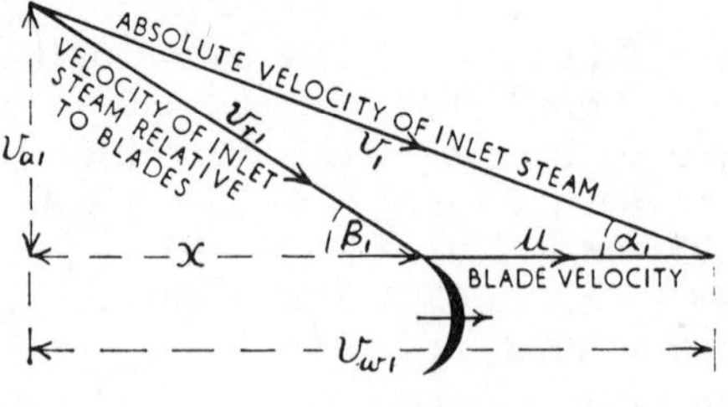

Fig. 87

absolute velocity of the steam directed towards the blades at an angle α_1 (as close as practicable) to the direction of their movement. The linear velocity of the blades is represented by u.

Drawing the vectors of the steam velocity and blade velocity towards a common point, the vector joining these to form a closed figure is the velocity of the steam relative to the moving blades, represented by v_{r_1}.

Velocity vector diagrams and relative velocity are explained in Volume 2 of this series (Applied Mechanics) to which the student should refer if revision is necessary.

The relative direction of the steam to the blades is β_1 and in order that the steam should glide on to the blades without shock, the entrance edge of the blades must be in this direction. Therefore β_1 is the *entrance angle of the blades.*

$v_{w1} = v_1 \cos \alpha_1$ is the component of the velocity of the steam jet in the direction of the blade movement, and is referred to as the *velocity of whirl at entrance.*

$v_{a1} = v_1 \sin \alpha_1$ this is the axial component of the steam jet, that is, the component in the direction of the axis of the turbine.

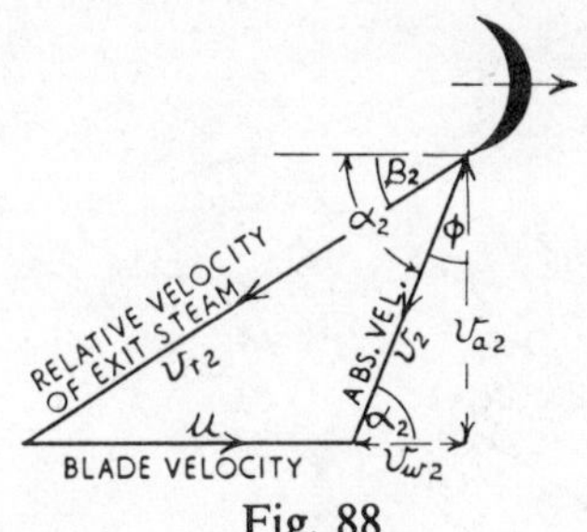

Fig. 88

Fig. 88 is the vector diagram of velocities at the exit side of the moving blades. β_2 is the *exit angle of the blades* and the relative velocity of the exit steam v_{r2} is in this direction. In impulse turbines, since there is no fall in steam pressure as it passes over the rotor blades, if friction is neglected, the relative velocity at exit is the same magnitude as the relative velocity at entrance, that is, $v_{r2} = v_{r1}$. Friction between the steam and the blade surface reduces the velocity and, to take friction into account, $v_{r2} = k v_{r1}$ where the velocity coefficient k is in the region of 0·8 to 0·95. Loss of kinetic energy in the steam due to friction over the blade surfaces is converted into heat energy.

In the simple impulse turbine the blades are often symmetrical, whence $\beta_1 = \beta_2$.

u is the vector of the blade velocity, v_2 is the vector of the absolute velocity of the exit steam. The direction of the exit steam is at α_2 to the direction of blade movement, or at angle ϕ to the axis of the turbine.

$v_{w2} = v_2 \cos \alpha_2$ this is the component of the exit steam velocity in the direction of blade movement and is referred to as the *velocity of whirl at exit*.

The axial component of the exit steam is $v_{a2} = v_2 \sin \alpha_2$

Example. Steam at a velocity of 600 m/s from a nozzle is directed on to the blades at 20° to the direction of blade movement. Calculate the inlet angle of the blades so that the steam will enter without shock when the linear velocity of the blades is 240 m/s. If the exit angle of the blades is the same as the inlet angle, find, neglecting blade friction, the magnitude and direction of the steam leaving the blades.

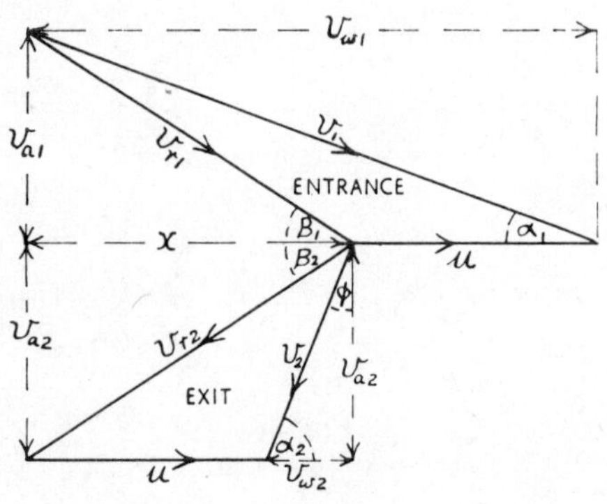

Fig. 89

Referring to Fig. 89:
$$v_{a1} = v_1 \sin \alpha_1 = 600 \times \sin 20° = 205\cdot2 \text{ m/s}$$
$$v_{w1} = v_1 \cos \alpha_1 = 600 \times \cos 20° = 563\cdot9 \text{ m/s}$$
$$x = v_{w1} - u = 563\cdot9 - 240 = 323\cdot9 \text{ m/s}$$

$$\tan \beta_1 = \frac{v_{a1}}{x} = \frac{205\cdot2}{323\cdot9} = 0\cdot6336$$

Inlet angle of blades $\beta_1 = 32° \, 22'$.   Ans. (i)

Neglecting friction across the blades, $v_{r2} = v_{r1}$ and since $\beta_2 = \beta_1$ then x is common to both entrance and exit diagrams, and $v_{a2} = v_{a1}$

$$v_{w2} = x - u$$
$$= 323 \cdot 9 - 240 = 83 \cdot 9 \text{ m/s}$$

$$\tan \phi = \frac{v_{w2}}{v_{a2}} = \frac{83 \cdot 9}{205 \cdot 2} = 0 \cdot 4089$$

$$\phi = 22° \ 14', \text{ and } \alpha_2 = 90° - 22° \ 14' = 67° \ 46'$$

$$v_2 = \frac{v_{w2}}{\sin \phi} = \frac{83 \cdot 9}{0 \cdot 3783} = 221 \cdot 8 \text{ m/s}$$

$$\left. \begin{array}{l} \text{Abs. velocity of steam at exit} = 221 \cdot 8 \text{ m/s} \\ \qquad \text{at } 67° \ 46' \text{ to blade movement} \\ \qquad \text{or } 22° \ 14' \text{ to turbine axis} \end{array} \right\} \quad \text{Ans. (ii)}$$

Both entrance and exit velocity diagrams contain the vector of the blade velocity u therefore they can be combined together by using the blade velocity as a common base, as shown in Fig. 90. This is a convenient diagram to solve either graphically or by calculation and will be used for all future problems of this type. A diagram is essential for reference in the calculations, students are advised to draw the diagram to scale rather than just a rough sketch, it will then serve as a check on the calculated results.

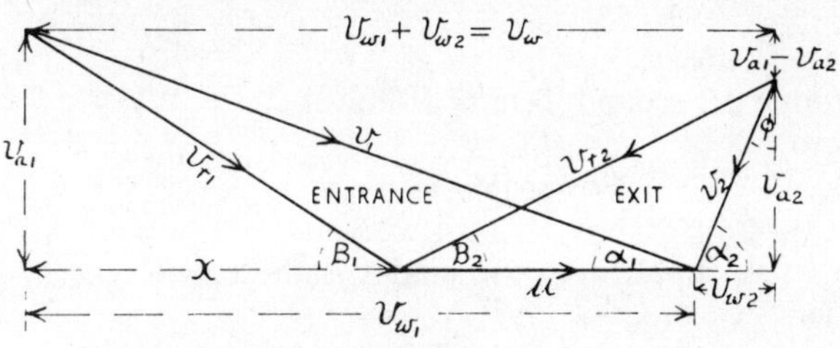

Fig. 90

FORCE ON BLADES

The effective velocity of the steam causing motion of the rotor blades is the component in the direction of movement of the blades. This is the velocity of whirl. The velocity of whirl at entrance is $v_{w1} = v_1 \cos \alpha_1$. The velocity of whirl at exit is $v_{w2} = v_2 \cos \alpha_2$. The effective change of velocity of the steam is the algebraic difference between v_{w1} and v_{w2}, let this be represented by

v_w. In the previous example, and in most cases, v_{w2} is in the opposite direction to v_{w1} then

$$\text{effective change of velocity} = v_{w1} - (- v_{w2})$$
$$\therefore v_w = v_{w1} + v_{w2}$$

If v_{w2} had been in the same direction as v_{w1} then the effective change of velocity would be $v_{w1} - v_{w2}$.

Since the direction of motion of the steam is changed as it passes over the blades, then the *blades* must exert a force *on the steam* to cause this change. From Newton's third law of motion which states that action and re-action are equal in magnitude and opposite in direction, this force is also the magnitude of the force exerted by the *steam* on the *blades*.

$$\text{Force [N]} = \text{mass [kg]} \times \text{acceleration [m/s}^2\text{]}$$
$$= \text{mass} \times \text{change of velocity per second}$$
$$= \text{change of momentum per second}$$

Therefore, if the mass rate of flow [kg/s] of the steam is represented by $\dot{m}$, then:

$$\text{Force on blades [N]} = \dot{m}\text{[kg/s]} \times \text{effective change of velocity}$$
$$= \dot{m}v_w.$$

$$\text{Work done} = \text{force} \times \text{distance}$$

Since the distance through which the force acts on the blades is the linear distance moved by the blades, we have:

$$\text{Work done per second [N m/s]} = \text{force on blades [N]} \times \text{blade velocity [m/s].}$$

Also,

$$\text{Work done per second [N m/s]} = \text{Power [J/s = W]}$$

therefore,

$$\text{Power [W]} = \dot{m}v_w u$$

The work supplied to the blades is the kinetic energy of the steam jet,

$$\text{work supplied per second} = \tfrac{1}{2}\dot{m}v_1^2$$

hence,

$$\text{Blade efficiency} = \frac{\text{work done on blades [J/s]}}{\text{work supplied [J/s]}}$$

$$= \frac{\dot{m}v_w u}{\tfrac{1}{2}\dot{m}v_1^2} = \frac{2u v_w}{v_1^2}$$

The *axial* force of the steam on the blades is due to the difference between the axial components of the steam velocities at entrance and exit, thus,

$$\text{axial thrust} = \dot{m}(v_{a1} - v_{a2})$$

ƒExample. Steam leaves the nozzles of a single stage impulse turbine at a velocity of 670 m/s at 19° to the plane of the wheel, and the steam consumption is 0·34 kg/s. The mean diameter of the blade ring is 1070 mm. Find (i) the inlet angle of the blades to suit a rotor speed of 83·3 rev/s. If the velocity coefficient of the steam across the blades is 0·9 and the blade exit angle is 32°, find (ii) the force on the blades, (iii) the power given to the wheel, and (iv) the blade efficiency.

Referring to Fig. 90:

Linear velocity of blades $=$ mean circumference $\times$ rev/s

$$u = \pi \times 1·07 \times 83·3 = 280 \text{ m/s}$$
$$v_{a1} = v_1 \sin \alpha_1 = 670 \times \sin 19° = 218·1 \text{ m/s}$$
$$v_{w1} = v_1 \cos \alpha_1 = 670 \times \cos 19° = 633·6 \text{ m/s}$$
$$x = v_{w1} - u = 633·6 - 280 = 353·6 \text{ m/s}$$

$$\tan \beta_1 = \frac{v_{a1}}{x} = \frac{218·1}{353·6} = 0·6169$$

Entrance angle $= 31° \, 40'$ Ans. (i)

$$v_{r1} = \frac{v_{a1}}{\sin \beta_1} = \frac{218·1}{\sin 31° \, 40'} = 415·6 \text{ m/s}$$

$$v_{r2} = 0·9 v_{r1} = 0·9 \times 415·6 = 374 \text{ m/s}$$
$$v_{w2} = v_{r2} \cos \beta_2 - u = 374 \cos 32° - 280$$
$$= 317·1 - 280 = 37·1 \text{ m/s}$$

Effective change of velocity $= v_w = v_{w1} + v_{w2}$
$$= 633·6 + 37·1 = 670·7 \text{ m/s}$$

Force on blades [N] $= \dot{m}[\text{kg/s}] \times v_w[\text{m/s}]$
$$= 0·34 \times 670·7$$
$$= 228·1 \text{ N}   \text{ Ans. (ii)}$$

Power [W $=$ J/s $=$ N m/s] $=$ force [N] $\times$ linear velocity [m/s]
$$= 228·1 \times 280$$
$$= 63870 \text{ W or } 63·87 \text{ kW}   \text{ Ans. (iii)}$$

$$\text{Blade efficiency} = \frac{\text{work done [J/s]}}{\text{work supplied [J/s]}}$$

$$= \frac{\dot{m} v_w u}{\tfrac{1}{2}\dot{m} v_1^2} = \frac{2uv_w}{v_1^2}$$

$$= \frac{2 \times 280 \times 670 \cdot 7}{670^2}$$

$$= 0 \cdot 8368 \text{ or } 83 \cdot 68\% \quad \text{Ans. (iv)}$$

VECTOR DIAGRAMS FOR REACTION TURBINES

It has been seen that, in the impulse turbine, the steam expands whilst passing through fixed nozzles and there is no expansion on its passage through the channels between the moving blades. All generation of velocity takes place in the nozzles.

In the reaction turbine, expansion of the steam takes place during its passage through the fixed (guide) blades which take the place of nozzles, and it also expands as it passes through the moving blades. Therefore the velocity of the steam is increased as it passes through the fixed blades, and the relative velocity of the steam to the moving blades is increased as it passes through the moving blades, so that v_{r_2} is greater than v_{r_1}.

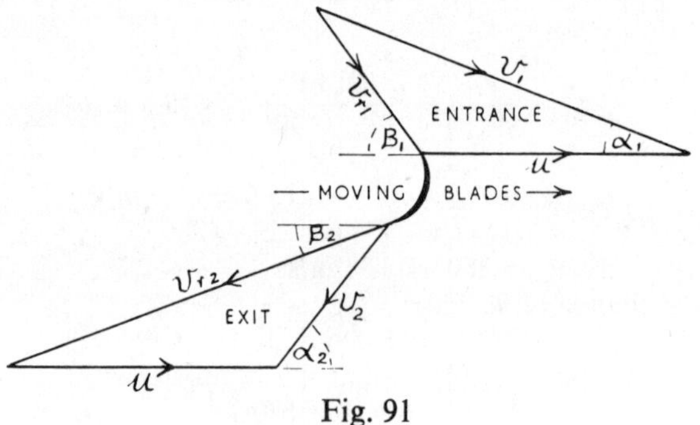

Fig. 91

In the reaction turbine, the fixed and moving blades are often of the same section and reversed in direction. In such cases, the entrance and exit angles of the fixed blades are the same as those of the moving blades, and the velocity vector diagram at entrance is identical with the velocity vector diagram at exit, therefore the combined diagram is symmetrical. See Figs. 91 and 92. Thus the relative velocity of the steam at exit from the moving blades is equal to the absolute velocity at entrance, $v_{r_2} = v_1$ and the absolute velocity of the steam at exit from the moving blades is equal to the relative velocity at entrance, $v_2 = v_{r_1}$. Hence, $\beta_2 = \alpha_1$ and $\alpha_2 = \beta_1$.

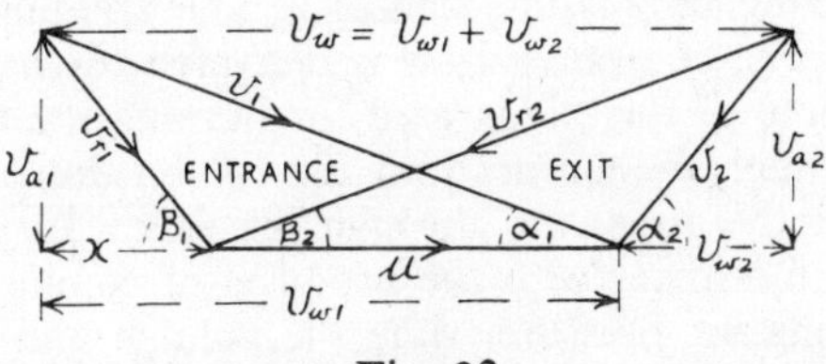

Fig. 92

Example. At one stage of a reaction turbine the velocity of the steam leaving the fixed blades is 90 m/s and the exit angle is 20°. The linear velocity of the moving blades is 60 m/s and the steam consumption is 0·55 kg/s. Assuming the fixed and moving blades to be of identical section, calculate (i) the entrance angle of the blades, (ii) the force on the blades, and (iii) the stage power.

$$v_{a1} = v_1 \sin \alpha_1 = 90 \times \sin 20° = 30 \cdot 78 \text{ m/s}$$
$$v_{w1} = v_1 \cos \alpha_1 = 90 \times \cos 20° = 84 \cdot 57 \text{ m/s}$$
$$x = v_{w1} - u = 84 \cdot 57 - 60 = 24 \cdot 57 \text{ m/s}$$

$$\tan \beta_1 = \frac{30 \cdot 78}{24 \cdot 57} = 1 \cdot 253$$

Entrance angle $= 51° \, 24'$ Ans. (i)

Effective change of velocity $= v_w = v_{w1} + v_{w2}$
(note that $v_{w2} = x$)
$$v_w = 84 \cdot 57 + 24 \cdot 57 = 109 \cdot 14 \text{ m/s}$$
$$\text{Force on blades [N]} = \dot{m}[\text{kg/s}] \times v_w[\text{m/s}]$$
$$= 0 \cdot 55 \times 109 \cdot 14$$
$$= 60 \cdot 02 \text{ N}   \text{Ans. (ii)}$$

$$\text{Power [W = J/s = N m/s]} = \text{force [N]} \times \text{velocity [m/s]}$$
$$= 60 \cdot 02 \times 60$$
$$= 3601 \text{ W or } 3 \cdot 601 \text{ kW}   \text{Ans. (iii)}$$

ƒ GAS TURBINES

Marine type gas turbines work on the constant pressure (Joule) cycle. Fig. 93 is a diagrammatic sketch of a simple open cycle gas turbine plant which consists of three essential parts—air compressor, combustion chamber, and turbine. Referring to Figs. 93 and 94, air is drawn in from the atmosphere and compressed from

$p_1V_1T_1$ to the higher pressure, smaller volume and higher temperature $p_2V_2T_2$. The compressed air is delivered to the combustion chamber. Some of this air is used for burning the fuel which is admitted through the burner into the combustion chamber, the remainder of the air passes through the jacket surrounding the burner housing, mixes with the products of combustion, and is heated at constant pressure while the volume and temperature increases, the conditions now being $p_3V_3T_3$. The mixture of hot air and gases now passes through the turbine where it expands to $p_4V_4T_4$ as it does work in driving the rotor. Finally, the gases exhaust at constant pressure.

Of the power developed in the turbine, some is absorbed in driving the compressor, the remainder being available for external use such as for propulsion, or driving an electric generator. A starting motor is fitted at the opposite end of the shaft to that for the external drive.

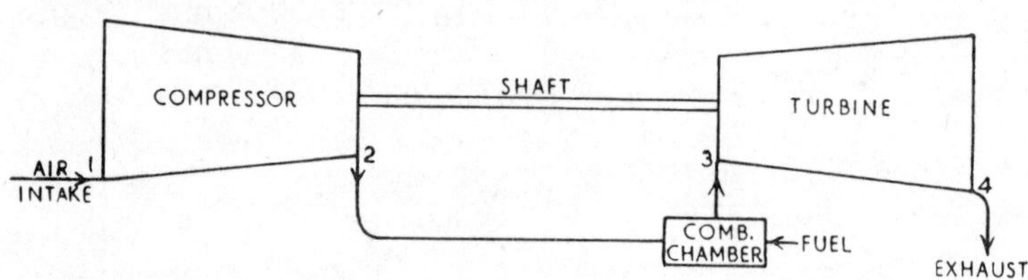

Fig. 93

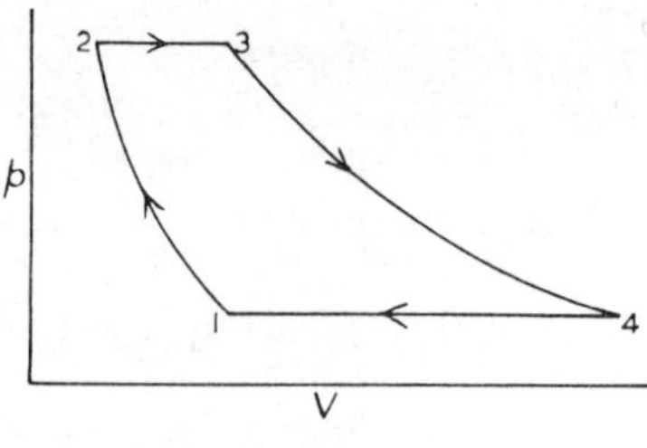

Fig. 94

In the ideal cycle, compression and expansion are isentropic between the same pressures p_2 and p_1 following the law $pV^\gamma = a$ constant. Referring to Fig. 94:

$$\text{Ideal thermal efficiency} = \frac{\text{heat energy converted into work}}{\text{heat energy supplied}}$$

$$= \frac{\text{heat supplied} - \text{heat rejected}}{\text{heat supplied}}$$

$$= 1 - \frac{\text{heat rejected}}{\text{heat supplied}}$$

$$= 1 - \frac{m \times c_P \times (T_4 - T_1)}{m \times c_P \times (T_3 - T_2)}$$

$$= 1 - \frac{T_4 - T_1}{T_3 - T_2} \qquad \dots \qquad \dots \qquad \text{(i)}$$

Let r_p = pressure ratio $= \dfrac{p_2}{p_1} = \dfrac{p_3}{p_4}$

$$\frac{T_2}{T_1} = \left\{ \frac{p_2}{p_1} \right\}^{\frac{\gamma-1}{\gamma}} \qquad \therefore T_2 = T_1 \times r_p^{(\gamma-1)/\gamma}$$

$$\frac{T_3}{T_4} = \left\{ \frac{p_3}{p_4} \right\}^{\frac{\gamma-1}{\gamma}} \qquad \therefore T_3 = T_4 \times r_p^{(\gamma-1)/\gamma}$$

$$T_3 - T_2 = r_p^{(\gamma-1)/\gamma} (T_4 - T_1)$$

Substituting this value of $T_3 - T_2$ into (i):

$$\text{Thermal efficiency} = 1 - \frac{1}{r_p^{(\gamma-1)/\gamma}} \dots \qquad \dots \qquad \dots \qquad \dots \qquad \text{(ii)}$$

Showing that the thermal efficiency depends upon the pressure ratio.

Example. In a simple gas turbine plant working on the ideal constant pressure cycle, air is taken into the compressor at 1 bar, 16°C, and delivered at 5·4 bar. If the temperature at turbine inlet is 700°C, calculate (i) the temperature at the end of compression, (ii) temperature at exit from the turbine, (ii) the ideal thermal efficiency. Take $\gamma = 1\cdot4$.

$$T_1 = 16 + 273 = 289\,\text{K}$$
$$T_3 = 700 + 273 = 973\,\text{K}$$

$$\text{Pressure ratio } r_p = \frac{p_2}{p_1} = \frac{p_3}{p_4} = \frac{5 \cdot 4}{1} = 5 \cdot 4$$

$$\frac{\gamma - 1}{\gamma} = \frac{0 \cdot 4}{1 \cdot 4} = \frac{2}{7}$$

$$r_p^{(\gamma-1)/\gamma} = 5 \cdot 4^{2/7} = 1 \cdot 619$$

$$\frac{T_2}{T_1} = \left\{ \frac{p_2}{p_1} \right\}^{\frac{\gamma-1}{\gamma}}$$

$$T_2 = 289 \times 1 \cdot 619 = 467 \cdot 8 \, \text{K}$$

Temperature at end of compression
$$= 467 \cdot 8 - 273 = 194 \cdot 8°\text{C.} \quad \text{Ans. (i)}$$

$$\frac{T_4}{T_3} = \left\{ \frac{p_4}{p_3} \right\}^{\frac{\gamma-1}{\gamma}}$$

$$T_4 = \frac{973}{1 \cdot 619} = 601 \, \text{K}$$

Temperature at end of expansion
$$= 601 - 273 = 328°\text{C.} \quad \text{Ans. (ii)}$$

$$\text{Thermal efficiency} = 1 - \frac{T_4 - T_1}{T_3 - T_2}$$

$$= 1 - \frac{601 - 289}{973 - 467 \cdot 8}$$

$$= 0 \cdot 3824 \text{ or } 38 \cdot 24\% \quad \text{Ans. (iii)}$$

Alternatively,

$$\text{Thermal efficiency} = 1 - \frac{1}{r_p^{(\gamma-1)/\gamma}}$$

$$= 1 - \frac{1}{1 \cdot 619} = 0 \cdot 3823$$

HEAT EXCHANGER. By including a heat exchanger, some of the heat energy in the exhaust gases can be utilized by transferring it to the air before it enters the combustion chamber, resulting in less fuel being needed to raise the temperature of the air to the required turbine inlet temperature and therefore increasing the thermal efficiency. A typical arrangement is shown diagrammatically in Fig. 95.

The lowest temperature entering the heat exchanger is that of the compressed air T_2 and the highest temperature is that of the exhaust gases from the turbine T_5. The difference $T_5 - T_2$ is the overall temperature range for the exchanger. The air, on its passage through the exchanger is increased in temperature from T_2 to T_3, that is, an increase of $T_3 - T_2$. The ratio of the increase in air temperature to the overall temperature range is termed the *thermal ratio* or *effectiveness* of the heat exchanger, thus:

$$\text{Thermal ratio of heat exchanger} = \frac{T_3 - T_2}{T_5 - T_2}$$

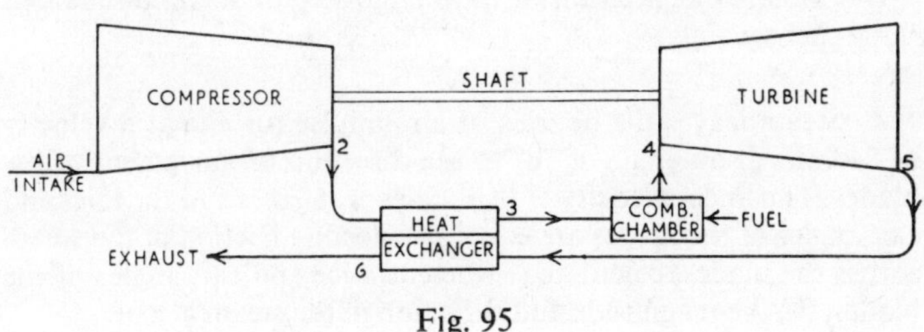

Fig. 95

TEST EXAMPLES 13

1. Steam is expanded through a nozzle and the enthalpy drop per kg of steam from the initial pressure to the final pressure is 60 kJ. Neglecting friction, find the velocity of discharge and the exit area of the nozzle to pass 0·2 kg/s if the specific volume of the steam at exit is 1·5 m³/kg.

2. Dry saturated steam at 8 bar is expanded in turbine nozzles to a pressure of 5 bar 0·97 dry. Find (i) the velocity at exit. If the area at exit is 14·5 cm² find (ii) the mass flow of the steam in kg/s.

f3. Dry saturated steam at 14 bar is expanded in a turbine nozzle to 10 bar, expansion following the law $pV^n = $ constant, where the value of n is 1·135, calculate:
    (i) the dryness fraction of the steam at exit,
    (ii) the enthalpy drop through the nozzle per kg of steam,
    (iii) the velocity of discharge,
    (iv) the area of nozzle exit in mm² per kg of steam discharged per second.

4. Steam leaves the nozzles of an impulse turbine at a velocity of 840 m/s at an angle of 18° to the direction of movement of the blades. The linear velocity of the blades is 360 m/s and the inlet and exit angles of the blades are equal. Neglecting friction of the steam across the blades, calculate (i) the entrance and exit angles of the blades, (ii) the magnitude and direction of the steam at exit.

5. The velocity of the steam from the nozzles of a turbine is 450 m/s, the angle of the nozzles to blade motion is 20° and the blade entrance angle is 33°. Find (i) the linear velocity of the blades so that the steam enters without shock, and (ii) the rotational speed of the rotor in rev/s if the mean diameter of the blade ring is 660 mm.

f6. Steam at a velocity of 750 m/s is directed on to the blades of an impulse turbine at an angle of 22° to their motion. The blade velocity is 225 m/s and the exit angle of the blades is 31°. Friction between the steam and blade surfaces reduces the relative velocity of the steam by 15%. Calculate (i) the entrance angle of the blades, and (ii) the magnitude and direction of the absolute velocity of the steam at exit from the blades.

*f*7. Steam enters the nozzles of an impulse turbine at 20 bar, 300°C, and leaves at 3 bar, 0·98 dry. The angle of the nozzles is 20° to the direction of motion of the blades and the blade velocity is 380 m/s. Calculate (i) the velocity of the steam leaving the nozzles, and (ii) the inlet angle of the blades.

*f*8. In a single-stage impulse turbine, the velocity of the steam from the nozzles is 900 m/s at 18° to the plane of the wheel. The linear velocity of the blades is 330 m/s and the steam consumption 0·095 kg/s. The blades are symmetrical in shape and the blade friction loss is 10%. Find (i) the tangential force on the blades, (ii) the power developed, and (iii) the axial thrust.

*f*9. In an impulse turbine the theoretical enthalpy drop of the steam through the nozzles is 312·5 kJ/kg and 10% of this is lost in friction in the nozzles. The nozzle angle is 20°, the inlet angle of the blades is 35°, and the absolute velocity of the steam leaving the blades is 204 m/s in the direction of the axis of the turbine. Calculate on the basis of one kg of steam supplied per second:

(i) blade velocity so that there is no shock at steam entry,
(ii) blade angle at exit,
(iii) energy lost due to friction of the steam across the blades,
(iv) axial thrust,
(v) power supplied,
(vi) efficiency of the blading.

10. At a certain stage of a reaction turbine, the steam leaves the guide blades and enters the moving blades at an absolute velocity of 243 m/s at an angle of 23° to the plane of rotation, and the blade velocity is 159 m/s. Fixed and moving blades have the same inlet and exit angles and the steam flow is 0·9 kg/s. Calculate (i) the inlet angle of the blades, (ii) the force on the blades, and (iii) the stage power.

*f*11. Air is drawn into a gas turbine working on the constant pressure cycle, at 1 bar 21°C and compressed to 5·7 bar. The temperature at the end of heat supply is 680°C. Taking expansion and compression to be adiabatic where $c_V = 0·718$ kJ/kg K, $c_P = 1·005$ kJ/kg K, calculate (i) temperature at end of compression, (ii) temperature at exhaust, (iii) heat energy supplied per kg at constant pressure, (iv) increase in internal energy per kg from inlet to exhaust, (v) ideal thermal efficiency.

CHAPTER 14

BOILERS AND COMBUSTION

There are two distinct types of boiler, (i) the fire-tube, in which the hot gases from the combustion of the fuel in the furnace pass through the tubes, while the water is around the outside of the tubes, and (ii) the water-tube boiler in which the water flows through the tubes while the hot gases pass around the outside of the tubes.

The Scotch boiler is the most popular of the fire-tube type but because of the advantages of the water-tube boiler, some of which are given below, very few Scotch boilers are now being manufactured. There are however, many still in regular use as main and auxiliary units.

The main advantages of water-tube boilers are (i) circulation of the water is natural and immediate upon lighting up the fires, thus steam can be raised from cold water in the matter of a few hours without the danger experienced in Scotch boilers of temperature differences between top and bottom parts of the boiler causing unequal expansion resulting in mechanical straining; (ii) the steam and water drums are small in diameter compared with the large shell of the Scotch boiler, therefore they are stronger and suitable for much higher steam pressures; (iii) less mass of water is carried in the boiler, hence a saving in weight; (iv) less space is taken up for the same output.

The water-tube boiler does however, require greater skill in operating and maintaining, especially with regard to the purity of the feed water, and its upkeep is more costly than the Scotch boiler.

There are various types of water-tube boilers in general use, such as the Babcock and Wilcox, Yarrow, and Foster-Wheeler, and each manufacturer has a variety of designs to offer depending upon requirements.

Fig. 96 shows diagrammatically the principle of the Foster-Wheeler D-type boiler. This consists of two horizontal cylindrical drums, one above the other, the top being the steam-and-water drum (referred to briefly as the steam drum) and the bottom the water drum, these are connected directly by vertical generating tubes, and by other tubes via headers. One set of tubes from the steam drum are bent at approximately the middle of their length so that the upper portion forms part of the roof of the combustion

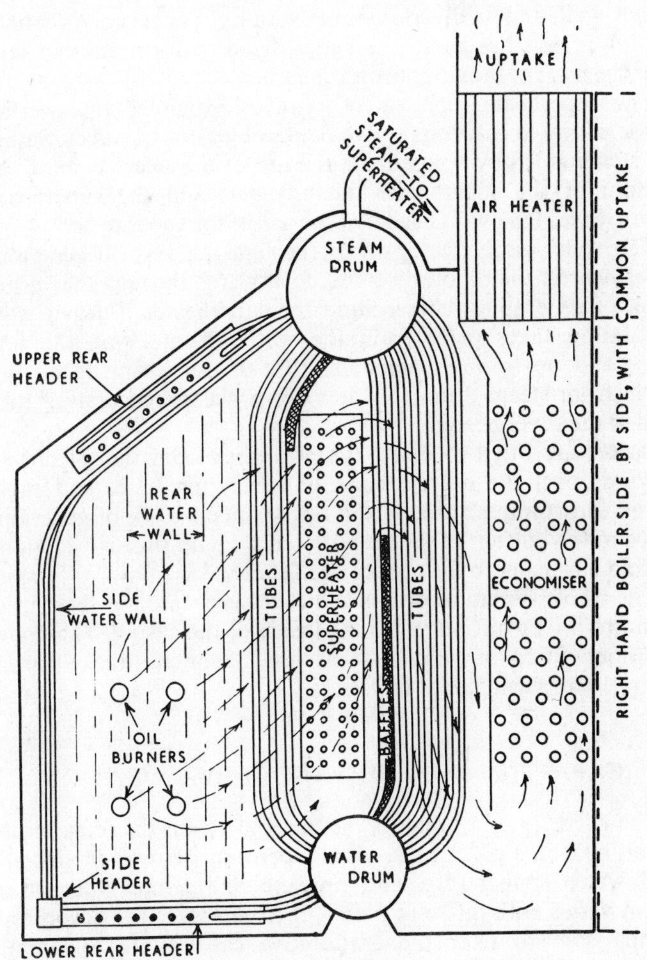

FOSTER WHEELER D TYPE BOILER
Fig. 96

chamber, and the lower portion forms the side wall, these tubes
are connected to a side header at combustion chamber floor level
which, in turn, is connected to the water drum by a bottom set
of tubes to form part of the floor of the combustion chamber.
Other sets of tubes, by being bent at right angles, connect the
steam drum to an upper rear header to form another roof layer,
vertical tubes from upper rear header to lower rear header form

a back wall, and finally more tubes with right angle bends connect the lower rear header to the water drum to form another layer over the floor of the combustion chamber.

The superheater consists of U-tubes expanded into vertical headers located between the two banks of generating tubes. Saturated steam is led by a pipe from the top of the steam drum to the bottom of the superheater inlet header, and the superheated steam leaves the outlet header at the top of the superheater.

The water circulates upwards through the vertical generating tubes nearest the oil burners and downwards through the vertical bank of generating tubes behind the superheater. Further water circulation takes place from the water drum along the floor tubes and up through the wall tubes to the steam drum, then down from steam drum to water drum via the generating tubes behind the super heater.

Baffles are fitted to direct the hot gases from the oil burners upwards over the nearest bank of generating tubes and superheater, downwards through the further main bank of generating tubes to the bottom of the economiser, then up through economiser and air-heater to the uptake and funnel.

The whole system of drums and tubes is encased in sheet metal with fire-brickwork bolted to the inside of the casing. The boilers are often fitted in pairs so that one common uptake can be arranged for each pair of boilers.

CAPACITY AND EQUIVALENT EVAPORATION

The *capacity* of a boiler is the mass of steam that can be produced by it in a given time, usually per hour. However, since the feed water temperature, and pressure and temperature of the steam varies with different plants, it is necessary for purposes of comparison, to refer the evaporative capacity to a common standard. This is the imaginary condition of assuming the feed water temperature as 100°C, to be converted into dry saturated steam at 100°C, and to reduce this to the mass of steam evaporated by unit mass of fuel burned in the furnace. This basis of comparison is termed the *equivalent evaporation, per kg of fuel, from and at* 100°C. For brevity, "per kg of fuel" is usually omitted.

Thus, if for example a boiler produces 60 Mg of steam per hour at 40 bar and 450°C, from feed water at 130°C, when 4800 kg of fuel are burned per hour, then:

       Tables page 7, steam 40 bar 450°C, $h = 3330$
          „   „   4, water at 130°C, $h = 546$

Heat energy transferred to each kg of steam
$$= 3330 - 546 = 2784 \text{ kJ}$$
Actual evaporative capacity is 60×10^3 kg/h therefore, total heat energy transferred to steam
$$= 2784 \times 60 \times 10^3 \text{ kJ/h}$$

Now suppose the temperature of the feed water was 100°C and dry saturated steam at 100°C was produced, the heat energy required per kg of steam would be h_{fg} at 100°C which (tables page 2) is 2256·7 kJ, therefore the mass of steam that would be produced under these conditions of "from and at 100°C" would be:

$$\frac{\text{total heat energy transferred to steam per hour}}{\text{heat energy required for each kg}}$$

$$= \frac{2784 \times 60 \times 10^3}{2256 \cdot 7} = 74 \times 10^3 \text{ kg/h or 74 Mg/h}$$

This is the capacity of the boiler from and at 100°C.

Since 4800 kg of fuel are burned per hour, the mass of steam that would be produced by each kg of fuel is,

$$\frac{74 \times 10^3}{4800} = 15 \cdot 42 \text{ kg steam/kg fuel}$$

This is the equivalent evaporation, per kg of fuel, from and at 100°C.

Summing up:

if m_s = actual mass of steam generated per unit mass of fuel burned

h_1 = enthalpy per kg of steam

h_w = enthalpy per kg of feed water

h_{fg} 100°C = enthalpy of evaporation per kg at 100°C then, equivalent evaporation from and at 100°C

$$= \frac{m_s(h_1 - h_w)}{h_{fg}100°C}$$

BOILER EFFICIENCY

The efficiency of a boiler is the ratio of the heat energy transferred to the feed water in converting it into steam, to the heat energy supplied to the boiler by the combustion of the fuel.

The heat energy transferred to the water to produce steam is the difference between the enthalpy of the steam leaving the boiler and the enthalpy of the feed water entering the boiler, thus:

$$\text{mass of steam generated} \times (h_1 - h_w)$$

The heat energy supplied to the boiler is the energy released during combustion of the fuel, which is the product of the mass of fuel burned and its calorific value. The calorific value, as previously explained (Chapter 7) is the heat energy given off during complete combustion of unit mass of the fuel. Therefore,

$$\text{Boiler efficiency} = \frac{\text{Heat energy transferred to water and steam}}{\text{Heat energy supplied by fuel}}$$

$$= \frac{\text{mass of steam} \times (h_1 - h_w)}{\text{mass of fuel} \times \text{calorific value}}$$

or, if we let m_s represent the mass of steam generated per unit mass of fuel burned, then,

$$\text{Boiler efficiency} = \frac{m_s(h_1 - h_w)}{\text{calorific value of the fuel}}$$

Example. An oil-fired boiler working at a pressure of 15 bar generates 14·5 kg of steam per kg of fuel burned. The feed water temperature is 95°C and the steam leaves the boiler 0·98 dry. If the calorific value of the oil is 42 MJ/kg, calculate the thermal efficiency of the boiler and the equivalent evaporation from and at 100°C.

$$\text{Tables page 4, steam 15 bar,} \quad h_f = 845 \quad h_{fg} = 1947$$
$$\text{,, \quad ,, \quad 2, water 95°C,} \quad h = 398$$
$$\text{Boiler steam,} \quad h_1 = h_f + xh_{fg}$$
$$= 845 + 0 \cdot 98 \times 1947 = 2753$$
Heat energy transferred to steam per kg of fuel burned
$$= m_s(h_1 - h_w)$$
$$= 14 \cdot 5 \,(2753 - 398)$$
$$= 14 \cdot 5 \times 2355 \text{ kJ}$$
Heat energy supplied to boiler per kg of fuel burned
$$= 42 \times 10^3 \text{ kJ}$$

$$\text{Boiler efficiency} = \frac{14 \cdot 5 \times 2355}{42 \times 10^3}$$

$$= 0 \cdot 8132 \quad \text{or} \quad 81 \cdot 32\%. \quad \text{Ans. (i)}$$

Equivalent evaporation, per kg of fuel, from and at $100°C$

$$= \frac{m_s(h_1 - h_w)}{h_{fg} \, 100°C}$$

$$= \frac{14 \cdot 5 \times 2355}{2256 \cdot 7}$$

$$= 15 \cdot 13 \text{ kg steam/kg fuel.} \quad \text{Ans. (ii)}$$

BOILER AND FEED WATER

Every precaution must be taken to maintain the boiler water as pure as possible. One of the most common causes of contamination of the feed water is a leakage of sea water into the condensate due to a leaking tube in the condenser.

There is always a certain amount of leakage in the steam circuit through the engines and this loss must be made up to maintain the working level of the water in the boiler. For Scotch boilers, make-up feed may be taken direct from the reserve fresh water tanks, or distilled water may be produced by evaporating sea water in evaporators. In water-tube boilers, because of their less robust construction, the higher pressures and temperatures, and the high rate of evaporation, the boiler water must be as near to pure as it is possible to maintain it and it is usual to pass all make-up feed water through evaporators before feeding into the boilers.

The quantity of dissolved solids present in water is expressed in parts of solids per million parts of water, abbreviated to *parts per million* and represented by p.p.m. This is the same ratio as the grammes of solids in one million grammes of water. Note that 10^6 g $= 10^3$ kg $= 1$ tonne, and this is the mass of one cubic metre of pure water. For example, if there are 80 g of dissolved solids in one m^3 of water, it is expressed as 80 p.p.m. As a further example, if a boiler contains 5 tonne of water having 150 p.p.m. dissolved solids, the total mass of dissolved solids in the boiler is

$$5 \times 150 = 750 \text{ g.}$$

Equations dealing with the amount of dissolved solids in evaporators and boilers are dealt with on this basis. Thus, in boilers with a contaminated feed, the equation is built upon the simple principle:

$$\begin{array}{ccc}\text{initial mass [g] of} \\ \text{solids in boiler}\end{array} + \begin{array}{c}\text{mass [g] of solids} \\ \text{put in with feed}\end{array} = \begin{array}{c}\text{final mass [g] of} \\ \text{solids in boiler}\end{array}$$

$$\begin{array}{c}\text{mass [t] of} \\ \text{water in} \\ \text{boiler}\end{array} \times \begin{array}{c}\text{initial} \\ \text{p.p.m.}\end{array} + \begin{array}{c}\text{mass [t]} \\ \text{of feed}\end{array} \times \begin{array}{c}\text{feed} \\ \text{p.p.m.}\end{array} = \begin{array}{c}\text{mass [t] of} \\ \text{water in} \\ \text{boiler}\end{array} \times \begin{array}{c}\text{final} \\ \text{p.p.m.}\end{array}$$

In evaporators where it is common to have a constant blow-down to maintain a steady pre-determined density of the water in the evaporator, the equation would be based upon

mass of solids put in = mass of solids blown out

mass of feed × feed p.p.m. = mass blown out × blow out p.p.m.

Example. A boiler initially contains 4 tonne of water at 80 p.p.m. If the evaporation rate is 500 kg/h and the feed water contains 150 p.p.m. of dissolved solids, calculate (i) the density (p.p.m.) of the boiler water after 12 hours, and (ii) the time for the water to reach 2000 p.p.m. from the time of initial condition.

$$\begin{aligned}\text{Amount of feed} &= \text{evaporation rate} \\ &= 500 \text{ kg/h} = 0.5 \text{ tonne/h} \\ &= 0.5 \times 12 \text{ tonne in 12 hours}\end{aligned}$$

solids in initially + solids put in = solids in finally

$$\begin{array}{c}\text{water in} \\ \text{boiler}\end{array} \times \begin{array}{c}\text{initial} \\ \text{p.p.m.}\end{array} + \begin{array}{c}\text{amount} \\ \text{of feed}\end{array} \times \begin{array}{c}\text{feed} \\ \text{p.p.m.}\end{array} = \begin{array}{c}\text{water in} \\ \text{boiler}\end{array} \times \begin{array}{c}\text{final} \\ \text{p.p.m.}\end{array}$$

$$4 \times 80 + 0.5 \times 12 \times 150 = 4 \times \text{final density}$$
$$320 + 900 = 4 \times \text{final density}$$

$$\text{Final density} = \frac{1220}{4} = 305 \text{ p.p.m. Ans. (i)}$$

Let t = time in hours for boiler water to reach 2000 p.p.m. then feed in t hours = $0.5t$ tonne:

$$\begin{array}{c}\text{water in} \\ \text{boiler}\end{array} \times \begin{array}{c}\text{initial} \\ \text{p.p.m.}\end{array} + \begin{array}{c}\text{amount} \\ \text{of feed}\end{array} \times \begin{array}{c}\text{feed} \\ \text{p.p.m.}\end{array} = \begin{array}{c}\text{water in} \\ \text{boiler}\end{array} \times \begin{array}{c}\text{final} \\ \text{p.p.m.}\end{array}$$

$$4 \times 80 + 0.5t \times 150 = 4 \times 2000$$
$$75t = 7680$$
$$t = 102.4 \text{ hours Ans. (ii)}$$

If an analysis of the feed water is given so that a distinction can be made with regard to the solids that remain in solution and those that precipitate to form scale when the water is heated, then the density in the evaporator or boiler is taken as that due to permanently soluble solids.

The density of sea water and the composition of the dissolved solids vary throughout different parts of the world. A typical sample could be taken as dissolved solids 32 280 p.p.m. and mass analysis of the solid matter:

Sodium chloride	79·3%
Magnesium chloride	10·2%
Magnesium sulphate	6·1%
Calcium sulphate	3·8%
Calcium bicarbonate	0·6%

Scale is formed from magnesium sulphate, calcium sulphate and calcium bicarbonate. All scales are bad heat conductors and therefore hinder transfer of heat, the result being overheating of the metal heating surfaces with consequent loss of strength and possibility of collapse. This is more serious in water-tube boilers with the added danger of partial blockage of the tubes. The remainder of the solids may be regarded as permanently soluble solids.

COMBUSTION

Combustion of fuel is the chemical combination, at high temperature, of the combustible elements in the fuel with oxygen, heat energy being released in the process.

In furnaces of boilers and cylinders of internal combustion engines, the oxygen is obtained from an air supply, air being composed of approximately 23% oxygen and 77% nitrogen, by mass. The oxygen is the active element. Nitrogen, being an inert gas, takes no active part, it acts as a moderator, dilutes the products of combustion and, as it absorbs some of the heat energy produced it reduces the temperature of combustion.

The principal combustible elements in fuels are carbon and hydrogen, others may be present in small quantities, such as sulphur.

The air must be intimately mixed with the fuel and the amount of fuel which can be burned depends upon the quantity of air supplied. An excess over the theoretical minimum quantity of air for complete combustion is always necessary, the amount of excess depending upon the design of the combustion space and conditions under which the fuel is burned. If there is an insufficient air supply, combustion will not be complete, one indication of this being black smoke. If two much air is supplied, an unnecessary amount of heat energy will be carried away to waste. Each case represents a loss of efficiency.

Taking the composition of air by mass as 23% oxygen and 77% nitrogen, then 23 kg of oxygen will be obtained from 100 kg of air. In the same proportion, to obtain 1 kg of oxygen, the supply of air must be $\dfrac{100}{23}$ kg. Thus, the theoretical minimum mass of air required is $\dfrac{100}{23}$ times the mass of oxygen needed for complete combustion.

To have a better understanding of combustion and enable the engineer to burn fuels correctly and efficiently, an elementary knowledge of the chemical equations involved is necessary.

Elements are substances in their simplest form and consist of molecules which are identical. They are represented by a symbol, usually the first letter of their names except when necessary to distinguish between two whose names have the same first letter. The molecules of some elements consist of single atoms, such as carbon (C) and sulphur (S). Other elements consist of molecules of two atoms each and these are distinguished by the subscript 2, examples of these are hydrogen (H_2), nitrogen (N_2) and oxygen (O_2).

Compounds are chemical combinations of different elements and denoted by placing together the symbols of the elements which constitute the compound. Thus, each molecule of water (and steam) is composed of two atoms of hydrogen and one atom of oxygen, and is therefore written H_2O. Carbon dioxide is represented by CO_2 which signifies that each molecule of carbon dioxide is composed of one atom of carbon and two atoms of oxygen. The number of molecules, when more than one, is represented by placing that number in front, thus $3CO_2$ represents three molecules of carbon dioxide.

Atomic weights are pure numbers representing the relative masses of the atoms. For instance, if the atomic weight of hydrogen is 1 and the atomic weight of sulphur is 32, it means that the mass of one atom of sulphur is 32 times greater than the mass of one atom of hydrogen. These numbers are purely relative and neither the actual weight nor actual mass.

The *molecular weight* is the sum of the atomic weights of the atoms of which the molecule is composed.

A list of the relative masses of the substances involved in calculations on the combustion of fuels is given below, to the nearest whole number which is sufficiently accurate for practical purposes.

Substance	Symbol	Atomic weight	Molecular weight
ELEMENTS:			
Hydrogen	H_2	1	$2 \times 1 = 2$
Carbon	C	12	$1 \times 12 = 12$
Nitrogen	N_2	14	$2 \times 14 = 28$
Oxygen	O_2	16	$2 \times 16 = 32$
Sulphur	S	32	$1 \times 32 = 32$
COMPOUNDS:			
Water, Steam	H_2O		$2 \times 1 + 1 \times 16 = 18$
Carbon dioxide	CO_2		$1 \times 12 + 2 \times 16 = 44$
Sulphur dioxide	SO_2		$1 \times 32 + 2 \times 16 = 64$
Carbon monoxide	CO		$1 \times 12 + 1 \times 16 = 28$

CALORIFIC VALUES. The calorific value (c.v.) of a substance is the amount of heat energy released during complete combustion of unit mass of that substance and, for fuels, it is usually expressed in megajoules of energy per kilogramme of mass [MJ/kg].

When hydrogen burns it combines with oxygen to form steam and the heat energy released is about 144 MJ per kg of hydrogen.

Carbon, if supplied with sufficient oxygen, will burn completely to carbon dioxide and in doing so, about 33·7 MJ of heat energy is released per kg of carbon. If there is a deficiency of oxygen, some or all of the carbon will burn to carbon monoxide and the heat energy released by each kg of carbon is then only about one-third of that when carbon dioxide is formed. Thus there is a great loss when carbon monoxide is produced due to an insufficient air supply to the fuel.

In the burning of sulphur, it chemically combines with oxygen to form sulphur dioxide and about 9·3 MJ of heat energy is released per kg of sulphur.

CHEMICAL EQUATIONS represent the proportions in which the elements combine and, by substituting the atomic weights, the relative mass of oxygen required for the burning of each combustible element can be calculated.

Combustion of hydrogen to steam,

$$2H_2 + O_2 = 2H_2O$$

inserting atomic weights,

$$2 \times (2 \times .1) + 2 \times 16 = 36$$
$$4 \text{ kg } H_2 + 32 \text{ kg } O_2 = 36 \text{ kg } H_2O$$
$$\text{reducing, } 1 \text{ kg } H_2 + 8 \text{ kg } O_2 = 9 \text{ kg } H_2O$$

That is, 1 kg of hydrogen requires 8 kg of oxygen to burn it completely, and this produces 9 kg of steam.

Combustion of carbon to carbon dioxide,

$$C + O_2 = CO_2$$
$$1 \times 12 + 2 \times 16 = 44$$
$$12 \text{ kg } C + 32 \text{ kg } O_2 = 44 \text{ kg } CO_2$$
$$\text{or, } 1 \text{ kg } C + 2\tfrac{2}{3} \text{ kg } O_2 = 3\tfrac{2}{3} \text{ kg } CO_2$$

Thus, 1 kg of carbon requires $2\tfrac{2}{3}$ kg of oxygen to burn it completely and this produces $3\tfrac{2}{3}$ kg of carbon dioxide.

Combustion of sulphur to sulphur dioxide,

$$S + O_2 = SO_2$$
$$1 \times 32 + 2 \times 16 = 64$$
$$32 \text{ kg } S + 32 \text{ kg } O_2 = 64 \text{ kg } SO_2$$
$$1 \text{ kg } S + 1 \text{ kg } O_2 = 2 \text{ kg } SO_2$$

Hence, 1 kg of sulphur requires 1 kg of oxygen to burn it and 2 kg of sulphur dioxide is produced.

The analysis of some fuels show that a little oxygen is present, in these cases it is usual to assume that this oxygen is combined with some of the hydrogen in the form of water and therefore this portion of hydrogen cannot be burned. We have seen above that hydrogen and oxygen combine in the proportion of 1 to 8, hence the amount of hydrogen which is not available for combustion is one-eighth of the mass of oxygen present in the fuel. The remaining hydrogen in the fuel is referred to as *available hydrogen*, thus,

$$\text{Available hydrogen} = H_2 - \frac{O_2}{8}$$

From the foregoing notes we can now write the expressions:

$$\text{Calorific value} = 33 \cdot 7\,C + 144 \left[H_2 - \frac{O_2}{8} \right] + 9 \cdot 3\,S \text{ MJ/kg}$$

$$\text{Oxygen required } = 2\tfrac{2}{3}\,C + 8\left(H_2 - \frac{O_2}{8}\right) + S \text{ kg } O_2/\text{kg fuel}$$
$$\text{or } 2\tfrac{2}{3}\,C + 8\,H_2 - O_2 + S$$

$$\text{Minimum air reqd. } = \frac{100}{23} \times \text{oxygen required}$$

$$= \frac{100}{23}\left\{2\tfrac{2}{3}\,C + 8\left(H_2 - \frac{O_2}{8}\right) + S\right\} \text{ kg air/kg fuel}$$

Example. An oil fuel is composed of 86% carbon, 11% hydrogen, 2% oxygen and 1% impurities. Calculate the calorific value and theoretical minimum mass of air required to burn 1 kg of the fuel, taking the calorific values of carbon and hydrogen as 33·7 and 144 MJ/kg respectively.

$$\text{Available hydrogen } = H_2 - \frac{O_2}{8}$$

$$= 0·11 - \frac{0·02}{8} = 0·1075 \text{ kg}$$

Heat energy from 0·86 kg of carbon
$$= 0·86 \times 33·7 = 28·98 \text{ MJ}$$
Heat energy from 0·1075 kg of hydrogen
$$= 0·1075 \times 144 = 15·48 \text{ MJ}$$
Total heat energy per kg of fuel
$$= 28·98 + 15·48 = 44·46 \text{ MJ/kg} \quad \text{Ans. (i)}$$
Oxygen required to burn the carbon
$$= 2\tfrac{2}{3} \times 0·86 = 2·293 \text{ kg}$$
Oxygen required to burn the available hydrogen
$$= 8 \times 0·1075 = 0·86 \text{ kg}$$
Total oxygen $= 2·293 + 0·86 = 3·153 \text{ kg}$

$$\text{Air required } = \frac{100}{23} \times 3·153 = 13·71 \text{ kg air/kg fuel} \quad \text{Ans. (ii)}$$

HIGHER AND LOWER CALORIFIC VALUES. H_2O formed by the hydrogen in the fuel cannot exist as water in the high temperatures of boiler flue gases or the exhaust gases from internal combustion engines, it exists in the form of steam, and this steam passing away in the waste gases carries enthalpy of evaporation which is not available as heat energy to the boiler or engine.

Therefore, the theoretical calorific value of a fuel containing hydrogen, as calculated in the above example, is termed the *higher* (or *gross*) *calorific value* and, when the unavailable heat energy is subtracted from this, it is termed the *lower* (or *net*) *calorific value*. It has been recommended that the amount of heat energy to be considered as not available should be 2·442 MJ per kg of steam in the products of combustion. The amount of steam, as shown above, is nine times the mass of hydrogen in the fuel.

In the previous example the mass of hydrogen in each kilogramme of fuel is 0·11 kg, therefore there will be $9 \times 0.11 = 0.99$ kg of steam formed. The unavailable heat energy is therefore to be taken as $0.99 \times 2.442 = 2.418$ MJ/kg fuel.

Hence the lower calorific value of this fuel is:
$$44.46 - 2.418 = 42.042 \text{ MJ/kg}$$

BOMB CALORIMETER

The calorific values of solid and liquid fuels can be found experimentally by means of a bomb calorimeter. There are various types but all are based on the same principle. The Darroch bomb calorimeter is shown diagrammatically in Fig. 97. It consists of a strong inner vessel known as the "bomb" which has a rubber-jointed cover held firmly down by a cover ring screwed to the top of the bomb. These parts are all made of monel metal. Fitted to the cover are (i) a support to carry the crucible into which a measured sample of fuel is placed, (ii) an oxygen inlet valve through which oxygen is supplied from an oxygen cylinder and shut off when the pressure is about 25 bar, (iii) an electrode (insulated from the cover) to which is attached one end of a fine fuse wire, this fuse dips into the fuel and its other end is connected to the crucible support. The bomb rests on supports inside the calorimeter and a known mass of water is poured into the calorimeter to completely surround and cover the bomb. A stirrer and finely graduated thermometer is immersed in this water jacket. The calorimeter with bomb is usually placed inside an insulated outer vessel (not shown here) to minimise loss of heat. The fuse is heated by a battery (usually 12 volts) which ignites the fuel. The heat energy given out during combustion of the fuel is transferred to the bomb and its fittings (of known water equivalent) and jacket water, and causes them to rise in temperature, uniform heating of the parts being obtained by continual stirring of the water during the experiment. Careful note is taken of the temperature rise and the calorific value of the fuel is calculated thus:

Heat energy given out by the fuel = Heat energy absorbed by water + Heat energy absorbed by bomb and its fittings

The result is the higher calorific value because the steam formed during combustion condenses inside the bomb and gives up its enthalpy of condensation to the water jacket and bomb fittings.

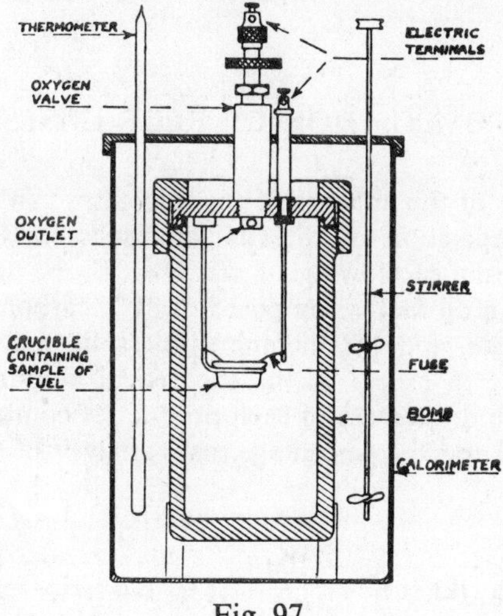

Fig. 97

Example. In an experiment to determine the calorific value of an oil fuel by means of a bomb calorimeter, the mass of the sample of oil was 0·7 gramme, and 1·9 litres of water were contained in the calorimeter. The temperature was observed to rise through 3·1°C. If the water equivalent of the bomb and its fittings is 420 grammes, calculate the calorific value of the oil.

Mass of 1·9 litres of water = 1·9 kg
Taking specific heat of water as 4·2 kJ/kg K,

Heat energy given out by fuel = Heat energy absorbed by water and bomb

Mass of oil $\times$ calorific value

$$= \left\{ \begin{matrix} \text{mass of} \\ \text{water} \end{matrix} + \begin{matrix} \text{water equiv.} \\ \text{of bomb} \end{matrix} \right\} \times \text{spec. heat} \times \text{temp. rise}$$

$$0.7 \times 10^{-3} \times \text{c.v.} = (1.9 + 0.42) \times 4.2 \times 3.1$$

$$\text{c.v.} = \frac{2.32 \times 4.2 \times 3.1 \times 10^3}{0.7}$$

$$= 43.15 \times 10^3 \text{ kJ/kg}$$
$$\text{or } 43.15 \text{ MJ/kg} \quad \text{Ans.}$$

COMPOSITION OF FLUE GASES

An estimate of the analysis of the flue gases can be calculated from the composition of the fuel and mass of supply air as demonstrated in the following.

Example. An oil fuel is composed of 85% carbon, 12% hydrogen, 2% oxygen, and 1% incombustible solid matter. If the air supply is 50% in excess of the theoretical amount for perfect combustion, find the mass of each product of combustion per kg of fuel burned and the percentage mass analysis of the flue gases.

$$\text{Theoretical air/kg fuel} = \frac{100}{23} \left\{ 2\tfrac{2}{3} C + 8 \left[H_2 - \frac{O_2}{8} \right] \right\}$$

$$= \frac{100}{23} \left\{ 2\tfrac{2}{3} C + 8 H_2 - O_2 \right\}$$

$$= \frac{100}{23} \left\{ 2\tfrac{2}{3} \times 0.85 + 8 \times 0.12 - 0.02 \right\}$$

$$= \frac{100}{23} \times 3.207 = 13.94 \text{ kg air/kg fuel}$$

$$\text{Excess air} = 0.5 \times 13.94 = 6.97 \text{ kg}$$
$$\text{Actual air supply} = 13.94 + 6.97 = 20.91 \text{ kg air/kg fuel}$$

Mass of each product of combustion,   Ans. (i):

Mass of CO_2 formed $= 3\tfrac{2}{3} \times 0{\cdot}85 = 3{\cdot}117$ kg

Mass of H_2O formed $= 9 \times 0{\cdot}12 = 1{\cdot}08$ kg

Mass of O_2 = excess oxygen from the excess air

$\qquad = 23\%$ of $6{\cdot}97 = 1{\cdot}603$ kg

Mass of N_2 = mass of nitrogen in all the air supply

$\qquad = 77\%$ of $20{\cdot}91 = 16{\cdot}1$ kg

Total products of combustion

$\qquad = 3{\cdot}117 + 1{\cdot}08 + 1{\cdot}603 + 16{\cdot}1 = 21{\cdot}9$ kg/kg fuel

Alternatively, the total mass of products of combustion per kg fuel = mass of air supplied + (mass of fuel — incombustibles)

$\qquad = 20{\cdot}91 + (1 - 0{\cdot}01)$

$\qquad = 20{\cdot}91 + 0{\cdot}99 = 21{\cdot}9$ kg/kg fuel

Composition expressed as percentages,   Ans. (ii):

$$CO_2 = \frac{3{\cdot}117}{21{\cdot}9} \times 100 = 14{\cdot}23\%$$

$$H_2O = \frac{1{\cdot}08}{21{\cdot}9} \times 100 = 4{\cdot}931\%$$

$$O_2 = \frac{1{\cdot}603}{21{\cdot}9} \times 100 = 7{\cdot}319\%$$

$$N_2 = \frac{16{\cdot}1}{21{\cdot}9} \times 100 = 73{\cdot}52\%$$

ORSAT APPARATUS

A volumetric analysis of the dry products of combustion can be made by the Orsat apparatus. This operates on the principle of passing a sample of the flue gases through a series of three bottles containing different solutions, each of which absorbs one of the constituents, and measuring each time the reduction in volume of the sample which is the volume of the constituent absorbed.

The solutions used for absorption are (i) caustic potash which can absorb CO_2, (ii) pyrogallic acid which can absorb CO_2 and O_2, (iii) cuprous chloride which can absorb CO_2, O_2 and CO. Therefore the first bottle through which the sample is passed must be that containing the caustic potash solution so that only CO_2 is absorbed, the remainder of the sample then contains O_2, CO and N_2. The sample is next passed through the bottle containing the pyrogallic acid solution which takes out the O_2, leaving CO and N_2. Finally, the sample is passed through the cuprous chloride solution which absorbs the CO, and the remainder is assumed to be all N_2. The temperature and pressure are maintained constant throughout the operation.

It is convenient to draw in a sample of exactly 100 ml to be tested so that the percentage volumetric analysis is simply the volumes in ml.

CONVERSION FROM VOLUMETRIC TO MASS ANALYSIS. The volumetric analysis obtained by the Orsat apparatus can now be converted into a mass analysis. By Avogadro's law, equal volumes of any gas, at the same temperature and pressure, contain the same number of molecules, therefore, although the *mass* of a molecule of one gas is different to the mass of a molecule of another gas, the *volume* of each molecule is the same at the same temperature and pressure.

Hence the ratio of the volumes of each gas in the mixture of flue gases can be converted into a ratio of masses by multiplying the volumetric ratio by the molecular weight of the gas, and from this the percentage composition by mass can be calculated.

Example. In a test on a sample of funnel gases the percentage composition by volume was found to be: $CO_2 = 8 \cdot 5\%$, $O_2 = 9 \cdot 5\%$, $CO = 3\%$ and the remainder 79% was assumed to be N_2. Convert these quantities into a mass analysis.

DRY FLUE GAS	RATIO OF VOLUMES	MOLECULAR WEIGHT	RATIO OF MASSES	RATIO OF MASSES AS PERCENTAGES
CO_2	8·5	44	$8 \cdot 5 \times 44 = 374$	$\dfrac{374}{2974} \times 100 = 12 \cdot 58\%$
O_2	9·5	32	$9 \cdot 5 \times 32 = 304$	$\dfrac{304}{2974} \times 100 = 10 \cdot 22\%$
CO	3·0	28	$3 \times 28 = 84$	$\dfrac{84}{2974} \times 100 = 2 \cdot 82\%$
N_2	79·0	28	$79 \times 28 = 2212$	$\dfrac{2212}{2974} \times 100 = 74 \cdot 38\%$
			Total $= \overline{2974}$	Total $= \overline{100 \cdot 00\%}$

CONVERSION FROM MASS TO VOLUMETRIC ANALYSIS is sometimes required and can be obtained by reversing the above procedure.

Example. The analysis of a sample of coal burned in the furnace of a boiler is 80% carbon, 5% hydrogen, 4% oxygen, and the remainder ash etc. Calculate (i) the theoretical mass of air required per kg of coal for perfect combustion, (ii) the actual mass of air if it is supplied with 70% excess, (iii) the percentage mass analysis of the products of combustion, (iv) the percentage analysis of the dry flue gases by volume.

$$\text{Theoretical air} = \frac{100}{23}\left\{ 2\tfrac{2}{3}\,C + 8\left[H_2 - \frac{O_2}{8}\right] \right\}$$

$$= \frac{100}{23}\left\{ 2\tfrac{2}{3}\,C + 8\,H_2 - O_2 \right\}$$

$$= \frac{100}{23}\left\{ 2\tfrac{2}{3} \times 0.8 + 8 \times 0.05 - 0.04 \right\}$$

$$= \frac{100}{23} \times 2.493 = 10.84 \text{ kg air/kg fuel} \quad \text{Ans. (i)}$$

$$\text{Excess air} = 0.7 \times 10.84 = 7.59$$
$$\text{Actual air} = 10.84 + 7.59 = 18.43 \text{ kg air/kg fuel} \quad \text{Ans. (ii)}$$

Mass products of combustion per kg of coal:
$$CO_2 = 3\tfrac{2}{3}\,C = 3\tfrac{2}{3} \times 0.8 = 2.933 \text{ kg}$$
$$H_2O = 9\,H_2 = 9 \times 0.05 = 0.45$$
$$O_2 = 23\% \text{ of excess air} = 0.23 \times 7.59 = 1.746$$
$$N_2 = 77\% \text{ of all air} = 0.77 \times 18.43 = 14.19$$
$$\text{Total} = 19.319 \text{ kg}$$

% mass analysis   Ans. (iii):

$$CO_2 = \frac{2.933}{19.319} \times 100 = 15.18\%$$

$$H_2O = \frac{0.45}{19.319} \times 100 = 2.33\%$$

$$O_2 = \frac{1 \cdot 746}{19 \cdot 319} \times 100 = 9 \cdot 04\%$$

$$N_2 = \frac{14 \cdot 19}{19 \cdot 319} \times 100 = 73 \cdot 45\%$$

Dry flue products are the total products less the amount of steam $\qquad = 19 \cdot 319 - 0 \cdot 45 = 18 \cdot 869\,kg$

% volumetric analysis of dry flue gases   Ans. (iv):

DRY FLUE GAS	RATIO OF MASSES	MOLECULAR WEIGHT	RATIO OF VOLUMES	RATIO OF VOLUMES AS PERCENTAGES
CO_2	15·18	44	$15 \cdot 18 \div 44$ $= 0 \cdot 345$	$\dfrac{0 \cdot 345}{3 \cdot 2505} \times 100 = 10 \cdot 61\%$
O_2	9·04	32	$9 \cdot 04 \div 32$ $= 0 \cdot 2825$	$\dfrac{0 \cdot 2825}{3 \cdot 2505} \times 100 = 8 \cdot 69\%$
N_2	73·45	28	$73 \cdot 45 \div 28$ $= 2 \cdot 623$ Total $= \overline{3 \cdot 2505}$	$\dfrac{2 \cdot 623}{3 \cdot 2505} \times 100 = 80 \cdot 7\%$

TEST EXAMPLES 14

1. An auxiliary boiler produces steam at 8 bar 0·98 dry from feed water at 43°C, and fuel of calorific value 42 MJ/kg is burned at the rate of 1·2 tonne per day. Assuming an efficiency of 70%, calculate the steam production per hour.

2. During a test on an oil-fired water-tube boiler the following data were recorded:

$$\begin{aligned}
\text{Pressure of steam} &= 25 \text{ bar} \\
\text{Temperature of steam} &= 300°C \\
\text{Mass of feed water entering boiler} &= 11\,350 \text{ kg/h} \\
\text{Temperature of feed water} &= 100°C \\
\text{Mass of fuel burned} &= 875 \text{ kg/h} \\
\text{Calorific value of fuel} &= 42·3 \text{ MJ/kg}
\end{aligned}$$

Calculate the efficiency of the boiler.

3. Fuel of calorific value 43·5 MJ/kg is burned in an auxiliary boiler at the rate of 81 kg/h. The feed water rate is 0·258 kg/s and its temperature is 48°C. If the efficiency of the boiler is 68% and the steam produced is dry saturated, estimate its pressure.

4. The equivalent evaporation of a boiler, from and at 100°C, is 15 kg steam/kg fuel, and the calorific value of the fuel burned is 41·9 MJ/kg. Find the efficiency of the boiler.

5. Steam at 30 bar, 375°C, is generated in a boiler at the rate of 30 000 kg/h from feed water at 130°C. The fuel has a calorific value of 42 MJ/kg and the daily consumption is 53 tonne. Calculate (i) the boiler efficiency, (ii) the equivalent engine power if the overall efficiency of the plant is 0·13, (iii) the evaporative capacity of the boiler in kg/h from and at 100°C, (iv) the equivalent evaporation per kg of fuel from and at 100°C.

6. A boiler contains 3·5 tonne of water initially having 40 p.p.m. dissolved solids, and after 24 hours the dissolved solids in the water is 2500 p.p.m. If the feed rate is 875 kg/h, find the p.p.m. of dissolved solids contained in the feed water.

For the following questions take the values:

	ATOMIC WEIGHT	CALORIFIC VALUE MJ/kg
Hydrogen	1	144
Carbon	12	33·7
Sulphur	32	9·3
Nitrogen	14	
Oxygen	16	

Air contains 23% oxygen and 77% nitrogen by mass.

7. The analysis of a sample of a coal is 81% carbon, 5% hydrogen, 5·6% oxygen, 1% sulphur, and the remainder is ash. Calculate the calorific value.

8. An oil fuel is composed of 85·2% carbon, 12% hydrogen, 1·6% oxygen, and 1·2% impurities. Calculate the calorific value and the minimum mass of air required per kg of fuel for perfect combustion.

9. The analysis of a petrol is 85·5% carbon, 14·4% hydrogen, and 0·1% sulphur. Calculate the volume of air at 1·0 bar and 15°C required for perfect combustion of 1 kg of the fuel. Take R for air $= 0·287\ kJ/kg\ K$.

10. A boiler fuel contains 86·1% carbon, 12·5% hydrogen, 0·4% oxygen and 1% sulphur. 40% excess air is supplied to the furnace and the fuel rate is 400 kg/h. Calculate (i) the mass of air supplied per hour, (ii) the quantity of heat energy transferred to the air per hour if it enters the air heater at 18°C and leaves at 130°C. Take c_p for air as $1·005\ kJ/kg\ K$.

11. The constituents of a fuel are 85% carbon, 13% hydrogen, and 2% oxygen. When burning this fuel in a boiler furnace the air supply is 50% in excess of the theoretical minimum required for complete combustion, the inlet temperature of the air being 31°C and funnel temperature 280°C. Calculate (i) the calorific value of the fuel, (ii) mass of air supplied per kg of fuel, and (iii) the heat energy carried away to waste in the funnel gases expressed as a percentage of the heat energy supplied, taking the specific heat of the flue gases as $1·005\ kJ/kg\ K$.

12. In an experiment to determine the calorific value of an oil fuel by means of a bomb calorimeter, the mass of the sample of fuel was 0·75 gramme, mass of water surrounding the bomb 1·8 kg, water equivalent of bomb and fittings 470 grammes, and the rise in temperature was 3·3°C. Calculate the calorific value of this oil in MJ/kg. Take specific heat of water = 4·2 kJ/kg K.

13. A fuel consists of 84% carbon, 13% hydrogen, 2% oxygen, and the remainder incombustible solid matter. Calculate (i) the calorific value, (ii) the theoretical mass of air required per kg of fuel, and (iii) an estimate of the mass analysis of the flue gases if 22 kg of air are supplied per kg of fuel burned.

14. The analysis of a fuel oil is 85·5% carbon, 11·9% hydrogen, 1·6% oxygen, and 1% impurities. Calculate the percentage CO_2 in the flue gases when (i) the quantity of air supplied is the minimum for complete combustion, and when the excess air over the minimum is (ii) 25%, (iii) 50%, and (iv) 75%.

CHAPTER 15

REFRIGERATION

The natural transfer of heat energy is from a hot body to a colder body. The function of a refrigeration plant is to act as a heat pump and reverse this process so that rooms can be maintained at low temperatures for the preservation of foodstuffs.

Refrigerating machines can be divided into two classes, (i) those which require a supply of mechanical work which is the vapour compression system, and (ii) those which require a heat supply and work on the absorption system. The former is more efficient and in general use on board ship, the latter is more suited to domestic use, therefore only the vapour compression system will be described here.

The refrigerating agent used in the circuit is a substance which will evaporate at low temperatures. The boiling and condensation points of a liquid depend upon the pressure exerted upon it, for example, if water is under atmospheric pressure it will vaporise at 100°C, if the pressure is 7 bar the water will not change into steam until its temperature is 165°C, at 14 bar the boiling point is 195°C and so on. The refrigerant used must vaporise at very low temperatures. The boiling point of carbon dioxide at atmospheric pressure is about —78°C (note *minus* 78), by increasing the pressure the temperature at which liquid CO_2 will vaporise (or CO_2 vapour will condense) is raised accordingly so that any desired vaporisation and condensation temperature can be attained, within certain limits, by subjecting it to the appropriate pressure.

Some of the agents employed as refrigerants and their more important characteristics are as follows:

Carbon dioxide (CO_2). This is an inert vapour, non-poisonous, odourless, and has no corrosive action on the metals. Its natural boiling point is very low which means that it must be run at very high pressures to bring it to the conditions where it will vaporise and condense at the normal temperatures of a refrigerating machine. A further disadvantage is that its critical temperature is about 31°C which falls within the range of sea water temperatures. At its critical temperature and above, it is impossible to liquefy the vapour no matter to what pressure it is subjected and, as part of the circuit depends upon condensing the gas in a condenser circulated by sea water, great difficulty is experienced, even with special additional devices, when in high temperature tropical waters.

Ammonia (NH_3). This is a poisonous vapour and therefore an ammonia machine should not be open to the engine room but have a compartment of its own so that it can be sealed off in the case of a serious leakage; water will absorb ammonia and therefore a water spray is a good combatant against a leakage. Ammonia will corrode copper and copper alloys and therefore parts in contact with it should be made of such metals as nickel steel and monel metal. Its natural boiling point is about —39°C therefore the pressures required throughout the system to obtain the necessary evaporation and liquefaction temperatures are much lower than those required in a CO_2 machine.

Methyl chloride (CH_3Cl). This vapour is slightly poisonous, inflammable and, under certain conditions, corrosive. It requires comparatively low pressures throughout the system.

Freon 12 (CCl_2F_2) or Arcton 6, the former is the American name and the latter the British name for the same refrigerant. This also requires only low pressures but has the advantage over methyl chloride in that it is non-inflammable, non-poisonous and non-corrosive.

WORKING CYCLE

Fig. 98 illustrates diagrammatically the essential components of the vapour compression system, which consists of the Compressor driven by an electric motor, the Condenser which is circulated by sea water, the Regulator (sometimes referred to as the expansion valve), and the Evaporator which is circulated by brine. It is a completely enclosed circuit, the same quantity of refrigerant passes continually through the system and it only requires to be charged when there are losses due to leakage. Any of the above agents may be used as the refrigerant, the difference in their working being only the pressures throughout the system, therefore in the following description the words "refrigerant", "the liquid" and "the vapour", etc. will be used.

The refrigerant is drawn as a vapour at low pressure from the evaporator into the compressor where it is compressed to a high pressure and delivered into the coils of the condenser in the state of a superheated vapour. As it passes through the condenser coils, the vapour is cooled and condensed into a liquid at approximately sea water temperature, the latent heat energy given up being absorbed by the sea water surrounding the coils and pumped overboard. The liquid, still at a high pressure, passes along to the regulator, this is a valve just partially open to limit the flow through it. The pipe-line from the discharge side of the compressor to the regulating valve is under high pressure but from the regulating valve to the suction side of the compressor is at low pressure due to the regulating valve being just a little way open. As the liquid passes through the regulator from a region of high pressure to a region of low pressure, throttling occurs, and some of the liquid automatically changes itself into a vapour absorbing the required amount of heat energy to do so from the remainder of the liquid and causing it to fall to a low temperature, this temperature being regulated by the

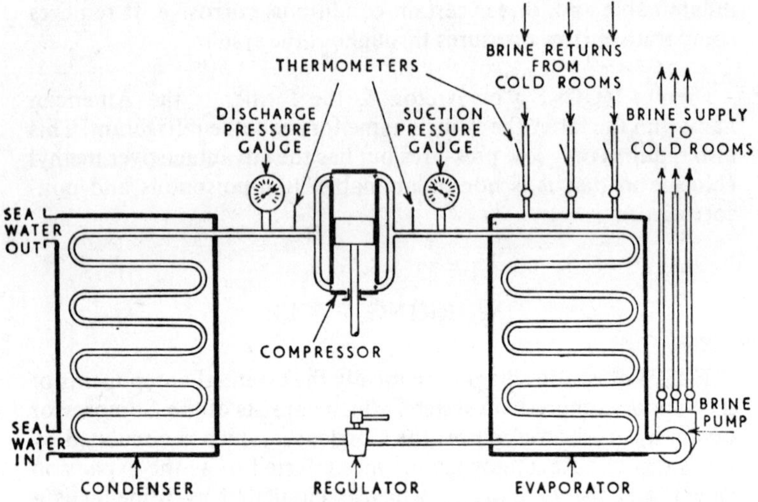

VAPOUR COMPRESSION REFRIGERATING MACHINE

Fig. 98

pressure, so that the refrigerant enters the evaporator at a temperature lower than that of the brine. The liquid (more correctly a mixture of liquid and vapour) now passes through the coils of the evaporator where it receives heat to evaporate it before being taken into the compressor to go through the cycle again. The required heat absorbed by the liquid refrigerant in the coils of the evaporator to cause evaporation is extracted from the brine surrounding the coils, resulting in the brine being cooled to a low temperature.

BRINE CIRCULATION

The cold brine from the evaporator is pumped through pipes led around the top of the walls and/or ceiling of the cold rooms where it extracts heat from the rooms before it is returned to the evaporator.

The quantity of brine flowing through the pipes of any one room determines the temperature of that room, therefore the control of the room temperature is by regulating the opening of the valve on the brine return at the evaporator (assuming that the brine supply to the pipes is at a sufficiently low temperature). Each room is provided with a thermometer so that the temperature inside can be taken regularly and recorded, there is also a thermometer at each brine return which gives a good indication to the experienced engineer of the room temperature.

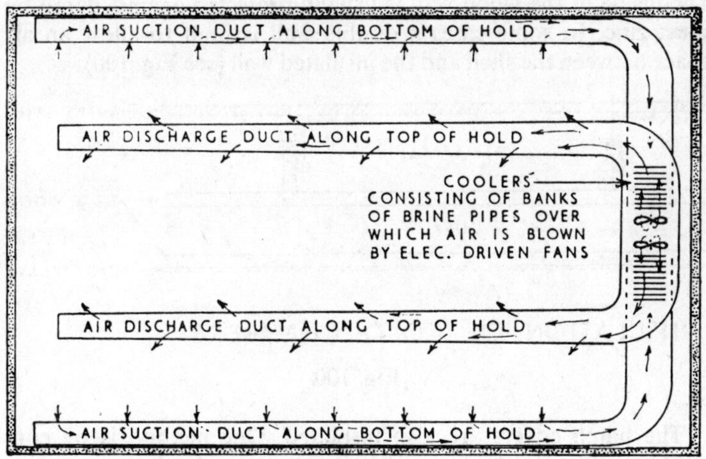

COLD AIR SYSTEM OF COOLING
Fig. 99

In fruit carrying rooms the air must be continually circulated to prevent stagnant pockets of CO_2 given off by and forming around the fruit, therefore the cold air system of cooling is employed. Instead of wrapping the brine pipes around the walls and ceiling inside the rooms, the brine is led into grid-boxes through which the air, drawn from the bottom of the rooms by fans or blowers, is passed over the brine grids and blown back into the rooms via ducting along the ceiling, see Fig. 99.

The brine is made by dissolving calcium chloride in fresh water, the freezing point depending upon the density, but the liquid should not be too thick as to impede its free flow through the pipes. A satisfactory mixture is obtained by 0·25 kg of calcium chloride per litre of fresh water. This will produce a brine with a freezing point well below any usual working temperature.

INSULATION

All cold rooms must be insulated to minimise transfer of heat into them from the outside, (see Chapter 4) this is often done by building a wood wall around the room leaving a space between it and the steel plating, and filling up this space with some light-weight heat-insulating material such as cork or silicate of cotton. The inside of the wood wall is usually protected against damp by sheet zinc. In some cases a double wall is built to allow an air space between the shell and the insulated wall (see Fig. 100).

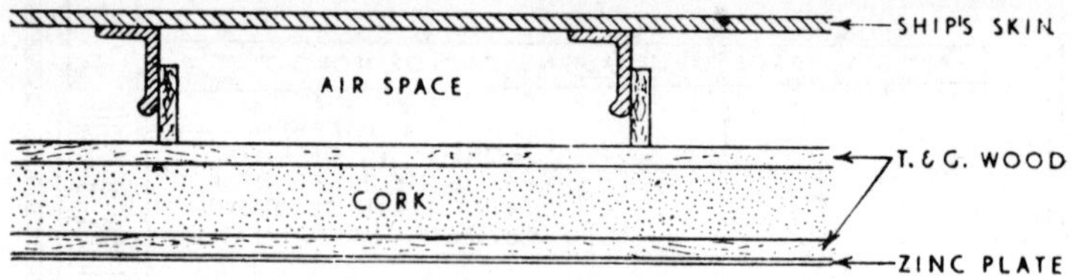

INSULATION OF COLD CHAMBERS

Fig. 100

The hatch covers are also insulated. One method is by constructing them in the form of hollow rectangular boxes or plugs filled with cork. The sides are inclined so that one interlocks another, necessitating the removal of the end plugs (keys) before lifting the others.

THE CIRCUIT OF THE REFRIGERANT

It is necessary to have a clear picture in mind when performing calculations involving the changes that take place on the refrigerant as it flows through the components of the circuit,

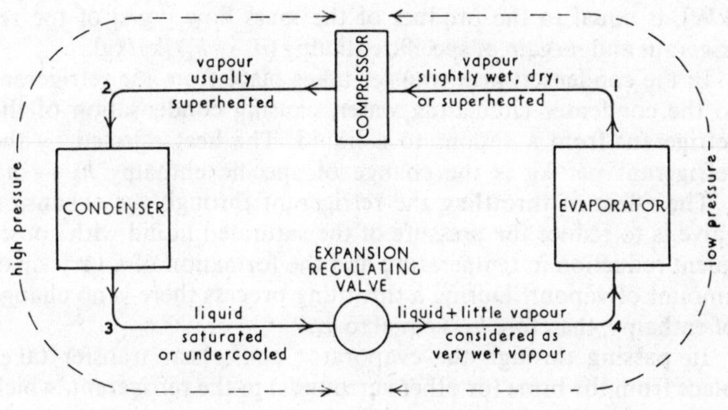

Fig. 101

therefore a simple diagrammatic sketch such as Fig. 101 should be made with points in the circuit suitably marked. Reference should be made to Fig. 101 when reading the following notes and examples.

Properties of some refrigerants are given in "Thermodynamic and Transport Properties of Fluids", SI Units, arranged by Y. R. Mayhew and G. F. C. Rogers, this booklet was referred to in previous Chapters dealing with properties of steam. In examples given here, extracts from these tables will be taken when appropriate, stating the page number to enable the student to follow the figures.

Principles and calculations involving enthalpy of the refrigerant, dryness fraction, volume, effect of throttling, etc. follow the same pattern as for other vapours. However, whereas the enthalpy and entropy of steam are measured from the datum of water at 0°C, it is more convenient to choose a lower datum temperature for refrigerants to avoid negative quantities within the normal range of refrigeration. This datum is —40°C, thus the enthalpy and entropy of the saturated liquid refrigerant are taken as zero at this temperature.

Note that when the value of h_{fg} is not given in the tables, it is found by subtraction: $h_{fg} = h_g - h_f$. Similarly for the value of s_{fg}.

Referring to Fig. 101:

The work done on the vapour in the compressor increases the specific enthalpy of the refrigerant from h_1 to h_2. Therefore the effective work done per second [kJ/s], which is the effective power [kW], is equal to the product of the mass flow [kg/s] of the refrigerant and its gain of specific enthalpy $(h_2 - h_1)$ [kJ/kg].

In the condenser, heat transfer takes place from the refrigerant to the condenser circulating water, causing condensation of the refrigerant from a vapour to a liquid. The heat rejected by the refrigerant per kg is the change of specific enthalpy $h_2 - h_3$.

The effect of throttling the refrigerant through the expansion valve is to reduce the pressure of the saturated liquid with consequent reduction in temperature and the formation of a very small amount of vapour. During a throttling process there is no change of enthalpy, therefore h_4 is equal to h_3.

In passing through the evaporator coils, heat transfer takes place from the brine (or other surrounds) to the refrigerant, which cools the brine and causes the refrigerant to evaporate. The gain in specific enthalpy of the refrigerant is $h_1 - h_4$ and this is the *refrigerating effect* or *cooling effect* per kg of the refrigerant flowing through the circuit.

Example. In an NH_3 refrigerator, the ammonia leaves the evaporator and enters the compressor as dry saturated vapour at 2·68 bar, it leaves the compressor and enters the condenser at 8·57 bar with 50°C of superheat. It is condensed at constant pressure and leaves the condenser as saturated liquid. If the rate of flow of the refrigerant through the circuit is 0·45 kg/min calculate (i) the compressor power, (ii) the heat rejected to the condenser cooling water in kJ/s, and (iii) the refrigerating effect in kJ/s.

From tables page 12, NH_3:

$$2\text{·}68 \text{ bar,} \quad h_g = 1430\text{·}5$$
$$8\text{·}57 \text{ bar,} \quad h_f = 275\text{·}1 \qquad h \text{ supht } 50° = 1597\text{·}2$$

Mass flow of refrigerant

$$= \frac{0\text{·}45}{60} = 0\text{·}0075 \text{ kg/s}$$

Enthalpy gain per kg of refrigerant in compressor
$$= h_2 - h_1$$
$$= 1597 \cdot 2 - 1430 \cdot 5 = 166 \cdot 7 \text{ kJ/kg}$$
Enthalpy gain per second
$$= \text{mass flow [kg/s]} \times 166 \cdot 7 \text{ [kJ/kg]}$$
$$= 0 \cdot 0075 \times 166 \cdot 7 = 1 \cdot 25 \text{ kJ/s}$$
kJ/s = kW, therefore useful compressor power
$$= 1 \cdot 25 \text{ kW} \quad \text{Ans. (i)}$$
Enthalpy drop per kg of refrigerant through condenser
$$= h_2 - h_3$$
$$= 1597 \cdot 2 - 275 \cdot 1 = 1322 \cdot 1 \text{ kJ/kg}$$
$$\text{Heat rejected} = 0 \cdot 0075 \times 1322 \cdot 1$$
$$= 9 \cdot 915 \text{ kJ/s} \quad \text{Ans. (ii)}$$
Enthalpy gain per kg of refrigerant through evaporator
$$= h_1 - h_4$$
$$= 1430 \cdot 5 - 275 \cdot 1 = 1155 \cdot 4 \text{ kJ/kg}$$

(Note that the value of h_4 is the same as h_3 because there is no change of enthalpy in the throttling process through the expansion valve between the condenser exit and the evaporator inlet).

Refrigerating effect $= 0 \cdot 0075 \times 1155 \cdot 4$
$$= 8 \cdot 666 \text{ kJ/s} \quad \text{Ans. (iii)}$$

Example. In a Freon-12 refrigerator, the freon leaves the condenser as a saturated liquid at 20°C, the evaporator temperature is —10°C and the freon leaves the evaporator as a vapour 0·97 dry. Calculate (i) the dryness fraction at the evaporator inlet, (ii) the cooling effect per kg of refrigerant, and (iii) the volume flow of refrigerant entering the compressor if the mass flow is 0·1 kg/s.

From tables page 13, Freon-12:

$$20°C, \quad h_f = 54 \cdot 87$$
$$-10°C, \quad h_f = 26 \cdot 87 \quad h_g = 183 \cdot 19 \quad v_g = 0 \cdot 0766$$
$$h_{fg} = h_g - h_f = 183 \cdot 19 - 26 \cdot 87 = 156 \cdot 32$$

Throttling process through expansion valve:
Enthalpy after throttling (h_4) = Enthalpy before (h_3)
$$h_{f4} + x_4 h_{fg4} = h_{f3}$$
$$26 \cdot 87 + x_4 \times 156 \cdot 32 = 54 \cdot 87$$
$$x_4 \times 156 \cdot 32 = 28$$
$$x_4 = 0 \cdot 1792 \quad \text{Ans. (i)}$$

Enthalpy gain of refrigerant through evaporator

$$= \text{cooling effect} = h_1 - h_4$$
$$= (26 \cdot 87 + 0 \cdot 97 \times 156 \cdot 32) - (26 \cdot 87 + 0 \cdot 1792 \times 156 \cdot 32)$$
$$= (0 \cdot 97 - 0 \cdot 1792) \times 156 \cdot 32$$
$$= 0 \cdot 7908 \times 156 \cdot 32 = 123 \cdot 6 \, \text{kJ/kg} \quad \text{Ans. (ii)}$$

Alternatively, since $h_4 = h_3$:

cooling effect

$$= (26 \cdot 87 + 0 \cdot 97 \times 156 \cdot 32) - 54 \cdot 87$$
$$= 123 \cdot 6 \, \text{kJ/kg}$$

Specific volume of refrigerant leaving evaporator

$$= 0 \cdot 97 \times 0 \cdot 0766 \, \text{m}^3/\text{kg}$$

Volume of refrigerant entering compressor per second [m³/s]

$$= \text{mass flow [kg/s]} \times \text{spec. vol [m}^3\text{/kg]}$$
$$= 0 \cdot 1 \times 0 \cdot 97 \times 0 \cdot 0766$$
$$= 7 \cdot 43 \times 10^{-3} \, \text{m}^3/\text{s}$$
$$\text{or} \quad 7 \cdot 43 \, \text{litre/s} \quad \text{Ans. (iii)}$$

CAPACITY AND PERFORMANCE

The capacity of a refrigerating plant is usually expressed in terms of the tonnes of ice at 0°C that could be made in 24 hours from water at 0°C.

The *coefficient of performance* is the ratio of the refrigerating effect to the net work done on the refrigerant in the compressor:

$$\text{Coeff. of performance} = \frac{\text{heat extracted by refrigerant}}{\text{work done on refrigerant}}$$

$$= \frac{h_1 - h_4}{h_2 - h_1}$$

ƒ In Chapter 8, the reversed Carnot cycle was explained as the ideal cycle for a refrigerator and the coefficient of performance of this cycle given as:

$$\frac{T_2}{T_1 - T_2}$$

where T_1 and T_2 are the highest and lowest absolute temperatures in the cycle. This is referred to as the *Carnot coefficient of performance*. Note however, that the notation is different to that used in the practical circuit of Fig. 101, to avoid confusion, the student may perhaps prefer to write subscripts H for highest and L for lowest:

$$\frac{T_\mathrm{L}}{T_\mathrm{H} - T_\mathrm{L}}$$

ƒExample. A Freon-12 refrigerating machine operates on the ideal vapour compression cycle between the limits —15°C and 25°C. The vapour is dry saturated at the end of isentropic compression and there is no undercooling of the condensate in the condenser. Calculate (i) the dryness fraction at the suction of the compressor, (ii) the refrigerating effect per kg of refrigerant, (iii) the coefficient of performance, (iv) the Carnot coefficient of performance.

Tables Freon-12 page 13:

$$\begin{aligned}
25°\mathrm{C}, \quad & h_f = 59{\cdot}7 & h_g = 197{\cdot}73 & \quad s_g = 0{\cdot}6869 \\
-15°\mathrm{C}, \quad & h_f = 22{\cdot}33 & h_g = 180{\cdot}97 & \\
& h_{fg} = 180{\cdot}97 - 22{\cdot}33 = 158{\cdot}64 \\
& s_f = 0{\cdot}0906 \quad s_g = 0{\cdot}7051 \\
& s_{fg} = 0{\cdot}7051 - 0{\cdot}0906 = 0{\cdot}6145
\end{aligned}$$

Isentropic compression in compressor:

$$\begin{aligned}
\text{Entropy before } (s_1) &= \text{Entropy after } (s_2) \\
0{\cdot}0906 + x_1 \times 0{\cdot}6145 &= 0{\cdot}6869 \\
x_1 &= 0{\cdot}9701 \quad \text{Ans. (i)} \\
h_1 &= 22{\cdot}33 + 0{\cdot}9701 \times 158{\cdot}64 \\
&= 153{\cdot}9
\end{aligned}$$

$$\text{Refrigerating effect/kg} = h_1 - h_4$$

Since there is no undercooling, the condensate leaves the condenser as saturated liquid, hence $h_3 = 59{\cdot}7$, and $h_4 = h_3$ because there is no change of enthalpy during the throttling process, therefore:

$$\begin{aligned}
\text{Refrigerating effect} &= 153{\cdot}9 - 59{\cdot}7 \\
&= 94{\cdot}2 \text{ kJ/kg} \quad \text{Ans. (ii)} \\
\text{Work done in compressor kg} &= h_2 - h_1 \\
= 197{\cdot}73 - 153{\cdot}9 &= 43{\cdot}83 \text{ kJ/kg}
\end{aligned}$$

$$\begin{aligned}
\text{Coefficient of performance} &= \frac{\text{refrigerating effect}}{\text{work transfer}} \\
&= \frac{h_1 - h_4}{h_2 - h_1} = \frac{94{\cdot}2}{43{\cdot}83} \\
&= 2{\cdot}149 \quad \text{Ans. (iii)}
\end{aligned}$$

$$\text{Carnot coeff. of performance} = \frac{T_L}{T_H - T_L}$$

$$\text{where } T_H = 25 + 273 = 298\,\text{K}$$
$$T_L = -15 + 273 = 258\,\text{K}$$

$$\therefore \text{Carnot c.o.p.} = \frac{258}{298 - 258} = 6{\cdot}45 \quad \text{Ans. (iv)}$$

TEST EXAMPLES 15

1. In a CO_2 refrigerating plant, the specific enthalpy of the refrigerant as it leaves the condenser is 135 kJ/kg, and as it leaves the evaporator it is 320 kJ/kg. If the mass flow of the refrigerant is 5 kg/min calculate the refrigerating effect per hour.

2. Freon-12 leaves the condenser of a refrigerating plant as a saturated liquid at 5·673 bar. The evaporator pressure is 1·509 bar and the refrigerant leaves the evaporator at this pressure and at a temperature of —5°C. Calculate (i) the dryness fraction of the refrigerant as it enters the evaporator, (ii) the refrigerating effect per kg.

3. A refrigerating machine uses ammonia as the working fluid. It leaves the compressor as dry saturated vapour at 8·57 bar, passes through the condenser at this pressure and leaves as saturated liquid. The pressure in the evaporator is 1·902 bar and the ammonia leaves the evaporator 0·96 dry. If the rate of flow of the refrigerant through the circuit is 2 kg/min, calculate (i) the heat rejected in the condenser in kJ/min, (ii) the refrigerating effect in kJ/min, (iii) the volume taken into the compressor in m^3/min.

4. The dryness fractions of the CO_2 entering and leaving the evaporator of a refrigerating plant are 0·28 and 0·92 respectively. If the specific enthalpy of evaporation (h_{fg}) of CO_2 at the evaporator pressure is 290·7 kJ/kg, calculate (i) the refrigerating effect per kg, (ii) the mass of ice at —5°C that would theoretically be made per day from water at 14°C when the mass flow of CO_2 through the machine is 0·5 kg/s. Take the values:

$$\text{Specific heat of water} = 4\text{·}2 \text{ kJ/kg K}$$
$$\text{Specific heat of ice} = 2\text{·}04 \text{ kJ/kg K}$$
$$\text{Enthalpy of fusion of ice} = 335 \text{ kJ/kg}$$

5. A refrigerating machine is driven by a motor of output power 2·25 kW and 2·5 tonne of ice at —7°C is made per day from water at 18°C. Calculate the coefficient of performance of the machine and express its capacity in terms of tonnes of ice per 24 hours from and at 0°C, taking the following values:

$$\text{Specific heat of water} = 4\text{·}2 \text{ kJ/kg K}$$
$$\text{Specific heat of ice} = 2\text{·}04 \text{ kJ/kg K}$$
$$\text{Enthalpy of fusion of ice} = 335 \text{ kJ/kg}$$

6. The refrigerant leaves the compressor and enters the condenser of a freon-12 refrigerating plant at 5·673 bar and 50°C, and leaves the condenser as saturated liquid at the same pressure. At compressor suction the pressure is 1·826 bar and temperature 0°C. Calculate the coefficient of performance.

*f*7. In an ammonia refrigerating plant the discharge from the compressor is 14·7 bar 63°C, the refrigerant leaves the condenser at this pressure as liquid with no undercooling. The suction pressure at the compressor is 2·077 bar and compression is isentropic. Calculate, for a mass flow of refrigerant of 0·15 kg/s, (i) the refrigerating effect, (ii) the work transfer in the compressor, (iii) the coefficient of performance.

SOLUTIONS TO TEST EXAMPLES 1

1. $\qquad$ Mass of water lifted $= 50 \times 10^3$ kg/h

Weight of water lifted $= 50 \times 10^3 \times 9\cdot81$ N/h

Power [W $= $ J/s $= $ N m/s]

$\qquad = $ weight lifted per second [N/s] $\times$ height [m]

$$= \frac{50 \times 10^3 \times 9\cdot81 \times 8}{3600}$$

$$= 1090 \text{ W} \qquad = 1\cdot09 \text{ kW} \quad \text{Ans. (i)}$$

$$\text{Input power} = \frac{\text{output power}}{\text{efficiency}}$$

$$= \frac{1\cdot09}{0\cdot69} = 1\cdot58 \text{ kW} \quad \text{Ans. (ii)}$$

Energy [kW h] $= $ power [kW] $\times$ time [h]

$\qquad = 1\cdot58 \times 2 = 3\cdot16$ kW h $\quad$ Ans. (iii)

Energy [MJ] $= $ energy [kW h] $\times 3\cdot6$ [MJ/kW h]

$\qquad = 3\cdot16 \times 3\cdot6 = 11\cdot38$ MJ $\quad$ Ans. (iv)

2. A column of water 1 mm high exerts a pressure of $9\cdot81$ N/m², therefore 20 mm water is equivalent to:

$$20 \times 9\cdot81 = 196\cdot2 \text{ N/m}^2 \quad \text{Ans. (ia)}$$

One mbar $= 100$ N/m²

$\therefore$ $196\cdot2$ N/m² $= 1\cdot962$ mbar $\quad$ Ans. (iia)

A column of mercury 1 mm high exerts a pressure of $133\cdot3$ N/m², therefore 750 mm Hg is equivalent to:

$$750 \times 133\cdot3 = 1 \times 10^5 \text{ N/m}^2$$

$$= 100 \text{ kN/m}^2 \quad \text{Ans. (ib)}$$

$$= 1 \text{ bar} \quad \text{Ans. (iib)}$$

3. $\qquad$ Abs. press. $= $ (barometer mm Hg — vac.

$\qquad$ gauge mm Hg) $\times 133\cdot3$

$\qquad = (757 - 715) \times 133\cdot3$

$\qquad = 42 \times 133\cdot3$

$\qquad = 5\cdot6 \times 10^3$ N/m² $= 5\cdot6$ kN/m²

$\qquad = 0\cdot056$ bar $\quad$ Ans.

4. Temp. interval °C $=$ temp. interval °F $\times \dfrac{5}{9}$

$$= 162 \times \dfrac{5}{9} = 90°C \quad \text{Ans. (i)}$$

Temp. interval °C $= 63 \times \dfrac{5}{9} = 35°C \quad \text{Ans. (ii)}$

5. $C = (140 - 32) \times \dfrac{5}{9}$ $\qquad\qquad$ $C = (5 - 32) \times \dfrac{5}{9}$

$\qquad = 108 \times \dfrac{5}{9}$ $\qquad\qquad\qquad\quad = -27 \times \dfrac{5}{9}$

$\qquad = 60°C \quad \text{Ans. (i)}$ $\qquad\qquad\quad = -15°C \quad \text{Ans. (ii)}$

$C = (-31 - 32) \times \dfrac{5}{9}$ $\qquad\qquad$ $C = (-40 - 32) \times \dfrac{5}{9}$

$\qquad = -63 \times \dfrac{5}{9}$ $\qquad\qquad\qquad\quad = -72 \times \dfrac{5}{9}$

$\qquad = -35°C \quad \text{Ans. (iii)}$ $\qquad\qquad = -40°C \quad \text{Ans. (iv)}$

6. Area of tube bore $= 0.7854 \times 0.03^2 \, m^2$
$\qquad$ Area of 16 tubes $= 0.7854 \times 0.03^2 \times 16 \, m^2$
$\qquad$ Volume flow [m³/s] $=$ area [m²] $\times$ velocity [m/s]
$\qquad\qquad\qquad\qquad\quad = 0.7854 \times 0.03^2 \times 16 \times 2$
$\qquad\qquad\qquad\qquad\quad = 0.02261 \, m^3/s$
$1 \, m^3 = 10^3$ litre
$\therefore$ Volume flow [l/s] $= 0.02261 \times 10^3$
$\qquad\qquad\qquad\qquad = 22.61 \, l/s \quad \text{Ans. (i)}$
density $= 0.85 \, g/ml = 0.85 \, kg/litre$
Mass flow [kg/min] $=$ vol. flow [litre/min] $\times$ density [kg/litre]
$\qquad\qquad\qquad\qquad = 22.61 \times 60 \times 0.85$
$\qquad\qquad\qquad\qquad = 1153 \, kg/min \quad \text{Ans. (ii)}$

SOLUTIONS TO TEST EXAMPLES 2

1. Normal force between surfaces of casting and floor
$$= \text{weight of casting} = mg$$
$$= 5 \times 9 \cdot 81 = 49 \cdot 05 \text{ N}$$

Force to overcome friction
$$= \mu \times \text{normal force between surfaces}$$
$$= 0 \cdot 4 \times 49 \cdot 05 = 19 \cdot 62 \text{ N}$$

Mechanical energy expended:
$$\text{work done [J = N m]} = \text{force [N]} \times \text{distance [m]}$$
$$= 19 \cdot 62 \times 10$$
$$= 196 \cdot 2 \text{ J}$$
$$\therefore \text{ Heat energy generated} = 196 \cdot 2 \text{ J} \quad \text{Ans.}$$

2. Mechanical energy converted into heat energy per minute:
$$\text{Energy [kJ]} = \text{power [kW = kJ/s]} \times \text{time [s]}$$
$$= 70 \times 60$$
$$= 4200 \text{ kJ} = 4 \cdot 2 \text{ MJ} \quad \text{Ans. (i)}$$

All the heat energy being transferred to the water:
$$Q \text{ [kJ]} = m \text{ [kg]} \times c \text{ [kJ/kg K]} \times (T_2 - T_1) \text{ [K]}$$
$$4200 = m \times 4 \cdot 2 \times 10$$
$$m = 100 \text{ kg/min} \quad \text{Ans. (ii)}$$

3. Friction force at surface of journals
$$= \mu \times \text{force between surfaces}$$
$$= 0 \cdot 02 \times 200 = 4 \text{ kN}$$

Work done to overcome friction per revolution
$$= \text{friction force [kN]} \times \text{circumference [m]}$$
$$= 4 \times \pi \times 0 \cdot 38 \text{ kN m} = \text{kJ}$$
$$\text{Power loss [kW = kJ/s]} = \text{work per rev [kJ]} \times \text{rev/s}$$

$$= 4 \times \pi \times 0 \cdot 38 \times \frac{105}{60}$$

$$= 8 \cdot 358 \text{ kW} \quad \text{Ans. (i)}$$
$$\text{Energy [kJ]} = \text{power [kW = kJ/s]} \times \text{time [s]}$$
$$= 8 \cdot 358 \times 60$$
$$= 501 \cdot 5 \text{ kJ} \quad \text{Ans. (ii)}$$

4. Friction force at effective radius of pads
$$= \mu \times \text{force between surfaces}$$
$$= 0{\cdot}025 \times 240 = 6 \text{ kN}$$

Work to overcome friction in one revolution
$$= \text{friction force [kN]} \times \text{circumference [m]}$$
$$= 6 \times 2\pi \times 0{\cdot}23 \text{ kN m} = \text{kJ}$$
Power loss [kW = kJ/s] = work per rev [kJ] $\times$ rev/s

$$= 6 \times 2\pi \times 0{\cdot}23 \times \frac{93}{60}$$

$$= 13{\cdot}45 \text{ kW Ans. (i)}$$

Mechanical energy converted into heat energy per hour:
$$\text{Energy [kW h]} = \text{power [kW]} \times \text{time [h]}$$
$$= 13{\cdot}45 \times 1 = 13{\cdot}45 \text{ kW h}$$
$$13{\cdot}45 \text{ kW h} \times 3{\cdot}6 = 48{\cdot}42 \text{ MJ} \quad \text{Ans. (ii)}$$

Quantity of heat energy transferred to oil per hour:
$$Q \text{ [kJ]} = m \text{ [kg]} \times c \text{ [kJ/kg K]} \times (T_2 - T_1) \text{ [K]}$$
$$48{\cdot}42 \times 10^3 = m \times 2 \times 20$$
$$m = 1210{\cdot}5 \text{ kg/h} \quad \text{Ans. (iii)}$$

5. Let θ = initial temperature of the copper
$$= \text{temperature of the flue gases}$$
Heat lost by copper = Heat gained by water
$$m_C \times c_C \times \text{temp. fall}_C = m_W \times c_W \times \text{temp. rise}_W$$
$$1{\cdot}8 \times 0{\cdot}395 \times (\theta - 37{\cdot}2) = 2{\cdot}27 \times 4{\cdot}2 \times (37{\cdot}2 - 20)$$
$$0{\cdot}711\,\theta - 26{\cdot}44 = 164$$
$$0{\cdot}711\,\theta = 190{\cdot}44$$
$$\theta = 267{\cdot}9°\text{C} \quad \text{Ans.}$$

6. Mass of 1·2 litre of water $= 1{\cdot}2$ kg
$$\text{Let } \theta = \text{resultant temperature}$$
Heat gained by water = Heat lost by brass
$$m_W \times c_W \times \text{temp. rise}_W = m_B \times c_B \times \text{temp. fall}_B$$
$$1{\cdot}2 \times 4{\cdot}2 \times (\theta - 14) = 2{\cdot}5 \times 0{\cdot}39 \times (176 - \theta)$$
$$5{\cdot}04\,\theta - 70{\cdot}56 = 171{\cdot}6 - 0{\cdot}975\,\theta$$
$$6{\cdot}015\,\theta = 242{\cdot}16$$
$$\theta = 40{\cdot}26°\text{C} \quad \text{Ans.}$$

7. Mass of 2·3 litre of water $= 2·3$ kg

Let c [kJ/kg K] $=$ specific heat of the iron

Heat lost by iron $=$ Heat gained by water and vessel

$$2·15 \times c \times (100 - 24·4) = (2·3 + 0·18) \times 4·2 \times (24·4 - 17)$$

$$2·15 \times c \times 75·6 = 2·48 \times 4·2 \times 7·4$$

$$c = \frac{2·48 \times 4·2 \times 7·4}{2·15 \times 75·6}$$

$$= 0·4742 \text{ kJ/kg K} \quad \text{Ans.}$$

8. Since all quantities are of the same liquid, the specific heat is the same in each case and, if included, would cancel from both sides of the following heat energy transfer equations. It is therefore omitted in each case.

Let the masses of the quantities be represented by m_A, m_B and m_C respectively.

Mixing of A and B:

Heat gained by A $=$ Heat lost by B

$$m_A \times (17 - 9) = m_B \times (21 - 17)$$
$$8m_A = 4m_B$$
$$m_A = 0·5m_B$$

Mixing of B and C:

Heat gained by B $=$ Heat lost by C

$$m_B \times (28 - 21) = m_C \times (38 - 28)$$
$$7m_B = 10 \, m_C$$
$$m_C = 0·7m_B$$

Mixing A and C, let resultant temperature $= \theta$:

Heat gained by A $=$ Heat lost by C

$$m_A \times (\theta - 9) = m_C \times (38 - \theta)$$

Substituting m_A and m_C in terms of m_B:

$$0·5m_B \times (\theta - 9) = 0·7m_B \times (38 - \theta)$$

Cancelling m_B and evaluating:

$$0·5 \, \theta - 4·5 = 26·6 - 0·7 \, \theta$$
$$1·2 \, \theta = 31·1$$
$$\theta = 25·92°C \quad \text{Ans. (i)}$$

Mixing all three:

Heat gained by A + Heat gained by B = Heat lost by C
$$m_A\,(\theta - 9) + m_B \times (\theta - 21) = m_C \times (38 - \theta)$$

Substituting m_A and m_C in terms of m_B:
$$0.5m_B\,(\theta - 9) + m_B\,(\theta - 21) = 0.7m_B\,(38 - \theta)$$

Cancelling m_B and evaluating:
$$0.5\,\theta - 4.5 + \theta - 21 = 26.6 - 0.7\,\theta$$
$$2.2\,\theta = 52.1$$
$$\theta = 23.68°C \quad \text{Ans. (ii)}$$

9. Let θ = initial temperature of the ice

Heat gained by ice = Heat lost by water and calorimeter
$$0.9\{2.04(0 - \theta) + 335 + 4.2(13 - 0)\} = (10 + 0.2) \times$$
$$4.2\,(22 - 13)$$
$$0.9\{- 2.04\,\theta + 335 + 54.6\} = 10.2 \times 4.2 \times 9$$
$$-2.04\,\theta + 389.6 = 10.2 \times 4.2 \times 10$$
$$-2.04\,\theta = 38.8$$
$$\theta = -19.02°C \quad \text{Ans.}$$

10. Heat transferred to the ice to raise all of it in temperature from $-5°C$ to $0°C$
$$= 0.5 \times 2.04 \times 5 = 5.1 \text{ kJ}$$

Heat transferred from the water and vessel in cooling from $17°C$ to $0°C$
$$= (1.8 + 0.148) \times 4.2 \times 17$$
$$= 1.948 \times 4.2 \times 17 = 139.1 \text{ kJ}$$

Therefore, heat from water available to melt some of the ice
$$= 139.1 - 5.1 = 134 \text{ kJ}$$

Each kg of ice requires 335 kJ to melt it, therefore mass of ice melted by 134 kJ
$$= \frac{134}{335} = 0.4 \text{ kg}$$

Hence,

Final mass of water $= 1.8 + 0.4 = 2.2$ kg $\Big\}$ Ans.
Final mass of ice $\quad = 0.5 - 0.4 = 0.1$ kg

SOLUTIONS TO TEST EXAMPLES 3

1. Free expansion $= \alpha \times l \times (\theta_2 - \theta_1)$
    $\qquad\qquad\quad = 1{\cdot}25 \times 10^{-5} \times 3{\cdot}85 \times (260 - 18)$
    $\qquad\qquad\quad = 0{\cdot}01165 \text{ m}$
    $\qquad\qquad\quad = 11{\cdot}65 \text{ mm} \quad \text{Ans.}$

2. Vol. of sphere $= \dfrac{\pi}{6} d^3 = \dfrac{\pi}{6} \times 0{\cdot}15^3 \text{ m}^3$

    Mass [kg] $=$ volume [m³] $\times$ density [kg/m³]

    $\qquad\qquad = \dfrac{\pi}{6} \times 0{\cdot}15^3 \times 7{\cdot}21 \times 10^3$

    $\qquad\qquad = 12{\cdot}75 \text{ kg}$

    $Q\text{[kJ]} = m \text{ [kg]} \times c \text{ [kJ/kg K]} \times (T_2 - T_1) \text{ [K]}$
    $2110 = 12{\cdot}75 \times 0{\cdot}54 \times (T_2 - T_1)$

    temp. rise $= \dfrac{2110}{12{\cdot}75 \times 0{\cdot}54} = 306{\cdot}7 \text{ K}$

    $\qquad\qquad = 306{\cdot}7°\text{C}$

    Increase in diameter $= \alpha \times d \times (\theta_2 - \theta_1)$
    $\qquad\qquad\qquad\qquad\quad = 1{\cdot}12 \times 10^{-5} \times 150 \times 306{\cdot}7$
    $\qquad\qquad\qquad\qquad\quad = 0{\cdot}5152 \text{ mm} \quad \text{Ans.}$

3.

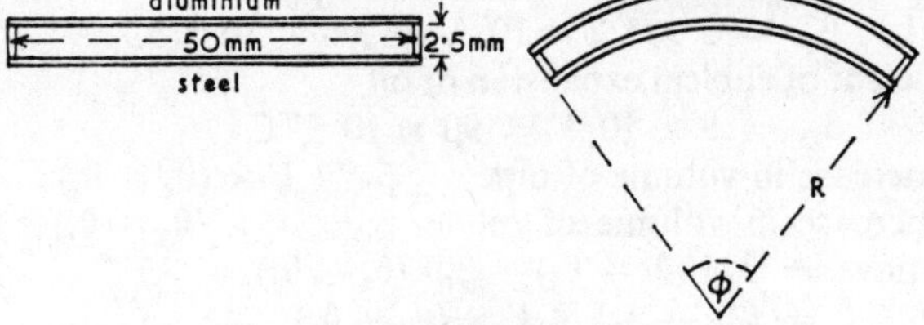

Fig. 102

Free expansion $= \alpha \times l \times (\theta_2 - \theta_1)$
Heated length $= l + \alpha l\,(\theta_2 - \theta_1) = l\{1 + \alpha\,(\theta_2 - \theta_1)\}$

Heated length of aluminium strip
$$= 50 (1 + 2 \cdot 5 \times 10^{-5} \times 200)$$
$$= 50 \times 1 \cdot 005 = 50 \cdot 25 \text{ mm}$$
Heated length of steel strip
$$= 50 (1 + 1 \cdot 2 \times 10^{-5} \times 200)$$
$$= 50 \times 1 \cdot 0024 = 50 \cdot 12 \text{ mm}$$
Heated length of brass distance pieces
$$= 2 \cdot 5 (1 + 2 \times 10^{-5} \times 200)$$
$$= 2 \cdot 5 \times 1 \cdot 004 = 2 \cdot 51 \text{ mm}$$

Let R = radius of curvature of steel
$R + 2 \cdot 51 =$,,   ,,   ,,   ,, aluminium
Subtended angle φ is common to both

$$\varphi[\text{rad}] = \frac{\text{arc length [mm]}}{\text{radius mm]}}$$

$$\therefore \frac{50 \cdot 25}{R + 2 \cdot 51} = \frac{50 \cdot 12}{R}$$

$$50 \cdot 25R = 50 \cdot 12 (R + 2 \cdot 51)$$
$$50 \cdot 25R = 50 \cdot 12R + 125 \cdot 8$$
$$0 \cdot 13R = 125 \cdot 8$$
$$R = 967 \cdot 7 \text{ mm}$$
$$\text{Radius of steel} = 967 \cdot 7 \text{ mm} \quad \text{Ans.}$$
$$\text{Radius of aluminium} = 967 \cdot 7 + 2 \cdot 51$$
$$= 970 \cdot 21 \text{ mm} \quad \text{Ans.}$$

4. Internal volume of pipe ($=$ vol. of oil in pipe)
$$= 0 \cdot 7854 \times 0 \cdot 03^2 \times 13 \cdot 7 \text{ m}^3$$
Coefficient of cubical expansion of steel pipe
$$\beta_P = 3 \times 1 \cdot 2 \times 10^{-5} = 3 \cdot 6 \times 10^{-5}/°C$$
Coefficient of cubical expansion of oil
$$\beta_O = 9 \times 10^{-4} = 90 \times 10^{-5}/°C$$
Increase in volume of pipe $= \beta_P \times V \times (\theta_2 - \theta_1)$
Increase in volume of oil $= \beta_O \times V \times (\theta_2 - \theta_1)$
Oil overflow $= \beta_O V(\theta_2 - \theta_1) - \beta_P V(\theta_2 - \theta_1)$
$$= (\beta_O - \beta_P) \times V \times (\theta_2 - \theta_1)$$
$$= (90 - 3 \cdot 6) \times 10^{-5} \times 0 \cdot 7854 \times 0 \cdot 03^2 \times 13 \cdot 7 \times 27$$
$$= 2 \cdot 259 \times 10^{-4} \text{ m}^3$$
one cubic metre $= 10^3$ litres
$$\therefore 2 \cdot 259 \times 10^{-4} \times 10^3 = 0 \cdot 2259 \text{ litre} \quad \text{Ans.}$$

5. Coefficient of cubical expansion of glass

$$\beta_G = 3 \times 8\cdot5 \times 10^{-6} = 2\cdot55 \times 10^{-5}/°C$$

Coeff. of cubical expansion of oil $= \beta_O$

$$\text{Oil overflow} = \beta_O V(\theta_2 - \theta_1) - \beta_G V(\theta_2 - \theta_1)$$

$$= (\beta_O - \beta_G) \times V \times (\theta_2 - \theta_1)$$

$$1\cdot57 = (\beta_O - 2\cdot55 \times 10^{-5}) \times 40 \times (65 - 15)$$

$$\beta_O - 2\cdot55 \times 10^{-5} = \frac{1\cdot57}{40 \times 50}$$

$$\beta_O = 0\cdot000\,785 + 0\cdot000\,0255$$
$$= 0\cdot000\,8105$$
$$= 8\cdot105 \times 10^{-4}/°C \quad \text{Ans.}$$

6. $\qquad \text{Free expansion} = \alpha \times l \times (\theta_2 - \theta_1)$

If fully restricted from returning to original length on cooling to original temperature:

$$\text{Strain} = \frac{\text{change of length}}{\text{original length}}$$

$$= \frac{\alpha l\,(\theta_2 - \theta_1)}{l} = \alpha\,(\theta_2 - \theta_1)$$

$$= 1\cdot2 \times 10^{-5} \times (113 - 16)$$
$$= 1\cdot2 \times 10^{-5} \times 97$$
$$\text{Stress} = \text{strain} \times E$$
$$= 1\cdot2 \times 10^{-5} \times 97 \times 207 \times 10^{9}$$
$$= 2\cdot41 \times 10^{8} \text{ N/m}^2$$
$$= 241 \text{ MN/m}^2 \text{ or } 24\cdot1 \text{ hbar} \quad \text{Ans. (i)}$$
$$\text{Force [N]} = \text{stress [N/m}^2] \times \text{area [m}^2]$$
$$= 2\cdot41 \times 10^{8} \times 0\cdot7854 \times 2\cdot5^2 \times 10^{-6}$$
$$= 1\cdot183 \times 10^{3} \text{ N}$$
$$= 1\cdot183 \text{ kN} \quad \text{Ans. (ii)}$$

7. $\qquad \text{Free expansion} = \alpha \times l \times (\theta_2 - \theta_1)$

If fully restricted,

$$\text{Strain} = \frac{\text{change of length}}{\text{original length}} = \frac{\alpha l\,(\theta_2 - \theta_1)}{l} = \alpha(\theta_2 - \theta_1)$$

$$\text{Stress} = \text{strain} \times E$$
$$= 1 \cdot 12 \times 10^{-5} \times (220 - 17) \times 206 \times 10^9$$
$$= 1 \cdot 12 \times 203 \times 206 \times 10^4$$
$$= 4 \cdot 684 \times 10^8 \, \text{N/m}^2 = 468 \cdot 4 \, \text{MN/m}^2$$

This compressive stress is to be relieved by exerting an initial tensile stress so that the compressive stress does not exceed $350 \, \text{MN/m}^2$:

$$\text{Initial tensile stress} = 468 \cdot 4 - 350$$
$$= 118 \cdot 4 \, \text{MN/m}^2$$
$$\text{or } 11 \cdot 84 \, \text{hbar} \quad \text{Ans.}$$

SOLUTIONS TO TEST EXAMPLES 4

1. $Q\,[\text{J}] = \dfrac{\lambda\,[\text{W/mK}] \times A\,[\text{m}^2] \times t\,[\text{s}] \times (T_1 - T_2)\,[\text{K}]}{S\,[\text{m}]}$

$$= \frac{150 \times 0.7854 \times 0.127^2 \times 60 \times 5}{0.019}$$

$$= 3 \times 10^4\,\text{J} = 30\,\text{kJ} \quad \text{Ans.}$$

2. Total surface area of walls, ceiling and floor
$$= 2\,(4.5 \times 2.5 + 4 \times 2.5 + 4.5 \times 4)$$
$$= 78.5\,\text{m}^2$$

$$Q = \frac{\lambda A t\,(T_1 - T_2)}{S}$$

$$= \frac{5.8 \times 10^{-2} \times 78.5 \times 3600 \times 20}{0.15}$$

$$= 2.185 \times 10^6\,\text{J}$$
$$= 2.185\,\text{MJ or } 2185\,\text{kJ} \quad \text{Ans.}$$

3. For each thickness:

$$Q = \frac{\lambda A t \times \text{temp. diff.}}{S} \qquad\qquad \therefore \text{ temp. diff.} = \frac{QS}{\lambda A t}$$

For the three thicknesses:

$$T_\text{D} = \frac{Q}{At}\left\{\frac{S_1}{\lambda_1} + \frac{S_2}{\lambda_2} + \frac{S_3}{\lambda_3}\right\}$$

where $T_\text{D} = 20 - (-3) = 23°\text{C temp. diff.} = 23\,\text{K}$

$$\Sigma\left\{\frac{S}{\lambda}\right\} = \frac{0.03}{0.2} + \frac{0.168}{0.042} + \frac{0.035}{0.2}$$

$$= 0.15 + 4 + 0.175$$
$$= 4.325$$

Heat transfer per second per metre^2 of area:

$$T_\mathrm{D} = \frac{Q}{At}\Sigma\left\{\frac{S}{\lambda}\right\}$$

$$23 = \frac{Q}{1 \times 1} \times 4{\cdot}325$$

$$Q = \frac{23}{4{\cdot}325} = 5{\cdot}318 \text{ J} \quad \text{Ans. (i)}$$

Heat transfer per hour through whole side:
$$\begin{aligned}
Q &= 5{\cdot}318 \times 6 \times 3{\cdot}7 \times 3600 \\
&= 4{\cdot}25 \times 10^5 \text{ J} \\
&= 425 \text{ kJ} \quad \text{Ans. (ii)}
\end{aligned}$$
Temperature drop across outer wood wall:

$$= \frac{QS_1}{\lambda_1 At}$$

$$= \frac{5{\cdot}318 \times 0{\cdot}03}{0{\cdot}2}$$

$$= 0{\cdot}7977 \text{ K} \quad \text{say } 0{\cdot}8°\text{C}$$
Temperature at interface of cork and outer wood wall
$$= 20 - 0{\cdot}8 = 19{\cdot}2°\text{C} \quad \text{Ans. (iiia)}$$
Temperature drop across cork:

$$= \frac{QS_2}{\lambda_2 At}$$

$$= \frac{5{\cdot}318 \times 0{\cdot}168}{0{\cdot}042} = 21{\cdot}27 \text{ K} = 21{\cdot}27°\text{C}$$

Temperature at interface of cork and inner wood wall
$$= 19{\cdot}2 - 21{\cdot}27 = -2{\cdot}07°\text{C} \quad \text{Ans. (iiib)}$$

$$\text{4.} \qquad \frac{1}{U} = \frac{1}{h_\mathrm{G}} + \frac{S_\mathrm{P}}{\lambda_\mathrm{P}} + \frac{1}{h_\mathrm{A}}$$

$$= \frac{1}{31{\cdot}5} + \frac{0{\cdot}01}{50} + \frac{1}{32}$$

$$= 0{\cdot}03175 + 0{\cdot}0002 + 0{\cdot}03125$$
$$= 0{\cdot}0632$$
$$U = 1/0{\cdot}0632 = 15{\cdot}82 \, \text{W/m}^2\text{K} \quad \text{Ans. (i)}$$
$$Q = UAtT_\text{D}$$
$$= 15{\cdot}82 \times 1 \times 60 \times (280 - 35)$$
$$= 2{\cdot}327 \times 10^5 \, \text{J} = 232{\cdot}7 \, \text{kJ} \quad \text{Ans. (ii)}$$

5. Heat flow from outside atmosphere to exposed surface of outer wood wall, and also from exposed surface of inner wood wall to room atmosphere:

$$Q = hAt \times \text{temp. difference}$$

per m² of surface area per second,

$$42 = 15 \times 1 \times 1 \times \text{temp. difference}$$

$$\text{temp. difference} = \frac{42}{15} = 2{\cdot}8^\circ\text{C}$$

Temperature of exposed surface of outer wood
$$= 25 - 2{\cdot}8 = 22{\cdot}2^\circ\text{C} \quad \text{Ans. (ia)}$$
Temperature of exposed surface of inner wood
$$= -20 + 2{\cdot}8 = -17{\cdot}2^\circ\text{C} \quad \text{Ans. (ib)}$$
Heat flow through each wood wall:

$$Q = \frac{\lambda At \times \text{temp. difference}}{S}$$

$$42 = \frac{0{\cdot}2 \times 1 \times 1 \times \text{temp. difference}}{0{\cdot}03}$$

Temperature difference across each wood wall

$$= \frac{42 \times 0{\cdot}03}{0{\cdot}2} = 6{\cdot}3^\circ\text{C}$$

Interface temperature of outer wood and cork
$$= 22{\cdot}2 - 6{\cdot}3 = 15{\cdot}9^\circ\text{C} \quad \text{Ans. (iia)}$$
Interface temperature of inner wood and cork
$$= -17{\cdot}2 + 6{\cdot}3 = -10{\cdot}9^\circ\text{C} \quad \text{Ans. (iib)}$$
Temperature difference across cork
$$= 15{\cdot}9 - (-10{\cdot}9) = 26{\cdot}8^\circ\text{C}$$
Heat flow through cork:

$$Q = \frac{\lambda At \times \text{temp. difference}}{S}$$

$$S = \frac{0{\cdot}05 \times 1 \times 1 \times 26{\cdot}8}{42}$$

$$= 0{\cdot}0319\,\text{m} = 31{\cdot}9\,\text{mm} \quad \text{Ans. (iii)}$$

6.
$$Q = 5{\cdot}67 \times 10^{-11}\,At\,(T_1{}^4 - T_2{}^4)$$
$$T_1 = 215 + 273 = 488\,\text{K}$$
$$T_2 = 45 + 273 = 318\,\text{K}$$
$$T_1{}^4 - T_2{}^4 = (488^2 + 318^2)(488^2 - 318^2)$$
$$= (238\,100 + 101\,100)(238\,100 - 101\,100)$$
$$= 339\,200 \times 137\,000$$
$$= 3{\cdot}392 \times 1{\cdot}37 \times 10^{10}$$
$$Q = 5{\cdot}67 \times 10^{-11} \times 0{\cdot}7854 \times 0{\cdot}5^2 \times 3600 \times 3{\cdot}392 \times 1{\cdot}37 \times 10^{10}$$
$$= 1862\,\text{kJ} \quad \text{Ans.}$$

7. Let T_1 = abs. temp. of shell before lagging
$$= 230 + 273 = 503\,\text{K}$$
     T_2 = abs. temp. of surrounds before lagging
$$= 51 + 273 = 324\,\text{K}$$
     T_3 = abs. temp. of cleading after lagging
$$= 69 + 273 = 342\,\text{K}$$
     T_4 = abs. temp. of surrounds after lagging
$$= 27 + 273 = 300\,\text{K}$$
Heat radiated from shell before lagging
$$= 5{\cdot}67 \times 10^{-11}(T_1{}^4 - T_2{}^4)\,\text{kJ/m}^2\text{s}$$
Heat radiated from cleading after lagging
$$= 5{\cdot}67 \times 10^{-11}(T_3{}^4 - T_4{}^4)\,\text{kJ/m}^2\text{s}$$
Heat saved by lagging
$$= 5{\cdot}67 \times 10^{-11}\{(T_1{}^4 - T_2{}^4) - (T_3{}^4 - T_4{}^4)\}\,\text{kJ/m}^2\text{s}$$
Working out quantity inside brackets:
$$T_1{}^2 = 503^2 = 253\,000$$
$$T_2{}^2 = 324^2 = 105\,000$$
$$T_1{}^2 + T_2{}^2 = 358\,000 = 3{\cdot}58 \times 10^5$$
$$T_1{}^2 - T_2{}^2 = 148\,000 = 1{\cdot}48 \times 10^5$$
$$T_3{}^2 = 342^2 = 117\,000$$
$$T_4{}^2 = 300^2 = 90\,000$$
$$T_3{}^2 + T_4{}^2 = 207\,000 = 2{\cdot}07 \times 10^5$$
$$T_3{}^2 - T_4{}^2 = 27\,000 = 0{\cdot}27 \times 10^5$$
$$(T_1{}^4 - T_2{}^4) - (T_3{}^4 - T_4{}^4)$$

$$= (3.58 \times 10^5 \times 1.48 \times 10^5) - (2.07 \times 10^5 \times 0.27 \times 10^5)$$
$$= (5.299 - 0.559) \times 10^{10}$$
$$= 4.74 \times 10^{10}$$

$$
\begin{aligned}
\text{Total surface area} &= \text{area of ends} + \text{area of cylinder} \\
&= \pi d^2 + \pi dl \\
&= \pi d\,(d + l) \\
&= \pi \times 1.22\,(1.22 + 4.78) \\
&= \pi \times 1.22 \times 6 \\
\text{Lagged area} &= 0.75 \times \pi \times 1.22 \times 6 \\
&= 17.26 \text{ m}^2
\end{aligned}
$$

Total heat saved by lagging per hour

$$
\begin{aligned}
&= 5.67 \times 10^{-11}\, At\,\{(T_1{}^4 - T_2{}^4) - (T_3{}^4 - T_4{}^4)\} \\
&= 5.67 \times 10^{-11} \times 17.26 \times 3600 \times 4.74 \times 10^{10} \\
&= 1.67 \times 10^5 \text{ kJ} \\
&= 167 \text{ MJ} \quad \text{Ans.}
\end{aligned}
$$

SOLUTIONS TO TEST EXAMPLES 5

1. Pressure $[N/m^2]$ = mm Hg $\times$ 133·3

$$= 760 \times 133\cdot3$$
$$= 1\cdot013 \times 10^5 \text{ N/m}^2$$
$$= 101\cdot3 \text{ kN/m}^2 \text{ or } 1\cdot013 \text{ bar} \quad \text{Ans.}$$

2.
$$p_1 V_1 = p_2 V_2 \text{ (Boyle's law)}$$
$$1500 \times 0\cdot14 = p_2 \times 0\cdot35$$
$$p_2 = \frac{1500 \times 0\cdot14}{0\cdot35} = 600 \text{ kN/m}^2 \quad \text{Ans.}$$

3.
$$p_1 V_1 = p_2 V_2$$
$$120 \times 1 = 960 \times V_2$$
$$V_2 = \frac{120 \times 1}{960} = 0\cdot125 \text{ m}^3 \quad \text{Ans. (i)}$$

$$p_1 V_1 = p_2 V_2$$
$$1\cdot05 \times V_1 = 42 \times 5\cdot6$$
$$V_1 = \frac{42 \times 5\cdot6}{1\cdot05} = 224 \text{ m}^3 \quad \text{Ans. (ii)}$$

4. Vol. of vessel = $\dfrac{\text{volume of two}}{\text{hemispherical ends}} + \dfrac{\text{volume of}}{\text{cylindrical part}}$

$$\text{Vol. of 2 vessels} = 2\left(\frac{\pi}{6}d^3 + \frac{\pi}{4}d^2 \times l\right)$$

$$= 2\pi d^2\left\{\frac{d}{6} + \frac{l}{4}\right\}$$

$$= 2\pi \times 0\cdot3^2\left\{\frac{0\cdot3}{6} + \frac{1\cdot2}{4}\right\}$$

$$= 2\pi \times 0\cdot09 \times 0\cdot35$$
$$= 0\cdot1979 \text{ m}^3$$

absolute pressure $= 2800 + 100 = 2900 \text{ kN/m}^2$

Since temperature is constant:

$$p_1 V_1 = p_2 V_2$$
$$100 \times V_1 = 2900 \times 0.1979$$

$$V_1 = \frac{2900 \times 0.1979}{100} = 5.74 \text{ m}^3$$

5.74 m³ is the volume of air at 100 kN/m² which will compress to 0.1979 m³ at 2900 kN/m². This would be the volume of atmospheric air required if the vessels initially contained no air, however, as they initially contain their own volume of 0.1979 m³ of air at 100 kN/m² then the volume to be taken from the atmosphere is
$$5.74 - 0.1979 = 5.5421 \text{ m}^3 \quad \text{Ans.}$$

5. For constant temp., $p_1 V_1 = p_2 V_2 = p_3 V_3$ etc.
On each suction stroke the volume is increased from the pipe-line volume of 14 m³ to the combined volume of pipe-line + stroke volume of pump, which is $14 + 0.08 = 14.08$ m³

At beginning of 1st stroke, $p_o = 101.3$ and $V_o = 14$
„   end   „ „   „ $p_1 = \quad ?$ and $V_1 = 14.08$

$$p_o V_o = p_1 V_1$$

$$p_1 = \frac{101.3 \times 14}{14.08} \text{ kN/m}^2$$

and this is the pressure at the beginning of the 2nd suction stroke when it begins with a volume of 14 m³.
At the end of the 2nd suction stroke the volume is 14.08 m³ and let the pressure then be p_2

$$p_1 V_1 = p_2 V_2$$

$$p_2 = \frac{101.3 \times 14 \times 14}{14.08 \times 14.08}$$

$$= 101.3 \times \left\{ \frac{14}{14.08} \right\}^2 \text{ kN/m}^2$$

p_2 is also the pressure at the beginning of the 3rd suction stroke when the volume is 14 m³. Let the pressure at the end of the 3rd stroke be p_3 when the volume is 14.08 m³,

$$p_2 V_2 = p_3 V_3$$

$$p_3 = \frac{101 \cdot 3 \times 14^2 \times 14}{14 \cdot 08^2 \times 14 \cdot 08}$$

$$= 101 \cdot 3 \times \left\{ \frac{14}{14 \cdot 08} \right\}^3 \text{kN/m}^2$$

and so on. It can therefore be seen that at the end of the nth stroke the pressure will be

$$101 \cdot 3 \times \left\{ \frac{14}{14 \cdot 08} \right\}^n \text{kN/m}^2$$

and this is to be 35 kN/m², hence:

$$101 \cdot 3 \times \left\{ \frac{14}{14 \cdot 08} \right\}^n = 35$$

$$\left\{ \frac{14}{14 \cdot 08} \right\}^n = \frac{35}{101 \cdot 3}$$

or, $$\left\{ \frac{14 \cdot 08}{14} \right\}^n = \frac{101 \cdot 3}{35}$$

$$1 \cdot 006^n = 2 \cdot 894$$
$$(\log 1 \cdot 006) \times n = \log 2 \cdot 894$$
$$0 \cdot 0025 \times n = 0 \cdot 4615$$

$$n = \frac{0 \cdot 4615}{0 \cdot 0025}$$

$$= 184 \cdot 6 \text{ suction strokes} \quad \text{Ans.}$$

6. Absolute temperatures:
$$T_1 = 21°C + 273 = 294 \text{ K}$$
$$T_2 = 315°C + 273 = 588 \text{ K}$$
By Charles' law, when pressure is constant,

$$\frac{V_1}{V_2} = \frac{T_1}{T_2}$$

$$V_2 = \frac{0 \cdot 25 \times 588}{294} = 0 \cdot 5 \text{ m}^3 \quad \text{Ans.}$$

$$\text{Atmospheric pressure} = 759 \times 133{\cdot}3 = 101{\cdot}2 \text{ kN/m}^2$$
$$\text{Initial abs. press.} = 140 + 101{\cdot}2 = 241{\cdot}2 \text{ kN/m}^2$$
$$\text{Initial abs. temp.} = 20 + 273 = 293 \text{ K}$$
$$\text{Final abs. temp.} = 40 + 273 = 313 \text{ K}$$

By Charles' law, when volume is constant,

$$\frac{p_1}{p_2} = \frac{T_1}{T_2}$$

$$p_2 = \frac{241{\cdot}2 \times 313}{293} = 257{\cdot}6 \text{ kN/m}^2 \text{ abs.}$$

Final gauge pressure
$$= 257{\cdot}6 - 101{\cdot}2 = 156{\cdot}4 \text{ kN/m}^2 \text{ gauge} \quad \text{Ans.}$$

8.
$$\frac{p_1 V_1}{T_1} = \frac{p_2 V_2}{T_2}$$

$$\frac{1350 \times 0{\cdot}2}{(177 + 273)} = \frac{250 \times 0{\cdot}9}{T_2}$$

$$T_2 = \frac{450 \times 250 \times 0{\cdot}9}{1350 \times 0{\cdot}2} = 375 \text{K}$$

$$375 - 273 = 102°\text{C} \quad \text{Ans.}$$

9.
$$pV = mRT$$

$$m = \frac{550 \times 0{\cdot}05}{0{\cdot}297 \times (28 + 273)}$$

$$= 0{\cdot}3076 \text{ kg} \quad \text{Ans. (i)}$$
$$pV = mRT$$

$$V = \frac{1 \times 0{\cdot}297 \times 273}{1000}$$

$$= 0{\cdot}0811 \text{ m}^3 \quad \text{Ans. (ii)}$$

10. Vol. of bottle $=$ volume of hemispherical end $+$ volume of cylindrical part

$$= \frac{1}{2} \times \frac{\pi}{6} d^3 + \frac{\pi}{4} d^2 \times l$$

$$= \pi d^2 \left\{ \frac{d}{12} + \frac{l}{4} \right\}$$

$$= \pi \times 0.24^2 \left\{ \frac{0.24}{12} + \frac{1.12}{4} \right\}$$

$$= \pi \times 0.24^2 \times 0.3 = 0.054\,29 \text{ m}^3$$

(i) $\quad pV = mRT$

$$m = \frac{(700 + 101.3) \times 0.054\,29}{0.26 \times (20 + 273)}$$

$$= \frac{801.3 \times 0.054\,29}{0.26 \times 293} = 0.571 \text{ kg} \quad \text{Ans. (i)}$$

(ii) $\quad pV = mRT$

$$m = \frac{(500 + 101.3) \times 0.054\,29}{0.26 \times (15 + 273)}$$

$$= \frac{601.3 \times 0.054\,29}{0.26 \times 288} = 0.4359 \text{ kg}$$

$$\text{mass drawn off} = 0.571 - 0.4359$$
$$= 0.1351 \text{ kg} \quad \text{Ans. (ii)}$$

11. By Charles' law, when volume is constant,

$$\frac{p_1}{p_2} = \frac{T_1}{T_2}$$

$$p_2 = \frac{(3200 + 100) \times (35 + 273)}{(16 + 273)}$$

$$= \frac{3300 \times 308}{289} = 3518 \text{ kN/m}^2 \text{ abs.}$$

or, $3518 - 100 = 3418 \text{ kN/m}^2$ gauge Ans. (i)

$$Q \, [\text{kJ}] = m \, [\text{kg}] \times c \, [\text{kJ/kg K}] \times (T_2 - T_1 \, [\text{K}]$$
$$= 20 \times 0 \cdot 718 \times (35 - 16)$$
$$= 272 \cdot 9 \, \text{kJ} \quad \text{Ans. (ii)}$$

12. Volume of saloon $= 12 \times 16 \cdot 5 \times 4 = 792 \, \text{m}^3$
Volume of air to be supplied per hour
$$= 2 \times 792 = 1584 \, \text{m}^3$$

Volume varies as the absolute temperature when the pressure is constant, therefore density varies inversely as the absolute temperature.

Density at $0°\text{C}$ and atmos. press. $= 1 \cdot 293 \, \text{kg/m}^3$

$\therefore$ density at $21°\text{C}$ and atmos. press.

$$= 1 \cdot 293 \times \frac{(0 + 273)}{(21 + 273)} = 1 \cdot 2 \, \text{kg/m}^3$$

Mass of air to be supplied per hour
$$= \text{volume} \, [\text{m}^3/\text{h}] \times \text{density} \, [\text{kg/m}^3]$$
$$= 1584 \times 1 \cdot 2 = 1901 \, \text{kg/h}$$
Heat extracted per hour
$$Q \, [\text{kJ/h}] = m \, [\text{kg/h}] \times c \, [\text{kJ/kg K}] \times (T_1 - T_2) \, [\text{K}]$$
$$= 1901 \times 1 \cdot 005 \times (30 - 21)$$
$$= 17190 \, \text{kJ/h or } 17 \cdot 19 \, \text{MJ/h} \quad \text{Ans. (i)}$$
Power $[\text{kW}] = \text{kilojoules per second}$

$$= \frac{17190}{3600} = 4 \cdot 775 \, \text{kW} \quad \text{Ans. (ii)}$$

13. (i) External work done $= p \times (V_2 - V_1)$
$$= 34 \cdot 5 \times 10^2 \times (0 \cdot 0935 - 0 \cdot 0425)$$
$$= 176 \, \text{kJ} \quad \text{Ans. (i)}$$

(ii) External work done $= mR \, (T_2 - T_1)$
$$= 0 \cdot 63 \times 0 \cdot 287 \times (1784 - 811)$$
$$= 175 \cdot 9 \, \text{kJ} \quad \text{Ans. (ii)}$$

Note that T_2 and T_1 are the absolute temperatures, their difference, however, is the same as the Celcius temperature difference of $1511 - 538$.

14. At constant volume:
$$Q = mc_v (T_2 - T_1)$$
$$= 1.36 \times 0.718 \times (468 - 40)$$
$$= 417.8 \text{ kJ} \quad \text{Ans. (ai)}$$

External work done = nil   Ans. (aii)

From the energy equation,

$$\begin{array}{ccc} \text{Heat energy} \\ \text{supplied} \end{array} = \begin{array}{ccc} \text{Increase in} \\ \text{internal energy} \end{array} + \begin{array}{ccc} \text{External} \\ \text{work done} \end{array}$$

$$417.8 = \text{Increase in internal energy} + 0$$

$\therefore$ Increase in internal energy = 417.8 kJ   Ans. (aiii)

At constant pressure:
$$Q = mc_p (T_2 - T_1)$$
$$= 1.36 \times 1.005 \times (468 - 40)$$
$$= 584.8 \text{ kJ} \quad \text{Ans. (bi)}$$

Internal energy depends only upon the temperature and is independent of changes in pressure and volume. Therefore the increase in internal energy is the same as before,
$$= 417.8 \text{ kJ} \quad \text{Ans. (biii)}$$

$$\begin{array}{ccc} \text{Heat energy} \\ \text{supplied} \end{array} = \begin{array}{ccc} \text{Increase in} \\ \text{internal energy} \end{array} + \begin{array}{ccc} \text{External} \\ \text{work done} \end{array}$$

$$584.8 = 417.8 + \text{work done}$$

$\therefore$ Work done = 584.8 − 417.8
$$= 167 \text{ kJ} \quad \text{Ans. (bii)}$$

Note also:
$$R = c_p - c_v = 1.005 - 0.718 = 0.287 \text{ kJ/kg K}$$
$$\text{Work done} = mR (T_2 - T_1)$$
$$= 1.36 \times 0.287 \times 428$$
$$= 167 \text{ kJ (as above)}$$

15. Ratio of partial pressures is equal to the ratio of partial volumes.

Partial press. of CO_2 = 0.1 × 1.015 = 0.1015 bar   Ans. (ia)

,, ,, ,, O_2 = 0.08 × 1.015 = 0.0812 bar   Ans. (ib)

,, ,, ,, N_2 = 0.82 × 1.015 = 0.8323 bar   Ans. (ic)

$$pV = mRT \qquad \therefore m = \frac{pV}{RT}$$

where p = partial pressure, and V = full volume
(or p = total pressure, and V = partial volume)

$$\text{Mass of } CO_2 = \frac{0.1015 \times 10^2 \times 500 \times 10^{-6}}{0.189 \times (20 + 273)}$$

$$= 9.162 \times 10^{-5} \, kg = 0.091\,62 \, g \quad \text{Ans. (iia)}$$

$$\text{Mass of } O_2 = \frac{0.0812 \times 10^2 \times 500 \times 10^{-6}}{0.26 \times (20 + 273)}$$

$$= 5.33 \times 10^{-5} \, kg = 0.0533 \, g \quad \text{Ans. (iib)}$$

$$\text{Mass of } N_2 = \frac{0.8323 \times 10^2 \times 500 \times 10^{-6}}{0.297 \times (20 + 273)}$$

$$= 4.782 \times 10^{-4} \, kg = 0.4782 \, g \quad \text{Ans. (iic)}$$

SOLUTIONS TO TEST EXAMPLES 6

1. For isothermal expansion
$$p_1 V_1 = p_2 V_2$$
$$600 \times 0.015 = p_2 \times 0.075$$

$$p_2 = \frac{600 \times 0.015}{0.075} = 120 \text{ kN/m}^2 \quad \text{Ans.}$$

2. For adiabatic compression
$$p_1 V_1^{\gamma} = p_2 V_2^{\gamma}$$
$$101.5 \times 0.012^{1.4} = p_2 \times 0.002^{1.4}$$

$$p_2 = \frac{101.5 \times 0.012^{1.4}}{0.002^{1.4}}$$

$$= 101.5 \times \left\{ \frac{0.012}{0.002} \right\}^{1.4}$$

$$= 101.5 \times 6^{1.4} = 1248 \text{ kN/m}^2 \quad \text{Ans.}$$

3. For polytropic expansion
$$p_1 V_1^{n} = p_2 V_2^{n}$$
$$2550 \times V_1^{1.3} = 210 \times 0.75^{1.3}$$

$$V_1^{1.3} = \frac{210 \times 0.75^{1.3}}{2550}$$

$$V_1 = 0.75 \times \sqrt[1.3]{\frac{210}{2550}}$$

$$\text{or } 0.75 \div \sqrt[1.3]{\frac{2550}{210}} = 0.1099 \text{ m}^3 \quad \text{Ans.}$$

4. Ratio of compression $= \dfrac{8.6}{1} = \dfrac{\text{initial volume}}{\text{final volume}} = \dfrac{V_1}{V_2}$

$$p_1 V_1^{n} = p_2 V_2^{n}$$
$$98 \times 8.6^{1.36} = p_2 \times 1^{1.36}$$
$$p_2 = 98 \times 8.6^{1.36} = 1828 \text{ kN/m}^2 \quad \text{Ans. (i)}$$

$$\frac{p_1 V_1}{T_1} = \frac{p_2 V_2}{T_2}$$

$$\frac{98 \times 8\cdot6}{(28 + 273)} = \frac{1828 \times 1}{T_2}$$

$$T_2 = \frac{301 \times 1828}{98 \times 8\cdot6} = 653 \text{ K}$$

$$653 - 273 = 380°\text{C} \quad \text{Ans. (ii)}$$

5.
$$p_1 V_1{}^n = p_2 V_2{}^n$$
$$1750 \times 0\cdot05^n = 122\cdot5 \times 0\cdot375^n$$

$$\frac{1750}{122\cdot5} = \left\{\frac{0\cdot375}{0\cdot05}\right\}^n$$

$$\log\left\{\frac{1750}{122\cdot5}\right\} = n \times \log\left\{\frac{0\cdot375}{0\cdot05}\right\}$$

$$1\cdot1548 = n \times 0\cdot8751$$

$$n = \frac{1\cdot1548}{0\cdot8751} = 1\cdot32 \quad \text{Ans.}$$

6. Tabulating values of pressure in bars and volume in litres with their respective logs:

p [bar]	$\log p$	V [litre]	$\log V$
30	1·4771	20	1·3010
17·7	1·2480	30	1·4771
12·2	1·0864	40	1·6021
9·1	0·9590	50	1·6990
7·2	0·8573	60	1·7782
5·9	0·7709	70	1·8451

The graph of $\log p$ on a base of $\log V$ is plotted as in Fig. 103. Choosing two points on the graph as shown,

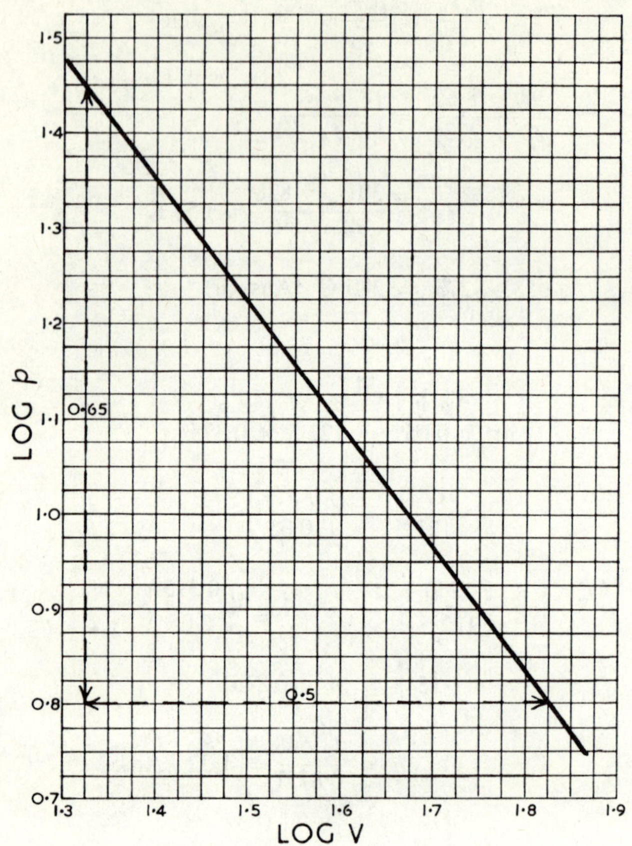

Fig. 103

$$n = \frac{\text{decrease in log } p}{\text{increase in log } V}$$

$$= \frac{1{\cdot}45 - 0{\cdot}8}{1{\cdot}825 - 1{\cdot}325}$$

$$= \frac{0{\cdot}65}{0{\cdot}5} = 1{\cdot}3 \quad \text{Ans.}$$

7.

$$\frac{T_1}{T_2} = \left\{\frac{p_1}{p_2}\right\}^{\frac{n-1}{n}}$$

$$\frac{n-1}{n} = \frac{1\cdot35 - 1}{1\cdot35} = \frac{0\cdot35}{1\cdot35} = \frac{0\cdot07}{0\cdot27} = \frac{7}{27}$$

$$\frac{(20 + 273)}{T_2} = \left\{\frac{101\cdot3}{1420}\right\}^{\frac{7}{27}}$$

$$T_2 = 293 \times \left\{\frac{1420}{101\cdot3}\right\}^{\frac{7}{27}} = 581\cdot1 \text{ K}$$

$$581\cdot1 - 273 = 308\cdot1°\text{C} \quad \text{Ans.}$$

8. For adiabatic expansion

$$\frac{T_1}{T_2} = \left\{\frac{V_2}{V_1}\right\}^{\gamma-1}$$

$$\text{where } \gamma = \frac{c_P}{c_V} = \frac{1\cdot005}{0\cdot718} = 1\cdot4$$

$$\gamma - 1 = 1\cdot4 - 1 = 0\cdot4$$

$$\frac{(66 + 273)}{(2 + 273)} = \left\{\frac{V_2}{0\cdot014}\right\}^{0\cdot4}$$

$$\left\{\frac{V_2}{0\cdot014}\right\}^{0\cdot4} = \frac{339}{275}$$

$$\frac{V_2}{0\cdot014} = \sqrt[0\cdot4]{\frac{339}{275}}$$

$$V_2 = 0\cdot014 \times \sqrt[0\cdot4]{\frac{339}{275}}$$

$$= 0\cdot023\ 62 \text{ m}^3 \quad \text{Ans.}$$

9.
$$\frac{T_1}{T_2} = \left\{\frac{V_2}{V_1}\right\}^{n-1}$$

$$\frac{(147 + 273)}{(21 + 273)} = \left\{\frac{0.21}{0.06}\right\}^{n-1}$$

$$1.429 = 3.5^{n-1}$$
$$\log 1.429 = (\log 3.5) \times (n - 1)$$
$$0.1549 = 0.5441(n - 1)$$
$$0.1549 = 0.5441n - 0.5441$$
$$0.1549 + 0.5441 = 0.5441n$$

$$n = \frac{0.6990}{0.5441} = 1.284 \quad \text{Ans.}$$

10.
$$\frac{T_1}{T_2} = \left\{\frac{p_1}{p_2}\right\}^{\frac{n-1}{n}}$$

$$\frac{(32 + 273)}{(500 + 273)} = \left\{\frac{117}{3655}\right\}^{\frac{n-1}{n}}$$

$$\frac{305}{773} = \left\{\frac{117}{3655}\right\}^{\frac{n-1}{n}}$$

$$\text{or,} \quad \frac{773}{305} = \left\{\frac{3655}{117}\right\}^{\frac{n-1}{n}}$$

$$\log\left\{\frac{773}{305}\right\} = \log\left\{\frac{3655}{117}\right\} \times \left\{\frac{n-1}{n}\right\}$$

$$0.4039 = 1.4947 \times \frac{n-1}{n}$$

$$0.4039n = 1.4947(n - 1)$$
$$0.4039n = 1.4947n - 1.4947$$
$$1.4947 = 1.4947n - 0.4039n$$
$$1.4947 = 1.0908n$$

$$n = \frac{1\cdot4947}{1\cdot0908} = 1\cdot37$$

$\therefore$ law of compression is:
$$pV^{1\cdot37} = C \quad \textbf{Ans.}$$

11. **Isothermal expansion:**

$$p_1V_1 = p_2V_2$$
$$8\cdot25 \times 0\cdot113 = p_2 \times 0\cdot331$$
$$p_2 = 2\cdot817 \text{ bar or } 281\cdot7 \text{ kN/m}^2 \quad \textbf{Ans. (ia)}$$
$$\text{Work done} = pV\ln r$$

$$r = \frac{0\cdot331}{0\cdot113} = 2\cdot929$$

$$\ln 2\cdot929 \text{ from tables} = 1\cdot0747$$
$$\text{Work done [kJ]} = 825\,[\text{kN/m}^2] \times 0\cdot113\,[\text{m}^3] \times 1\cdot0747$$
$$= 100\cdot2 \text{ kJ} \quad \textbf{Ans. (ib)}$$

Adiabatic expansion:
$$p_1V_1^{\gamma} = p_2V_2^{\gamma}$$
$$8\cdot25 \times 0\cdot113^{1\cdot4} = p_2 \times 0\cdot331^{1\cdot4}$$

$$p_2 = 8\cdot25 \times \left\{ \frac{0\cdot113}{0\cdot331} \right\}^{1\cdot4}$$

$$\text{or } 8\cdot25 \div \left\{ \frac{0\cdot331}{0\cdot113} \right\}^{1\cdot4}$$

$$p_2 = 1\cdot832 \text{ bar or } 183\cdot2 \text{ kN/m}^2 \quad \textbf{Ans. (iia)}$$

$$\text{Work done} = \frac{p_1V_1 - p_2V_2}{n - 1}$$

$$= \frac{825 \times 0\cdot113 - 183\cdot2 \times 0\cdot331}{1\cdot4 - 1}$$

$$= \frac{93\cdot24 - 60\cdot64}{0\cdot4}$$

$$= 81\cdot5 \text{ kJ} \quad \textbf{Ans. (iib)}$$

12.

$$T_1 = 510 + 273 = 783 \text{ K}$$
$$T_2 = 40 + 273 = 313 \text{ K}$$

$$\frac{p_1 V_1}{T_1} = \frac{p_2 V_2}{T_2}$$

$$\frac{36 \times 0.125}{783} = \frac{p_2 \times 1.5}{313}$$

$$p_2 = \frac{36 \times 0.125 \times 313}{783 \times 1.5}$$

$$= 1.199 \text{ bar or } 119.9 \text{ kN/m}^2 \quad \text{Ans. (i)}$$

$$p_1 V_1^n = p_2 V_2^n \quad \text{or} \quad \frac{T_1}{T_2} = \left\{ \frac{V_2}{V_1} \right\}^{n-1}$$

$$36 \times 0.125^n = 1.199 \times 1.5^n$$

$$\frac{36}{1.199} = \left\{ \frac{1.5}{0.125} \right\}^n$$

$$\log \left\{ \frac{36}{1.199} \right\} = n \times \log \left\{ \frac{1.5}{0.125} \right\}$$

$$1.4775 = n \times 1.0792$$

$$n = \frac{1.4775}{1.0792} = 1.37 \quad \text{Ans. (ii)}$$

$$p_1 V_1 = mRT_1$$

$$m = \frac{3600 \times 0.125}{0.284 \times 783} = 2.023 \text{ kg} \quad \text{Ans. (iii)}$$

Increase in internal energy:

$$E_2 - E_1 = mc_v (T_2 - T_1)$$
$$= 2.023 \times 0.71 \times (313 - 783)$$
$$= 2.023 \times 0.71 \times (-470)$$
$$= -675.1 \text{ kJ}$$

The minus sign indicates a *decrease* in internal energy.
Decrease in internal energy = 675.1 kJ   Ans. (iv)

$$\text{Work done} = \frac{p_1 V_1 - p_2 V_2}{n - 1} \text{ or } \frac{mR(T_1 - T_2)}{n - 1}$$

$$= \frac{3600 \times 0{\cdot}125 - 119{\cdot}9 \times 1{\cdot}5}{1{\cdot}37 - 1}$$

$$= \frac{450 - 179{\cdot}9}{0{\cdot}37} = 730{\cdot}2 \text{ kJ} \quad \text{Ans. (v)}$$

$$\text{Heat supplied} = \begin{array}{c}\text{Increase in}\\\text{internal energy}\end{array} + \begin{array}{c}\text{External}\\\text{work done}\end{array}$$

$$= -675{\cdot}1 + 730{\cdot}2$$
$$= 55{\cdot}1 \text{ kJ} \quad \text{Ans. (vi)}$$

SOLUTIONS TO TEST EXAMPLES 7

1. Indicated mean effective pressure [kN/m²]
    = mean height of diagram [mm] × spring scale [kN/m²mm]

$$= \frac{\text{area of diagram}}{\text{length of diagram}} \times \text{spring scale}$$

$$= \frac{390}{70} \times 0 \cdot 8 \times 10^2 = 445 \cdot 8 \text{ kN/m}^2$$

$$\text{ip} = p_m ALn \times 4 \text{ (for 4 cylinders)}$$

$$= 445 \cdot 8 \times 0 \cdot 7854 \times 0 \cdot 15^2 \times 0 \cdot 2 \times \frac{5 \cdot 5}{2} \times 4$$

$$= 17 \cdot 33 \text{ kW} \quad \text{Ans.}$$

2. $\text{ip} = p_m ALn \times 8$ (for 8 cylinders)

$$= 586 \times 0 \cdot 7854 \times 0 \cdot 75^2 \times 1 \cdot 125 \times \frac{110}{60 \times 2} \times 8$$

$$= 2136 \text{ kW} \quad \text{Ans. (i)}$$

$$\begin{aligned}
\text{bp} &= \text{ip} \times \text{mech. efficiency} \\
&= 2136 \times 0 \cdot 86 \\
&= 1838 \text{ kW} \quad \text{Ans. (ii)}
\end{aligned}$$

3. $\text{ip} = \dfrac{\text{bp}}{\text{mech. effic.}} = \dfrac{1125}{0 \cdot 84} = 1340 \text{ kW}$

$$\begin{aligned}
\text{Let } d &= \text{diameter of cylinders in meters} \\
\text{then } L &= 1 \cdot 25d \text{ metres} \\
\text{ip} &= p_m ALn \times 6 \\
1340 &= 5 \times 10^2 \times 0 \cdot 7854 \times d^2 \times 1 \cdot 25d \times 2 \times 6
\end{aligned}$$

$$d = \sqrt[3]{\frac{1340}{500 \times 0 \cdot 7854 \times 1 \cdot 25 \times 2 \times 6}}$$

$$\left.\begin{aligned}
&= 0 \cdot 6103 \text{ m} &= 610 \cdot 3 \text{ mm} \\
L &= 1 \cdot 25 \times 610 \cdot 3 &= 763 \text{ mm}
\end{aligned}\right\} \quad \text{Ans.}$$

4. Ratio of strokes, top piston : bottom piston
$$= \quad 6 : 7.5$$

Strokes of pistons:

$$\text{Top} = \frac{6}{6 + 7.5} \times 2430 = 1080 \text{ mm}$$

$$\text{Bottom} = \frac{7.5}{6 + 7.5} \times 2430 = 1350 \text{ mm}$$

Ans. (i)

$$\begin{aligned}
\text{ip} &= p_m ALn \times 6 \\
&= 6.5 \times 10^2 \times 0.7854 \times 0.625^2 \times 2.43 \times 1.75 \times 6 \\
&= 5089 \text{ kW} \quad \text{Ans. (ii)}
\end{aligned}$$

$$\begin{aligned}
\text{bp} &= \text{ip} \times \text{mech. efficiency} \\
&= 5089 \times 0.9 = 4580 \text{ kW} \quad \text{Ans. (iii)}
\end{aligned}$$

5. $\text{ip} = p_m ALn$

For top side of piston in each cylinder:

$$\begin{aligned}
\text{ip} &= 5.8 \times 10^2 \times 0.7854 \times 0.7^2 \times 1.35 \times 1.8 \\
&= 542.4 \text{ kW} \quad \text{Ans. (i)}
\end{aligned}$$

For bottom side of piston in each cylinder:

$$\begin{aligned}
\text{ip} &= 4.9 \times 10^2 \times 0.7854 \,(0.7^2 - 0.25^2) \times 1.35 \times 1.8 \\
&= 399.7 \text{ kW} \quad \text{Ans. (ii)}
\end{aligned}$$

Total engine power for eight cylinders:

$$\text{ip} = (542.4 + 399.7) \times 8 = 7537 \text{ kW} \quad \text{Ans. (iii)}$$

$$\text{bp} = 7537 \times 0.8 = 6030 \text{ kW} \quad \text{Ans. (iv)}$$

6. Load on brake $= 480 - 84 = 396$ N

Effective radius $= \tfrac{1}{2}(1220 + 24) = 622$ mm $= 0.622$m

Brake power $= T\omega$

$$= 396 \times 0.622 \times \frac{250 \times 2\pi}{60}$$

$$= 6449 \text{ W} = 6.449 \text{ kW} \quad \text{Ans. (i)}$$

$$1 \text{ kWh} = 3.6 \text{ MJ}$$

Heat carried away by water in one hour
$$= 0.9 \times 6.449 \times 3.6 \times 10^3 \text{ kJ}$$

Heat received by water

$$= \text{mass} \times \text{spec. heat} \times \text{temp. rise}$$
$$Q[\text{kJ}] = m\,[\text{kg}] \times c\,[\text{kJ/kgK}] \times \theta\,[\text{K}]$$
$$= m \times 4{\cdot}2 \times 18$$
$$\therefore\ m \times 4{\cdot}2 \times 18 = 0{\cdot}9 \times 6{\cdot}449 \times 3{\cdot}6 \times 10^3$$
$$m = 276{\cdot}4\ \text{kg}$$

Mass of one litre of fresh water = one kg
$\therefore$ Quantity in litres = 276·4 litres/h   Ans. (ii)

7. $\text{bp} = \dfrac{W \times \text{rev/s}}{300} = \dfrac{180 \times 50}{300} = 30\ \text{kW}$

Heat generated at brake in one hour

$$= \text{heat equivalent of 30 kWh}$$
$$= 30 \times 3{\cdot}6 \times 10^3\ \text{kJ}$$

Heat carried away by water [kJ]

$$= m\,[\text{kg}] \times c\,[\text{kJ/kgK}] \times \theta\,[\text{K}]$$
$$= 1218 \times 4{\cdot}2 \times (36{\cdot}9 - 16{\cdot}4)\ \text{kJ}$$

% heat carried away

$$= \frac{1218 \times 4{\cdot}2 \times 20{\cdot}5}{30 \times 3{\cdot}6 \times 10^3} \times 100$$

$$= 97{\cdot}09\%\quad \text{Ans.}$$

8. Brake power [W] = Torque [Nm] $\times\ \omega$ [rad/s]
When speed is 24·5 rev/s
$\quad \text{bp} = T \times 2\pi \times 24{\cdot}5 = 153{\cdot}9\,T\ \text{watts} = 0{\cdot}1539\,T\ \text{kW}$

With all cylinders firing:
$\quad\quad \text{bp} = 0{\cdot}1539 \times 193{\cdot}8 = 29{\cdot}84\ \text{kW}\quad \text{Ans. (i)}$

$$
\begin{aligned}
\text{no. 1 cyl. cut out, bp} &= 0{\cdot}1539 \times 130{\cdot}8 = 20{\cdot}13\ \text{kW}\\
\text{ip of no. 1 cyl.} &= 29{\cdot}84 - 20{\cdot}13 = 9{\cdot}71\ \text{kW}\\
\text{no. 2 cyl. cut out, bp} &= 0{\cdot}1539 \times 130{\cdot}2 = 20{\cdot}04\ \text{kW}\\
\text{ip of no. 2 cyl.} &= 29{\cdot}84 - 20{\cdot}04 = 9{\cdot}8\ \text{kW}\\
\text{no. 3 cyl. cut out, bp} &= 0{\cdot}1539 \times 129{\cdot}9 = 20{\cdot}0\ \text{kW}\\
\text{ip of no. 3 cyl.} &= 29{\cdot}84 - 20 = 9{\cdot}84\ \text{kW}
\end{aligned}
$$

no. 4 cyl. cut out, bp $= 0.1539 \times 131.1 = 20.18$ kW
ip of no. 4 cyl. $= 29.84 - 20.18 = 9.66$ kW

Total ip of engine
$= 9.71 + 9.8 + 9.84 + 9.66 = 39.01$ kW   Ans. (ii)

$$\text{Mech. efficiency} = \frac{\text{brake power}}{\text{indicated power}}$$

$$= \frac{29.84}{39.01} = 0.7649 = 76.49\% \quad \text{Ans. (iii)}$$

9. Brake thermal effic. $=$

$$\frac{3.6 \ [\text{MJ/kWh}]}{\text{kg fuel/brake kW h} \times \text{cal. value [MJ/kg]}}$$

$$= \frac{3.6}{0.255 \times 43.5} = 0.3245 \text{ or } 32.45\% \quad \text{Ans. (ii)}$$

$$\text{Indicated thermal effiic.} = \frac{\text{brake thermal effic.}}{\text{mechanical effic.}}$$

$$= \frac{0.3245}{0.86} = 0.3774 \text{ or } 37.74\% \quad \text{Ans. (i)}$$

For each kg of fuel burned,
    total mass of gases $= 35$ kg air $+ 1$ kg fuel $= 36$ kg

Heat carried away in exhaust gases,
$$\begin{aligned}
Q[\text{kJ}] &= m\,[\text{kg}] \times c\,[\text{kJ/kgK}] \times \theta\,[\text{K}] \\
&= 36 \times 1.005 \times (393 - 26) \\
&= 1.327 \times 10^4 \text{ kJ} \\
&= 13.27 \text{ MJ/kg fuel}
\end{aligned}$$
Heat supplied $= 43.5$ MJ/kg fuel

$\therefore$ percentage heat in exhaust gases

$$= \frac{13.27}{43.5} \times 100 = 30.52\% \quad \text{Ans. (iii)}$$

10. Ind. thermal eff. $= 100\% - (31{\cdot}7 + 30{\cdot}8)$
$= 37{\cdot}5\%$ Ans. (i)

$$\text{Mech. eff.} = \frac{\text{bp}}{\text{ip}} = \frac{4060}{4960}$$

$$= 0{\cdot}8185 \text{ or } 81{\cdot}85\% \quad \text{Ans. (ii)}$$

Overall effic. $=$ ind. thermal effic. $\times$ mech. effic.
$= 0{\cdot}375 \times 0{\cdot}8185$
$= 0{\cdot}307$ or $30{\cdot}7\%$ Ans. (iii)

Specific fuel consumption (indicated)

$$= \frac{27 \times 10^3}{24 \times 4960} = 0{\cdot}2269 \text{ kg/ind. kWh} \quad \text{Ans. (iv)}$$

$$\text{Indicated thermal effic.} = \frac{3{\cdot}6 \,[\text{MJ/kWh}]}{\text{kg fuel/ind. kWh} \times \text{cal. val. } [\text{MJ/kg}]}$$

$$0{\cdot}375 = \frac{3{\cdot}6}{0{\cdot}2269 \times \text{cal. value}}$$

$$\text{Calorific value} = \frac{3{\cdot}6}{0{\cdot}375 \times 0{\cdot}2269}$$

$$= 42{\cdot}32 \text{ MJ/kg} \quad \text{Ans. (v)}$$

11. Heat supplied to engine [kJ/h]
$= 0{\cdot}243 \times 2600 \times 42 \times 10^3$

Heat carried away by lub. oil [kJ/h]
$= m\,[\text{kg}] \times c\,[\text{kJ/kgK}] \times \theta\,[\text{K}]$
$= 40 \times 10^3 \times 2{\cdot}1 \times (49 - 24)$

Heat carried away as a percentage of the heat supplied

$$= \frac{40 \times 10^3 \times 2{\cdot}1 \times 25}{0{\cdot}243 \times 2600 \times 42 \times 10^3} \times 100$$

$$= 7{\cdot}914\% \quad \text{Ans. (i)}$$

Through the lub. oil cooler, heat transfer per hour:

Heat gained by water $=$ Heat lost by oil

$$m_w \times c_w \times \theta_w = m_o \times c_o \times \theta_o$$

$$m_w[\text{t/h}] \times 10^3 \times 4{\cdot}2 \times (31-21) = 40 \times 10^3 \times 2{\cdot}1 \times (49-24)$$

$$m_w = \frac{40 \times 2{\cdot}1 \times 25}{4{\cdot}2 \times 10}$$

$$= 50 \text{ tonne/h} \quad \text{Ans. (ii)}$$

12. Due to the "hit and miss" governor cutting off the supply of gas, the number of power strokes per second is the number of explosions per second, which is $123 \div 60 = 2{\cdot}05$

$$\begin{aligned}
\text{ip} &= p_m A L n \\
&= 3{\cdot}93 \times 10^2 \times 0{\cdot}7854 \times 0{\cdot}18^2 \times 0{\cdot}3 \times 2{\cdot}05 \\
&= 6{\cdot}152 \text{ kW Ans. (i)}
\end{aligned}$$

$$\text{Mech. effic.} = \frac{\text{brake power}}{\text{indicated power}}$$

$$= \frac{4{\cdot}33}{6{\cdot}152} = 0{\cdot}7039 = 70{\cdot}39\% \quad \text{Ans. (ii)}$$

Indicated thermal efficiency

$$= \frac{\text{heat equivalent of work done in cyl. [kJ/h]}}{\text{heat supplied [kJ/h]}}$$

$$= \frac{6{\cdot}152 \times 3600}{3{\cdot}1 \times 17{\cdot}6 \times 10^3}$$

$$= 0{\cdot}4059 = 40{\cdot}59\% \quad \text{Ans. (iii)}$$

$$\begin{aligned}
\text{Brake thermal effic.} &= \text{ind. thermal effic.} \times \text{mech. effic.} \\
&= 0{\cdot}4059 \times 0{\cdot}7039 \\
&= 0{\cdot}2858 = 28{\cdot}58\% \quad \text{Ans. (iv)}
\end{aligned}$$

SOLUTIONS TO TEST EXAMPLES 8

1.
$$p_1 V_1^{1\cdot34} = p_2 V_2^{1\cdot34}$$
$$0\cdot98 \times 9^{1\cdot34} = p_2 \times 1^{1\cdot34}$$
$$p_2 = 0\cdot98 \times 9^{1\cdot34}$$
$$= 18\cdot61 \text{ bar} \quad \text{Ans. (i)}$$

$$\frac{p_1 V_1}{T_1} = \frac{p_2 V_2}{T_2}$$

$$\frac{0\cdot98 \times 9}{(38 + 273)} = \frac{18\cdot61 \times 1}{T_2}$$

$$T_2 = \frac{311 \times 18\cdot61}{0\cdot98 \times 9} = 656\cdot3\,\text{K}$$

$$656\cdot3 - 273 = 383\cdot3°\text{C} \quad \text{Ans. (ii)}$$

2. Clearance length [mm] $= \dfrac{\text{clearance volume [mm}^3]}{\text{area of cylinder [mm}^2]}$

$$= \frac{36 \times 10^3}{0\cdot7854 \times 70^2} = 9\cdot354\,\text{mm}$$

$$V_1 = \text{clearance} + \text{stroke}$$
$$= 9\cdot354 + 75 = 84\cdot354\,\text{mm}$$

$$V_2 = \text{clearance} = 9\cdot354\,\text{mm}$$

$$p_1 V_1^{1\cdot37} = p_2 V_2^{1\cdot37}$$
$$0\cdot97 \times 84\cdot35^{1\cdot37} = p_2 \times 9\cdot354^{1\cdot37}$$

$$p_2 = 0\cdot97 \times \left\{ \frac{84\cdot35}{9\cdot354} \right\}^{1\cdot37}$$

$$= 19\cdot73 \text{ bar} \quad \text{Ans.}$$

3.   Let stroke volume $= 100$
then clearance volume $= 7{\cdot}5\%$ of $100 = 7{\cdot}5$

$$\therefore V_1 = 100 + 7{\cdot}5 = 107{\cdot}5$$
$$V_2 = 7{\cdot}5$$
$$p_1 V_1^{1{\cdot}36} = p_2 V_2^{1{\cdot}36}$$
$$1{\cdot}1 \times 107{\cdot}5^{1{\cdot}36} = p_2 \times 7{\cdot}5^{1{\cdot}36}$$

$$p_2 = 1{\cdot}1 \times \left\{ \frac{107{\cdot}5}{7{\cdot}5} \right\}^{1{\cdot}36}$$

$$= 41{\cdot}11 \text{ bar} \quad \text{Ans. (i)}$$

$$\frac{p_1 V_1}{T_1} = \frac{p_2 V_2}{T_2}$$

$$\frac{1{\cdot}1 \times 107{\cdot}5}{(35 + 273)} = \frac{41{\cdot}11 \times 7{\cdot}5}{T_2}$$

$$T_2 = \frac{308 \times 41{\cdot}11 \times 7{\cdot}5}{1{\cdot}1 \times 107{\cdot}5} = 803{\cdot}4 \text{ K}$$

$$803{\cdot}4 - 273 = 530{\cdot}4°\text{C} \quad \text{Ans. (ii)}$$

4.   **BEFORE ALTERATION:**

$$V_1 = 87{\cdot}5 + 12{\cdot}5 = 100$$
$$V_2 = 12{\cdot}5$$
$$p_1 V_1^{1{\cdot}35} = p_2 V_2^{1{\cdot}35}$$
$$0{\cdot}97 \times 100^{1{\cdot}35} = p_2 \times 12{\cdot}5^{1{\cdot}35}$$

$$p_2 = 0{\cdot}97 \times \left\{ \frac{100}{12{\cdot}5} \right\}^{1{\cdot}35}$$

$$= 0{\cdot}97 \times 8^{1{\cdot}35} = 16{\cdot}07 \text{ bar} \quad \text{Ans. (i)}$$

AFTER ALTERATION:

$$V_1 = 87{\cdot}5 + 10 = 97{\cdot}5$$
$$V_2 = 10$$

$$p_1 V_1^{1\cdot 35} = p_2 V_2^{1\cdot 35}$$
$$0\cdot 97 \times 97\cdot 5^{1\cdot 35} = p_2 \times 10^{1\cdot 35}$$
$$p_2 = 0\cdot 97 \times 9\cdot 75^{1\cdot 35} = 20\cdot 99 \text{ bar} \quad \text{Ans. (ii)}$$

5.   **BEFORE ALTERATION:**

$$V_1 = 880 + 80 = 960$$
$$V_2 = 80$$
$$p_1 V_1^{1\cdot 38} = p_2 V_2^{1\cdot 38}$$
$$p_1 \times 960^{1\cdot 38} = 32 \times 80^{1\cdot 38}$$

$$p_1 = 32 \times \left\{ \frac{80}{960} \right\}^{1\cdot 38}$$

AFTER CLEARANCE IS REDUCED BY 5 mm:

$$V_1 = 880 + 75 = 955$$
$$V_2 = 75$$
$$p_1 V_1^{1\cdot 38} = p_2 V_2^{1\cdot 38}$$

$$32 \times \left\{ \frac{80}{960} \right\}^{1\cdot 38} \times 955^{1\cdot 38} = p_2 \times 75^{1\cdot 38}$$

$$p_2 = 32 \times \left\{ \frac{80 \times 955}{960 \times 75} \right\}^{1\cdot 38}$$

$$= 34\cdot 72 \text{ bar}$$

$\therefore$ Increase in final pressure

$$= 34\cdot 72 - 32 = 2\cdot 72 \text{ bar} \quad \text{Ans.}$$

6.

$$\frac{T_2}{T_1} = \left\{ \frac{V_1}{V_2} \right\}^{n-1}$$

$$T_1 = 49 + 273 = 322\,\text{K}$$
$$V_1 \div V_2 = 16$$
$$n - 1 = 1\cdot 34 - 1 = 0\cdot 34$$

$$\therefore T_2 = 322 \times 16^{0\cdot 34} = 826\cdot 6\,\text{K}$$
$$826\cdot 6 - 273 = 553\cdot 6°\text{C} \quad \text{Ans.}$$

7. Volume [m³] of air in cylinder at beginning of compression

$$= 0{\cdot}7854 \times 0{\cdot}65^2 \times (0{\cdot}675 + 0{\cdot}065)$$
$$= 0{\cdot}2456 \text{ m}^3$$

Absolute pressure $= 13{\cdot}8 + 101{\cdot}3 = 115{\cdot}1 \text{ kN/m}^2$

Absolute temperature $= 43 + 273 = 316\text{K}$

$$p_1 V_1 = mRT_1$$

$$m = \frac{115{\cdot}1 \times 0{\cdot}2456}{0{\cdot}287 \times 316} = 0{\cdot}3118 \text{ kg} \quad \text{Ans.}$$

8.
$$\frac{T_2}{T_1} = \left\{\frac{V_1}{V_2}\right\}^{n-1}$$

$$V_1 \div V_2 = \text{ratio of compression} = r$$

$$\frac{(382 + 273)}{(30 + 273)} = r^{0{\cdot}36}$$

$$r = \sqrt[0{\cdot}36]{\frac{655}{303}} = 8{\cdot}511 \quad \text{Ans.}$$

9.
$$p_1 V_1^{1{\cdot}35} = p_2 V_2^{1{\cdot}35}$$
$$1 \times 8{\cdot}5^{1{\cdot}35} = p_2 \times 1^{1{\cdot}35}$$
$$p_2 = 8{\cdot}5^{1{\cdot}35} = 17{\cdot}97 \text{ bar} \quad \text{Ans. (i)}$$

$$\frac{p_1 V_1}{T_1} = \frac{p_2 V_2}{T_2}$$

$$\frac{1 \times 8{\cdot}5}{(40 + 273)} = \frac{17{\cdot}97 \times 1}{T_2}$$

$$T_2 = \frac{313 \times 17{\cdot}97}{8{\cdot}5} = 661{\cdot}8\,\text{K}$$

$$661{\cdot}8 - 273 = 388{\cdot}8°\text{C} \quad \text{Ans. (ii)}$$

$$\frac{T_3}{T_2} = \frac{p_3}{p_2}$$

$$T_3 = \frac{661{\cdot}8 \times 31}{17{\cdot}97} = 1141 \text{ K}$$

$$1141 - 273 = 868°\text{C} \quad \text{Ans. (iii)}$$

10.
$$p_1 = 0.99 \times 10^2 = 99 \text{ kN/m}^2$$
$$V_1 = 113 \times 10^{-3} = 0.113 \text{ m}^3$$
$$T_1 = 48 + 273 = 321 \text{ K}$$
$$p_1 V_1 = mRT_1$$

$$m = \frac{99 \times 0.113}{0.29 \times 321} = 0.1201 \text{ kg} \quad \text{Ans. (i)}$$

$$p_1 V_1^{1.37} = p_2 V_2^{1.37}$$

$$p_2 = p_1 \times \left\{ \frac{V_1}{V_2} \right\}^{1.37}$$

$$= 0.99 \times 10^{1.37} = 23.2 \text{ bar} \quad \text{Ans. (ii a)}$$

$$\frac{p_1 V_1}{T_1} = \frac{p_2 V_2}{T_2}$$

$$T_2 = \frac{321 \times 23.2 \times 1}{0.99 \times 10} = 752.3 \text{ K}$$

$$752.3 - 273 = 479.3°\text{C} \quad \text{Ans. (ii b)}$$

Heat energy given to gas:
$$Q[\text{kJ}] = m\,[\text{kg}] \times c\,[\text{kJ/kg K}] \times \theta\,[\text{K}]$$
$$95 = 0.1201 \times 0.712 \times (T_3 - 752.3)$$

$$T_3 - 752.3 = \frac{95}{0.1201 \times 0.712}$$

$$T_3 = 1111 + 752.3 = 1863.3 \text{ K}$$
$$1863.3 - 273 = 1590.3 \text{ °C} \quad \text{Ans. (iii b)}$$

$$\frac{p_3}{p_2} = \frac{T_3}{T_2}$$

$$p_3 = \frac{23.2 \times 1863}{752.3} = 57.45 \text{ bar} \quad \text{Ans. (iii a)}$$

11.

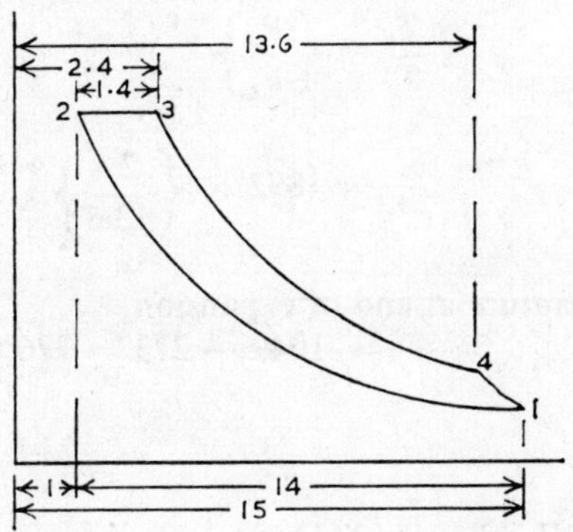

Let $V_1 = 15$, and $V_2 = 1$
then, stroke volume $= 15 - 1 = 14$
Fuel burning period $= \frac{1}{10} \times 14 = 1\cdot4$

$$\therefore \quad V_3 = 1\cdot4 + 1 = 2\cdot4$$
$$V_4 = 1 + \frac{9}{10} \times 14 = 13\cdot6$$
$$T_1 = 41 + 273 = 314 \text{ K}$$

COMPRESSION PERIOD:

$$\frac{T_2}{T_1} = \left\{\frac{V_1}{V_2}\right\}^{n-1}$$

$$T_2 = 314 \times 15^{0\cdot34} = 788\cdot4 \text{ K}$$
$\therefore$ Temperature at end of compression
$$= 788\cdot4 - 273 = 515\cdot4°C \quad \text{Ans. (i)}$$

BURNING PERIOD:

$$\frac{T_3}{T_2} = \frac{V_3}{V_2}$$

$$T_3 = 788\cdot4 \times 2\cdot4 = 1892 \text{ K}$$

$\therefore$ Temperature at end of combustion
$$= 1892 - 273 = 1619°C \quad \text{Ans. (ii)}$$

EXPANSION PERIOD:

$$\frac{T_4}{T_3} = \left\{\frac{V_3}{V_4}\right\}^{n-1}$$

$$T_4 = 1892 \times \left\{\frac{2\cdot4}{13\cdot6}\right\}^{0\cdot34} = 1049 \text{ K}$$

$\therefore$ Temperature at end of expansion
$$= 1049 - 273 = 776°C \quad \text{Ans. (iii)}$$

12. CONSTANT VOLUME CYCLE:

$$\text{A.S.E.} = 1 - \frac{1}{r^{\gamma-1}} = 1 - \frac{1}{7^{0\cdot4}}$$

$$= 1 - 0\cdot4592 = 0\cdot5408 \text{ or } 54\cdot08\% \quad \text{Ans. (i)}$$

DIESEL CYCLE:

Referring to Fig. 41, $V_1 = 14$, and $V_2 = 1$
$$\therefore \text{ stroke} = 14 - 1 = 13$$
$$\text{Fuel burning period} = 6\% \text{ of } 13 = 0\cdot78$$
$$V_3 = 1 + 0\cdot78 = 1\cdot78$$

$$\text{Fuel cut-off ratio} = \rho = \frac{V_3}{V_2} = 1\cdot78$$

$$\text{A.S.E.} = 1 - \frac{1}{\gamma} \times \frac{1}{r^{\gamma-1}} \left\{\frac{\rho^{\gamma} - 1}{\rho - 1}\right\}$$

$$= 1 - \frac{1}{1\cdot4} \times \frac{1}{14^{0\cdot4}} \left\{\frac{1\cdot78^{1\cdot4} - 1}{1\cdot78 - 1}\right\}$$

$$= 1 - \frac{1}{1\cdot4} \times \frac{1}{14^{0\cdot4}} \times \frac{1\cdot242}{0\cdot78}$$

$$= 1 - 0\cdot3959$$
$$= 0\cdot6041 \text{ or } 60\cdot41\% \quad \text{Ans. (ii)}$$

13. $p_1 V_1^{1\cdot35} = p_2 V_2^{1\cdot35}$ $1\cdot03 \times 14^{1\cdot35} = p_2 \times 1^{1\cdot35}$

$$p_2 = 1\cdot03 \times 14^{1\cdot35} = 36\cdot31 \text{ bar} \quad \text{Ans. (i)}$$

$$\frac{p_1 V_1}{T_1} = \frac{p_2 V_2}{T_2} \qquad \frac{1\cdot03 \times 14}{(51 + 273)} = \frac{36\cdot31 \times 1}{T_2}$$

$$T_2 = \frac{324 \times 36\cdot31}{1\cdot03 \times 14} = 815\cdot8 \text{ K}$$

$\therefore$ Temperature at end of compression
$$= 815\cdot8 - 273 = 542\cdot8°\text{C} \quad \text{Ans. (ii)}$$

$$\text{Stroke} = 1\cdot2 \times 500 = 600 \text{ mm}$$

Stroke volume,
$$V_1\text{-}V_2 = 0\cdot7854 \times 0\cdot5^2 \times 0\cdot6 = 0\cdot1179 \text{ m}^3$$

$$V_1/V_2 = \text{ratio of compression} = 14$$

$$\therefore V_1\text{-}V_2 = V_1 - \frac{V_1}{14} = \frac{13}{14} V_1 = 0\cdot1179$$

$$V_1 = \frac{14}{13} \times 0\cdot1179 = 0\cdot1269 \text{ m}^3$$

$$p_1 V_1 = mRT_1$$

$$m = \frac{1\cdot03 \times 10^2 \times 0\cdot1269}{0\cdot287 \times 324} = 0\cdot1406 \text{ kg} \quad \text{Ans (iii)}$$

$$\text{Work done} = \frac{MR(T_1 - T_2)}{n - 1}$$

$$= \frac{0\cdot1406 \times 0\cdot287 \times (324 - 815\cdot8)}{1\cdot35 - 1}$$

$$= -56\cdot67 \text{ kJ}$$

i.e. Work done ON the air $= 56\cdot67$ kJ   Ans. (iv)

Increase in internal energy $= m\, c_v(T_2 - T_1)$
$$= 0\cdot1406 \times 0\cdot718 \times (815\cdot8 - 324)$$
$$= 49\cdot63 \text{ kJ} \quad \text{Ans. (v)}$$

$$\begin{matrix} \text{Heat energy} \\ \text{supplied} \end{matrix} = \begin{matrix} \text{Increase in} \\ \text{internal energy} \end{matrix} + \begin{matrix} \text{External} \\ \text{work done} \end{matrix}$$

$$= 49{\cdot}63 - 56{\cdot}67$$
$$= -7{\cdot}04 \text{ kJ}$$

i.e. Heat energy transferred *from* air *to* cylinder walls during compression $= 7{\cdot}04$ kJ   Ans. (vi)

14. Referring to Fig. 42, data given:

$$\begin{aligned} p_1 &= 1 & p_3 = p_4 &= 41 \\ V_1 &= V_5 = 10{\cdot}7 & V_2 = V_3 &= 1 \\ T_1 &= 32 + 273 = 305 \\ T_4 &= 1593 + 273 = 1866 \end{aligned}$$

$$\gamma = \frac{c_P}{c_V} = \frac{1{\cdot}005}{0{\cdot}718} = 1{\cdot}4$$

COMPRESSION PERIOD:

$$p_1 V_1{}^\gamma = p_2 V_2{}^\gamma \qquad p_2 = \frac{1 \times 10{\cdot}7^{1{\cdot}4}}{1^{1{\cdot}4}}$$

$$= 27{\cdot}62 \text{ bar}$$

$$\frac{p_1 V_1}{T_1} = \frac{p_2 V_2}{T_2} \qquad T_2 = \frac{305 \times 27{\cdot}62 \times 1}{1 \times 10{\cdot}7}$$

$$= 787{\cdot}2 \text{ K}$$
$$787{\cdot}2 - 273 = 514{\cdot}2°\text{C}$$

PART COMBUSTION AT CONSTANT VOLUME:

Absolute temperature varies as absolute pressure,

$$\therefore T_3 = T_2 \times \frac{p_3}{p_2} = \frac{787{\cdot}2 \times 41}{27{\cdot}62} = 1169 \text{ K}$$

$$1169 - 273 = 896°\text{C}$$

PART COMBUSTION AT CONSTANT PRESSURE:

Volume varies as absolute temperature,

$$\therefore V_4 = V_3 \times \frac{T_4}{T_3} = \frac{1 \times 1866}{1169} = 1{\cdot}596$$

EXPANSION PERIOD:

$$p_4 V_4^{1\cdot4} = p_5 V_5^{1\cdot4}$$

$$p_5 = \frac{41 \times 1\cdot596^{1\cdot4}}{10\cdot7^{1\cdot4}} = 2\cdot858 \text{ bar}$$

Since $V_5 = V_1$, then $\dfrac{T_5}{T_1} = \dfrac{p_5}{p_1}$

$$T_5 = \frac{305 \times 2\cdot858}{1} = 871\cdot6 \text{ K}$$

$$871\cdot6 - 273 = 598\cdot6°\text{C}$$

Hence, pressures and temperatures required:

$$\left.\begin{array}{l} p_2 = 27\cdot62 \text{ bar}, \ \theta_2 = 514\cdot2°\text{C} \\ \phantom{p_2 = 27\cdot62 \text{ bar}, \ } \theta_3 = 896°\text{C} \\ p_5 = 2\cdot858 \text{ bar}, \ \theta_5 = 598\cdot6°\text{C} \end{array}\right\} \quad \text{Ans. (i)}$$

Ideal thermal efficiency

$$= 1 - \frac{\text{heat rejected}}{\text{heat supplied}}$$

$$= 1 - \frac{c_V (T_5 - T_1)}{c_V (T_3 - T_2) + c_P (T_4 - T_3)}$$

$$= 1 - \frac{(T_5 - T_1)}{(T_3 - T_2) + \gamma (T_4 - T_3)}$$

$$= 1 - \frac{871\cdot6 - 305}{(1169 - 787\cdot2) + 1\cdot4 (1866 - 1169)}$$

$$= 1 - 0\cdot4175$$
$$= 0\cdot5825 = 58\cdot25\% \quad \text{Ans. (ii)}$$

SOLUTIONS TO TEST EXAMPLES 9

1. $\text{Clearance length [mm]} = \dfrac{\text{clearance volume [mm}^3]}{\text{area of cylinder [mm}^2]}$

$$= \frac{900 \times 10^3}{0{\cdot}7854 \times 250^2} = 18{\cdot}33 \text{ mm}$$

$$V_1 = 350 + 18{\cdot}33 = 368{\cdot}33 \text{ mm}$$
$$p_1 V_1^{1{\cdot}25} = p_2 V_2^{1{\cdot}25}$$
$$0{\cdot}986 \times 368{\cdot}3^{1{\cdot}25} = 4{\cdot}1 \times V_2^{1{\cdot}25}$$

$$V_2 = 368{\cdot}3 \times \sqrt[1{\cdot}25]{\frac{0{\cdot}986}{4{\cdot}1}}$$

$$= 117{\cdot}8 \text{ mm}$$

$$\text{Compression period} = V_1 - V_2$$
$$= 368{\cdot}3 - 117{\cdot}8$$
$$= 250{\cdot}5 \text{ mm} \quad \text{Ans.}$$

2. Let c = clearance length, in millimetres
$$V_1 = \text{stroke} + \text{clearance} = 160 + c$$
$$V_2 = (160 + c) - 120 = 40 + c$$

$$p_1 V_1^{1{\cdot}3} = p_2 V_2^{1{\cdot}3}$$
$$1 \times (160 + c)^{1{\cdot}3} = 5 \times (40 + c)^{1{\cdot}3}$$

$$\left\{ \frac{160 + c}{40 + c} \right\}^{1{\cdot}3} = \frac{5}{1}$$

$$\frac{160 + c}{40 + c} = \sqrt[1{\cdot}3]{5} = 3{\cdot}449$$

$$160 + c = 3{\cdot}449 \,(40 + c)$$
$$160 + c = 137{\cdot}96 + 3{\cdot}449c$$
$$22{\cdot}04 = 2{\cdot}449c$$
$$c = 9 \text{ mm}$$

$$\text{Clearance volume [cm}^3] = \text{area [cm}^2] \times \text{length [cm]}$$
$$= 0{\cdot}7854 \times 8^2 \times 0{\cdot}9$$
$$= 45{\cdot}24 \text{ cm}^3 \quad \text{Ans.}$$

3. $pV = mRT$ $\therefore m = \dfrac{pV}{RT}$

Hence, mass varies directly as the absolute pressure and volume, and inversely as the absolute temperature.

$$\therefore \text{Mass in vessel} = 1{\cdot}293 \times \frac{40}{1} \times \frac{0{\cdot}566}{1} \times \frac{(0 + 273)}{(26 + 273)}$$

$$= 26{\cdot}73 \text{ kg} \quad \text{Ans.}$$

4. $pV = mRT$ $\therefore V = \dfrac{mRT}{p}$

Hence, volume varies directly as the mass and absolute temperature, and inversely as the absolute pressure.

$\therefore$ Volume of 0·9 kg at atmospheric pressure and 24°C

$$= 0{\cdot}7734 \times \frac{0{\cdot}9}{1} \times \frac{(24 + 273)}{(0 + 273)}$$

$$= 0{\cdot}7571 \text{ m}^3 \quad \text{Ans. (i)}$$

$$p_1 V_1^{1{\cdot}2} = p_2 V_2^{1{\cdot}2}$$
$$1 \times 0{\cdot}7571^{1{\cdot}2} = 10 \times V_2^{1{\cdot}2}$$

$$V_2 = \frac{0{\cdot}7571}{{}^{1{\cdot}2}\sqrt{10}}$$

$$= 0{\cdot}1111 \text{ m}^3 \quad \text{Ans. (ii)}$$

$$\frac{p_1 V_1}{T_1} = \frac{p_2 V_2}{T_2}$$

$$\frac{1 \times 0{\cdot}7571}{(24 + 273)} = \frac{10 \times 0{\cdot}1111}{T_2}$$

$$T_2 = \frac{297 \times 10 \times 0{\cdot}1111}{0{\cdot}7571}$$

$$= 436\,K$$

$\therefore$ Temperature of the compressed air
$$= 436 - 273 = 163°C \quad \text{Ans. (iii)}$$

5. Working in litres of volume and bars of pressure:
$$V_1 = 0.0424\,\text{m}^3 \times 10^3 = 42.4\,\text{litres}$$
$$\text{Clearance vol.} = 1230\,\text{cm}^3 \times 10^{-3} = 1.23\,\text{litres}$$

ISOTHERMAL COMPRESSION:
$$p_1 V_1 = p_2 V_2$$
$$1 \times 42.4 = 8 \times V_2$$
$$V_2 = 5.3\,\text{litres}$$
$$\text{Vol. delivered per stroke} = V_2 - \text{clearance vol.}$$
$$= 5.3 - 1.23 = 4.07\,\text{litres} \quad \text{Ans. (i)}$$

ADIABATIC COMPRESSION:
$$p_1 V_1^{1.4} = p_2 V_2^{1.4}$$
$$1 \times 42.4^{1.4} = 8 \times V_2^{1.4}$$

$$V_2^{1.4} = \frac{42.4^{1.4}}{8}$$

$$V_2 = \frac{42.4}{^{1.4}\sqrt{8}} = 9.601\,\text{litres}$$

Volume delivered per stroke
$$= 9.601 - 1.23 = 8.371\,\text{litres} \quad \text{Ans. (ii)}$$

6. Working in cm³ of volume and bars of pressure:
$$\text{Stroke volume} = 0.7854 \times 14^2 \times 20 = 3078\,\text{cm}^3$$

WHEN CLEARANCE IS 10% OF STROKE VOLUME:

$$\text{Clearance volume} = 10\% \text{ of } 3078 = 307.8\,\text{cm}^3$$

$$V_1 = \text{stroke volume} + \text{clearance volume}$$
$$= 3078 + 307.8 = 3385.8\,\text{cm}^3$$

$$p_1 V_1^{1.28} = p_2 V_2^{1.28}$$
$$0.98 \times 3386^{1.28} = 7 \times V_2^{1.28}$$

$$V_2 = 3386 \times \sqrt[1.28]{\frac{0.98}{7}}$$

$$= 728.8 \text{ cm}^3$$

Volume delivered per stroke
$$= 728.8 - 307.8 = 421 \text{ cm}^3 \quad \text{Ans. (i)}$$

WHEN CLEARANCE IS 5 % OF STROKE VOLUME:

$$\text{Clearance volume} = 5\% \text{ of } 3078 = 153.9 \text{ cm}^3$$
$$V_1 = 3078 + 153.9 = 3231.9$$
$$p_1 V_1^{1.28} = p_2 V_2^{1.28}$$
$$0.98 \times 3232^{1.28} = 7 \times V_2^{1.28}$$

$$V_2 = 3232 \times \sqrt[1.28]{\frac{0.98}{7}}$$

$$= 695.6 \text{ cm}^3$$

Volume delivered per stroke
$$= 695.6 - 153.9 = 541.7 \text{ cm}^3 \quad \text{Ans. (ii)}$$

7.
$$\text{Clearance length} = 7\% \text{ of } 380 = 26.6 \text{ mm}$$
$$V_1 = 380 + 26.6 = 406.6$$
$$V_2 = 406.6 - 260 = 146.6$$

$$p_1 V_1^n = p_2 V_2^n$$
$$0.99 \times 406.6^n = 4 \times 146.6^n$$

$$\left\{ \frac{406.6}{146.6} \right\}^n = \left\{ \frac{4}{0.99} \right\}$$

$$n \times \log \left\{ \frac{406.6}{146.6} \right\} = \log \left\{ \frac{4}{0.99} \right\}$$

$$n \times 0.4430 = 0.6065$$

$$n = \frac{0.6065}{0.443} = 1.37 \quad \text{Ans.}$$

8. Volume of free air (at 1·01 bar from the atmosphere) to make 22 m³ at 31·01 bar abs., at the same temperature

$$= 22 \times \frac{31 \cdot 01}{1 \cdot 01} \text{ m}^3$$

Volume of free air to make 22 m³ at 20·01 bar abs.

$$= 22 \times \frac{22 \cdot 01}{1 \cdot 01} \text{ m}^3$$

∴ Volume of free air to supply the difference

$$= \frac{22}{1 \cdot 01} \ (31 \cdot 01 - 20 \cdot 01)$$

$$= \frac{22 \times 11}{1 \cdot 01} = 239 \cdot 6 \text{ m}^3 \ \ldots \text{ (i)}$$

Volume of free air dealt with by compressor per minute
$$= 0 \cdot 7854 \ (0 \cdot 35^2 - 0 \cdot 075^2) \times 0 \cdot 3 \times 0 \cdot 92 \times 170$$
$$= 4 \cdot 306 \text{ m}^3/\text{min} \ \ldots \text{ (ii)}$$

$$\therefore \ \text{Time} = \frac{239 \cdot 6}{4 \cdot 306} = 55 \cdot 64 \text{ minutes} \quad \text{Ans.}$$

9. Volume of three air bottles

$$= 3 \left\{ \frac{\pi}{6} \times 0 \cdot 45^3 + \frac{\pi}{4} \times 0 \cdot 45^2 \times 1 \cdot 8 \right\}$$

$$= 3 \times \frac{\pi}{2} \times 0 \cdot 45^2 \left\{ \frac{0 \cdot 45}{3} + \frac{1 \cdot 8}{2} \right\}$$

$$= 1 \cdot 002 \text{ m}^3$$

To produce 1·002 m³ of air at an absolute pressure of 41 bar, volume of free air required (assuming same temperature)
$$= 1 \cdot 002 \times 41 \text{ m}^3$$

Volume of free air to be dealt with by compressor
$$= 1 \cdot 002 \times 41 - 1 \cdot 002$$
$$= 40 \cdot 08 \text{ m}^3 \ \ldots \text{ (i)}$$

Volume of free air taken into compressor per minute

$$= 0{\cdot}7854\,(0{\cdot}27^2 - 0{\cdot}08^2) \times 0{\cdot}25 \times 0{\cdot}9 \times 110$$
$$= 1{\cdot}293 \text{ m}^3 \; \ldots \text{(ii)}$$

$$\therefore \text{ Time} = \frac{40{\cdot}08}{1{\cdot}293} = 31 \text{ minutes} \quad \text{Ans.}$$

10.
$$\frac{T_2}{T_1} = \left\{\frac{V_1}{V_2}\right\}^{n-1}$$

$$T_2 = (25 + 273)\left\{\frac{22{\cdot}8}{5{\cdot}7}\right\}^{0{\cdot}2}$$

$$= 298 \times 4^{0{\cdot}2} = 393{\cdot}1 \text{ K}$$

$\therefore$ Temperature at end of compression
$$= 393{\cdot}1 - 273 = 120{\cdot}1°C \quad \text{Ans. (i)}$$

Volume delivered per second to reservoirs

$$= 5{\cdot}7 \times \frac{298}{393{\cdot}1} \times 5 = 21{\cdot}61 \text{ litres} \quad \text{Ans. (ii)}$$

From $pV = mRT$, $\qquad\qquad m = \dfrac{pV}{RT}$

Mass of air taken in per second

$$= \frac{(1 \times 10^2) \times (22{\cdot}8 \times 10^{-3})}{0{\cdot}287 \times 298} \times 5$$

$$= 0{\cdot}1332 \text{ kg} \quad \text{Ans. (iii)}$$

Heat gained by water $=$ Heat lost by air
$$m \times 4{\cdot}12 \times 11 = 0{\cdot}1332 \times 1{\cdot}005 \times (120{\cdot}1 - 25)$$
$$m = 0{\cdot}281 \text{ kg} \quad \text{Ans. (iv)}$$

11.

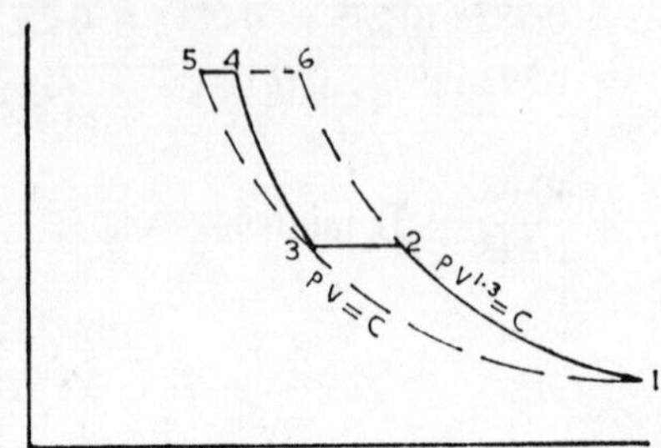

$$pV = mRT, \qquad V_1 = \frac{mRT}{p}$$

$$V_1 = \frac{0{\cdot}05 \times 0{\cdot}287 \times (17 + 273)}{0{\cdot}97 \times 10^2} = 0{\cdot}0429 \text{ m}^3$$

Working in litres of volume and bars of pressure:

FIRST STAGE COMPRESSION:

$$p_1 V_1^{1{\cdot}3} = p_2 V_2^{1{\cdot}3}$$
$$0{\cdot}97 \times 42{\cdot}9^{1{\cdot}3} = 2{\cdot}91 \times V_2^{1{\cdot}3}$$

$$V_2 = 42{\cdot}9 \times \sqrt[1{\cdot}3]{\frac{0{\cdot}97}{2{\cdot}91}} = 18{\cdot}43$$

Effect of cooling at constant pressure to the initial temperature is to reduce the volume to V_3, that is, to the same volume had it been compressed isothermally.

$$p_1 V_1 = p_3 V_3$$
$$0{\cdot}97 \times 42{\cdot}9 = 2{\cdot}91 \times V_3$$
$$V_3 = 14{\cdot}3$$

% decrease in volume at end of first stage

$$= \frac{18{\cdot}43 - 14{\cdot}3}{18{\cdot}43} \times 100 = 22{\cdot}42\% \quad \text{Ans. (ia)}$$

SECOND STAGE COMPRESSION:

$$p_3 V_3^{1\cdot3} = p_4 V_4^{1\cdot3}$$
$$2\cdot91 \times 14\cdot3^{1\cdot3} = 7\cdot3 \times V_4^{1\cdot3}$$

$$V_4 = 14\cdot3 \times \sqrt[1\cdot3]{\frac{2\cdot91}{7\cdot3}} = 7\cdot049$$

$$p_3 V_3 = p_5 V_5$$

$$2\cdot91 \times 14\cdot3 = 7\cdot3 \times V_5$$
$$V_5 = 5\cdot701$$

% decrease in volume at end of second stage

$$= \frac{7\cdot049 - 5\cdot701}{7\cdot049} \times 100 = 19\cdot12\% \quad \text{Ans. (ib)}$$

SINGLE STAGE COMPRESSION:

$$p_1 V_1^{1\cdot3} = p_6 V_6^{1\cdot3}$$
$$0\cdot97 \times 42\cdot9^{1\cdot3} = 7\cdot3 \times V_6^{1\cdot3}$$

$$V_6 = 42\cdot9 \times \sqrt[1\cdot3]{\frac{0\cdot97}{7\cdot3}} = 9\cdot084$$

% decrease in volume from V_6 to V_5

$$= \frac{9\cdot084 - 5\cdot701}{9\cdot084} \times 100 = 37\cdot24\% \quad \text{Ans. (ii)}$$

12. $V_1 = 0\cdot7854 \times 0\cdot25^2 \times 0\cdot4 = 0\cdot01963$ m³
Working in litres of volume and bars of pressure:
when $n = 1\cdot2$,
$$p_1 V_1^{1\cdot2} = p_2 V_2^{1\cdot2}$$
$$1\cdot01 \times 19\cdot63^{1\cdot2} = 5\cdot5 \times V_2^{1\cdot2}$$

$$V_2 = 19\cdot63 \times \sqrt[1\cdot2]{\frac{1\cdot01}{5\cdot5}} = 4\cdot78$$

To obtain work in kJ pressure must be in kN/m² ($= $ bars $\times 10^2$)
and volumes in cubic metres ($= $ litres $\times 10^{-3}$),

$$\text{Work/cycle} = \frac{n}{n-1}(p_2V_2 - p_1V_1)$$

$$= \frac{1\cdot2}{0\cdot2} \times 10^2 \times 10^{-3}(5\cdot5 \times 4\cdot78 - 1\cdot01 \times 19\cdot63)$$

$$= 3\cdot876 \text{ kJ}$$
$$\text{Power [kW]} = \text{work per second [kJ/s]}$$
$$= \text{work/cycle} \times \text{cycle/s}$$
$$= 3\cdot876 \times 2\cdot5 = 9\cdot69 \text{ kW} \quad \text{Ans. (i)}$$

$$\text{When } n = 1\cdot3,$$
$$p_1V_1^{1\cdot3} = p_2V_2^{1\cdot3}$$
$$1\cdot01 \times 19\cdot63^{1\cdot3} = 5\cdot5 \times V_2^{1\cdot3}$$

$$V_2 = 19\cdot63 \times \sqrt[1\cdot3]{\frac{1\cdot01}{5\cdot5}} = 5\cdot33$$

$$\text{Work/cycle} = \frac{n}{n-1}(p_2V_2 - p_1V_1)$$

$$= \frac{1\cdot3}{0\cdot3} \times 10^2 \times 10^{-3}(5\cdot5 \times 5\cdot33 - 1\cdot01 \times 19\cdot63)$$

$$= 4\cdot112 \text{ kJ}$$
$$\text{Power} = 4\cdot112 \times 2\cdot5 = 10\cdot28 \text{ kW}$$

$$\% \text{ extra power} = \frac{10\cdot28 - 9\cdot69}{9\cdot69} \times 100$$

$$= 6\cdot088\% \quad \text{Ans. (ii)}$$

13. Referring to Fig. 54,

$$\text{Work/cycle} = \frac{n}{n-1}\{(p_2V_2 - p_1V_1) - (p_3V_3 - p_4V_4)\}$$

$$= \frac{n}{n-1}\{(p_2V_2 - p_1V_1) - (p_2V_3 - p_1V_4)\}$$

$$= \frac{n}{n-1}\{p_2(V_2 - V_3) - p_1(V_1 - V_4)\}$$

Piston swept volume $(V_1 - V_3)$
$$= 0.7854 \times 0.2^2 \times 0.23 = 7.226 \times 10^{-3} \text{ m}^3$$

Working in litres of volume and bars of pressure:
$V_3 = 364 \text{ cm}^3 = 0.364$ litre
V_1 = piston swept vol. + clearance vol.
$\quad = 7.226 + 0.364 = 7.59$

$$p_1 V_1^n = p_2 V_2^n$$
$$1 \times 7.59^{1.28} = 5 \times V_2^{1.28}$$

$$V_2 = \frac{7.59}{\sqrt[1.28]{5}} = 2.158$$

$$p_3 V_3^n = p_4 V_4^n$$
$$5 \times 0.364^{1.28} = 1 \times V_4^{1.28}$$

$$V_4 = 0.364 \times \sqrt[1.28]{5} = 1.28$$

Expressing pressure in kN/m^2 ($1 \text{ bar} = 10^2 \text{ kN/m}^2$) and volumes in m^3 ($1 \text{ litre} = 10^{-3} \text{ m}^3$) to obtain work in kJ:

$$\text{Work/cycle} = \frac{n}{n-1} \{p_2(V_2 - V_3) - p_1(V_1 - V_4)\}$$

$$= \frac{1.28}{0.28} \times 10^2 \times 10^{-3} \{5(2.158 - 0.364) - 1(7.59 - 1.28)\}$$

$$= 1.216 \text{ kJ}$$

Indicated power [kW] = 1.216 [kJ/cycle] $\times$ 2 [cycle/s]
$$= 2.432 \text{ kW} \quad \text{Ans. (i)}$$

$$\text{Mean indicated press. [kN/m}^2] = \frac{\text{area of diagram [kJ]}}{\text{length of diagram [m}^3]}$$

$$= \frac{1.216}{7.226 \times 10^{-3}} = 168.3 \text{ kN/m}^2 = 1.683 \text{ bar} \quad \text{Ans. (ii)}$$

$$\text{Vol. effic.} = \frac{\text{volume drawn in per stroke}}{\text{piston swept volume}}$$

$$= \frac{V_1 - V_4}{V_1 - V_3} = \frac{7\cdot59 - 1\cdot28}{7\cdot226}$$

$$= 0\cdot8732 \text{ or } 87\cdot32\% \quad \text{Ans. (iii)}$$

14. Referring to Fig. 54:

From pV/T = constant, volume rate of air induced at suction pressure and temperature

$$= \frac{5 \times 1\cdot013 \times (24 + 273)}{60 \times 0\cdot98 \times (16 + 273)} = 0\cdot088\,53 \text{ m}^3/\text{s}$$

$$\text{Work/cycle} = \frac{n}{n-1} p_1(V_1 - V_4)\left[\left\{\frac{p_2}{p_1}\right\}^{\frac{n-1}{n}} - 1\right]$$

The above expression is the work per cycle when $(V_1 - V_4)$ is the volume drawn in per cycle. Similarly, if $(V_1 - V_4)$ is taken as the volume drawn in per second, the expression will give work per second, which is power, thus:

Power [kW] = work per second [kJ/s = kN m/s]

$$\frac{n}{n-1} = \frac{1\cdot25}{0\cdot25} = 5 \qquad\qquad \frac{n-1}{n} = \frac{1}{5}$$

$$V_1 - V_4 = 0\cdot088\,53 \text{ m}^3/\text{s}$$
$$p_1 = 0\cdot98 \times 10^2 \text{ kN/m}^2$$

$$\frac{p_2}{p_1} = \text{pressure ratio} = 4\cdot55$$

$$\text{Power} = \frac{n}{n-1} p_1(V_1 - V_4)\left[\left\{\frac{p_2}{p_1}\right\}^{\frac{n-1}{n}} - 1\right]$$

$$= 5 \times 0\cdot98 \times 10^2 \times 0\cdot088\,53\,(4\cdot55^{\frac{1}{5}} - 1)$$
$$= 15\cdot36 \text{ kW} \quad \text{Ans. (i)}$$

Let stroke volume $(V_1 - V_3)$ on Fig. 54 be represented by unity, then $V_3 = 0\cdot05$ and $V_1 = 1\cdot05$

From $p_3 V_3{}^n = p_4 V_4{}^n$

$$V_4 = 0\cdot05 \times {}^{1\cdot25}\sqrt{4\cdot55} = 0\cdot1681$$

$$\text{Suction period} = V_1 - V_4$$
$$= 1{\cdot}05 - 0{\cdot}1681 = 0{\cdot}8819$$

$$\text{Volumetric effic.} = \frac{\text{vol. drawn in per stroke}}{\text{stroke volume}}$$

$$= \frac{0{\cdot}8819}{1} = 0{\cdot}8819 \quad \text{Ans. (ii)}$$

Actual volume of air drawn in per stroke [m³]

$$= \frac{0{\cdot}088\,53 \; [\text{m}^3/\text{s}]}{2 \times 8 \; [\text{strokes/s}]}$$

$$\text{Piston swept vol.} = \frac{\text{induced volume}}{\text{volumetric effic.}}$$

$$= \frac{0{\cdot}088\,53}{2 \times 8 \times 0{\cdot}8819} \; \text{m}^3$$

$$\text{Let } d \,[\text{m}] = \text{cyl. diameter}, \qquad \text{stroke} = 1{\cdot}2d$$
$$\text{Piston swept vol.} = 0{\cdot}7854 d^2 \times 1{\cdot}2d$$

$$= \frac{0{\cdot}088\,53}{2 \times 8 \times 0{\cdot}8819}$$

$$d = \sqrt[3]{\frac{0{\cdot}088\,53}{0{\cdot}7854 \times 1{\cdot}2 \times 2 \times 8 \times 0{\cdot}8819}}$$

$$= 0{\cdot}1881 \; \text{m}$$

$$\left.\begin{array}{l}\text{Diameter of cylinder} = 188{\cdot}1 \; \text{mm} \\ \text{Stroke} = 1{\cdot}2 \times 188{\cdot}1 = 225{\cdot}7 \; \text{mm}\end{array}\right\}\text{Ans. (iii)}$$

SOLUTIONS TO TEST EXAMPLES 10

1. Tables page 2, water at 80°C, $h = 334.9$
    ,,   ,,   4, steam at 9 bar:

$$h = h_f + xh_{fg}$$
$$= 743 + 0.96 \times 2031 = 2693$$

Heat energy transferred $=$ change in enthalpy
$$= 2693 - 334.9$$
$$= 2358.1 \text{ kJ/kg} \quad \text{Ans.}$$

2. Tables page 4, 8 bar, $h_{fg} = 2048$

Enthalpy of wet steam $+ 82 =$ Enthalpy of dry steam

$$h_f + xh_{fg} + 82 = h_f + h_{fg}$$
$$xh_{fg} = h_{fg} - 82$$
$$x \times 2048 = 2048 - 82$$
$$x \times 2048 = 1966$$
$$x = 0.9601 \quad \text{Ans.}$$

3. Tables page 2, water 85°C, $h = 355.9$
    ,,   ,,   4, steam 17 bar:

$$h = h_f + xh_{fg}$$
$$= 872 + 0.97 \times 1923 = 2737$$

Heat energy supplied $=$ increase in enthalpy
$$= 2737 - 355.9$$
$$= 2381 \text{ kJ/kg} \quad \text{Ans. (i)}$$

h of water at 110°C is not listed in tables but temperatures close to this are given on page 4, by interpolation:

$$h \text{ at } 111.4°C = 467$$
$$h \text{ ,, } 109.3°C = 458$$
$$\text{difference for } 2.1°C = 9 \text{ kJ/kg}$$

110° is 0.7° more than 109.3°

$$\text{difference for } 0.7°C = \frac{0.7}{2.1} \times 9 = 3 \text{ kJ/kg}$$

$\therefore h$ at 110°C is 3 kJ/kg more than h at 109.3°C
$$= 3 + 458 = 461 \text{ kJ/kg}$$

Therefore, to produce steam at same pressure and dryness from feed water at 110°C,

$$\text{Change in enthalpy} = 2737 - 461 = 2276 \text{ kJ/kg}$$

$$\therefore \ \% \text{ less heat energy} = \frac{2381 - 2276}{2381} \times 100$$

$$= 4 \cdot 41 \% \quad \text{Ans. (ii)}$$

4. Tables page 7, 30 bar 350°C, $h = 3117$

   „   „   3, 0·06 bar dryness 0·88 :

$$h = h_f + x h_{fg}$$
$$= 152 + 0 \cdot 88 \times 2415 = 2277$$

$$\text{Enthalpy drop per kg} = 3117 - 2277$$
$$= 840 \text{ kJ/kg} \quad \text{Ans. (i)}$$

At 0·25 kg of steam per second, total change in enthalpy in the steam through the turbine per second

$$= 0 \cdot 25 \times 840 = 210 \text{ kJ/s}$$

$$\text{kilojoules per second} = \text{kilowatts}$$
$$\therefore \text{ Power equivalent} = 210 \text{ kW} \quad \text{Ans. (ii)}$$

5. Tables page 7, 20 bar 350°C, $h \cdot = 3138$, $v = 0 \cdot 1386$

   „   „   4, 20 bar 0·98 dry :

$$h = h_f + x h_{fg}$$
$$= 909 + 0 \cdot 98 \times 1890 = 2761$$
$$\text{kJ/kg}$$

$$v = x v_g$$
$$= 0 \cdot 98 \times 0 \cdot 099 \, 57$$
$$= 0 \cdot 097 \, 57 \text{ m}^3/\text{kg}$$

Heat energy supplied to steam in superheaters

$$= 3138 - 2761 = 377 \text{ kJ/kg}$$
$$\text{Ans. \ (i)}$$

% increase in specific volume

$$= \frac{0 \cdot 1386 - 0 \cdot 09757}{0 \cdot 09757} \times 100$$

$$= 42 \cdot 06 \% \quad \text{Ans (ii)}$$

6. Tables page 2, water 95°C, $h = 398$

   „   „   4, steam 17 bar, 0·95 dry,

$$h = h_f + x h_{fg}$$
$$= 872 + 0 \cdot 95 \times 1923 = 2699$$

Heat energy to produce wet steam
$$= 2699 - 398 = 2301 \text{ kJ/kg}$$

Extra heat energy to dry the steam
$$= 0.05 \times 1923 = 96.15 \text{ kJ/kg}$$
% extra heat energy to produce dry steam
$$= \frac{96.15}{2301} \times 100 = 4.178\% \quad \text{Ans. (i)}$$

Superheated steam at 17 bar, 350°C, tables page 7, by interpolation,

$$15 \text{ bar } 350°C, h = 3148$$
$$20 \text{ bar } 350°C, h = 3138$$
difference for 5 bar increase, $h = 10$ kJ/kg decrease
   ,,   ,, 2 bar   ,, $= \frac{2}{5} \times 10 = 4$ kJ/kg decrease
17 bar 350°C, $h = 3148 - 4 = 3144$

Heat energy to produce superheated steam
$$= 3144 - 398 = 2746 \text{ kJ/kg}$$
% extra heat energy to produce superheated steam

$$= \frac{2746 - 2301}{2301} \times 100 = 19.34\% \quad \text{Ans. (ii)}$$

7. Tables page 4, 4 bar, $v_g = 0.4623$
               8 bar, $v_g = 0.2403$

For 8 bar, dryness 0.94,
$$v = xv_g = 0.94 \times 0.2403 = 0.2258 \text{ m}^3/\text{kg}$$
$$p_1 v_1^n = p_2 v_2^n$$
$$8 \times 0.2258^{1.12} = 4 \times v_2^{1.12}$$
$$v_2 = 0.2258 \times {}^{1.12}\!\sqrt{2} = 0.4194 \text{ m}^3/\text{kg}$$
0.4194 m³/kg is the specific volume of wet steam at 4 bar, let its dryness fraction $= x$:
$$v = xv_g$$
$$0.4194 = x \times 0.4623$$
$$x = 0.9072 \quad \text{Ans.}$$

8. Tables page 7, 15 bar 250°C, $h = 2925$
   ,,   ,, 4, 15 bar,
$$h_f = 845, \quad h_{fg} = 1947, \quad h_g = 2792$$
Enthalpy before mixing = Enthalpy after

$$1 \text{ kg sup. steam} + 1 \text{ kg wet steam} = 2 \text{ kg dry sat. steam}$$
$$2925 + (845 + x \times 1947) = 2 \times 2792$$
$$x \times 1947 = 1814$$
$$x = 0.9315 \quad \text{Ans.}$$

9. Tables page 4, 7 bar, $v_g = 0.2728 \text{ m}^3/\text{kg}$
  „   „   „ 4·5 bar, $v_g = 0.4139 \text{ m}^3/\text{kg}$
Volume of 1 kg wet steam, 7 bar, 0·95 dry
$$= 0.95 \times 0.2728 = 0.2592 \text{ m}^3/\text{kg}$$

$$\text{Mass [kg] of steam expanded} = \frac{\text{volume [m}^3]}{\text{specific volume [m}^3/\text{kg]}}$$

$$= \frac{0.2}{0.2592} = 0.7716 \text{ kg}$$

Volume of 0·7716 kg of dry steam at 4·5 bar
$$= 0.7716 \times 0.4139 = 0.3194 \text{ m}^3$$

WHEN EXPANSION FOLLOWS THE LAW $pV = C$:
$$p_1 V_1 = p_2 V_2$$
$$7 \times 0.2 = 4.5 \times V_2 \qquad \therefore \ V_2 = 0.3111 \text{ m}^3$$

$$\text{Dryness fraction} = \frac{0.3111}{0.3194} = 0.974 \quad \text{Ans. (i)}$$

WHEN EXPANSOIN FOLLOWS THE LAW $pV^{1.14} = C$:
$$p_1 V_1^{1.14} = p_2 V_2^{1.14}$$
$$7 \times 0.2^{1.14} = 4.5 \times V_2^{1.14}$$

$$V_2 = 0.2 \times \sqrt[1.14]{\frac{7}{4.5}} = 0.2946 \text{ m}^3$$

$$\text{Dryness fraction} = \frac{0.2946}{0.3194} = 0.9226 \quad \text{Ans. (ii)}$$

10. Tables page 7, 20 bar 400°C,
$$h = 3248, \quad v = 0.1511$$
  „   „ 4, 12 bar,
$$h_f = 798, \quad h_{fg} = 1986,$$
$$v_g = 0.1632$$

The steam is cooled at *constant volume* therefore the volume of one kg of steam at 12 bar is the same as it was before cooling, that is 0·1511 m³. This is *less* than the volume of one kg of dry saturated steam at 12 bar, therefore it must now be wet and its dryness fraction is:

$$\text{Dryness fraction} = \frac{0\cdot1511}{0\cdot1632} = 0\cdot9258 \quad \text{Ans. (i)}$$

Enthalpy of steam at 12 bar,
$$\begin{aligned}
h &= h_f + x h_{fg} \\
&= 798 + 0\cdot9258 \times 1986 \\
&= 2637 \text{ kJ/kg}
\end{aligned}$$
$$\text{Heat transfer} = 3248 - 2637 = 611 \text{ kJ/kg} \quad \text{Ans. (ii)}$$

11. Tables page 7, by interpolation,
$$\begin{aligned}
20 \text{ bar } 400°C, \quad v &= 0\cdot1511 \\
30 \text{ bar } 400°C, \quad v &= 0\cdot0993
\end{aligned}$$
Therefore, 25 bar 400°C (mid-way between these two listed pressures at the same temperature)
$$= \tfrac{1}{2}(0\cdot1511 + 0\cdot0993) = 0\cdot1252 \text{ m}^3/\text{kg}$$
Mass of steam expanded, that is the mass [kg] of 0·0025 m³ (*i.e.* 2·5 litres $\times$ 10⁻³) at 0·1252 m³/kg

$$= \frac{0\cdot0025}{0\cdot1252} = 0\cdot019\,96 \text{ kg}$$

Tables page 3, 0·1 bar, $v_g = 14\cdot67$
Specific volume of steam 0·1 bar, dryness 0·9
$$v = 0\cdot9 \times 14\cdot67 = 13\cdot2 \text{ m}^3/\text{kg}$$
$\therefore$ Final volume of 0·019 96 kg of steam
$$\begin{aligned}
&= 0\cdot019\,96 \times 13\cdot2 \\
&= 0\cdot2636 \text{ m}^3 \text{ or } 263\cdot6 \text{ litres} \quad \text{Ans.}
\end{aligned}$$

12. Tables page 4, steam 5 bar, $h_f = 640$, $h_{fg} = 2109$
,,   ,,   2, water 12°C, $h = 50\cdot4$
Enthalpy before mixing $= H$ of 1·5 kg steam $+ H$ of 70 kg water
$$\begin{aligned}
&= 1\cdot5(640 + 0\cdot95 \times 2109) + 70 \times 50\cdot4 \\
&= 3965 + 3528 \\
&= 7493 \text{ kJ}
\end{aligned}$$

$$\text{Enthalpy after mixing} = H \text{ of } 71 \cdot 5 \text{ kg water}$$
$$\text{Enthalpy after} = \text{Enthalpy before}$$
$$\therefore 71 \cdot 5 \times h = 7493$$
$$h = 104 \cdot 8 \text{ kJ/kg}$$

Tables page 2, $104 \cdot 8$ is the value of h for water at $25°C$
$$\therefore \text{Temperature} = 25°C \quad \text{Ans.}$$

13. Tables page 3, steam $0 \cdot 12$ bar, $h_f = 207$, $h_{fg} = 2383$
    ,,   ,,   2, water $40°C$, $h = 167 \cdot 5$
         ,, $29°C$, $h = 121 \cdot 5$
         ,, $12°C$, $h = \;\; 50 \cdot 4$

Enthalpy of one kg of steam and m kg of circulating water entering condenser $= (207 + 0 \cdot 88 \times 2383) + (m \times 50 \cdot 4)$.

Enthalpy of one kg of condensate and m kg of circulating water leaving condenser $= 167 \cdot 5 + (m \times 121 \cdot 5)$.

$$\text{Enthalpy entering condenser} = \text{Enthalpy leaving}$$
$$(207 + 0 \cdot 88 \times 2383) + (m \times 50 \cdot 4) = 167 \cdot 5 + (m \times 121 \cdot 5)$$
$$(207 + 0 \cdot 88 \times 2383) - 167 \cdot 5 = m (121 \cdot 5 - 50 \cdot 4)$$

which is the same as:

$$\text{Heat energy lost by steam} = \text{Heat energy gained by water}$$
$$2136 \cdot 5 = m \times 71 \cdot 1$$
$$m = 30 \cdot 04 \text{ kg} \quad \text{Ans. (i)}$$

Each kg of circulating water carries away $71 \cdot 1$ kJ therefore energy carried away by 500 tonne
$$= 500 \times 10^3 \times 71 \cdot 1 \text{ kJ/h}$$

Equivalent power $[\text{kW} = \text{kJ/s}]$

$$= \frac{500 \times 10^3 \times 71 \cdot 1}{3600}$$

$$= 9877 \text{ kW} \quad \text{Ans. (ii)}$$

14. Tables page 4, steam $1 \cdot 1$ bar, $h_f = 429$, $h_{fg} = 2251$
    ,,   ,,   2, water $15°C$, $h = 62 \cdot 9$
         ,, $55°C$, $h = 230 \cdot 2$

Mass of steam sample $= 10 \cdot 75 - 10 = 0 \cdot 75$ kg

Taking into account the vessel of water equivalent $0 \cdot 45$ kg:

Initial mass of water at $15°C = 10 + 0 \cdot 45 = 10 \cdot 45$ kg
Final   ,,   ,,   ,,   ,, $55°C = 10 \cdot 75 + 0 \cdot 45 = 11 \cdot 2$ kg

$$\text{Enthalpy (steam and water) before} = \text{Enthalpy (water) after}$$
$$0 \cdot 75 (429 + x \times 2251) + 10 \cdot 45 \times 62 \cdot 9 = 11 \cdot 2 \times 230 \cdot 2$$
$$321 \cdot 8 + x \times 1689 + 657 \cdot 3 = 2578$$
$$x \times 1689 = 1598 \cdot 9$$
$$x = 0 \cdot 9469 \quad \text{Ans.}$$

15. Tables page 4, steam 2·4 bar, $\quad h_g = 2715$
   „   „   2, water 42°C, $\quad h = 175·8$
   „   „   3,   „   99·6°C, $\quad h = 417$

See Fig. 58. Consider one kg of steam from boiler, let x kg be tapped off low pressure turbine to heater, then $(1 - x)$ kg passes through condenser and as water to hotwell.

$$\frac{\text{Enthalpy of heating steam and water entering heater}}{} = \frac{\text{Enthalpy of water leaving heater}}{}$$

$$x \times 2715 + (1 - x) \times 175·8 = 1 \times 417$$
$$2715x + 175·8 - 175·8x = 417$$
$$2539·2x = 241·2$$
$$x = 0·095$$
$$\therefore \% \text{ of steam tapped off} = 9·5\% \quad \text{Ans.}$$

16. Tables page 4, steam 2·9 bar, $\quad h_f = 556, \quad h_{fg} = 2168$
   „   „   2, water 40°C, $\quad h = 167·5$
   „   „   2,   „   100°C, $\quad h = 419·1$

Treating heater and hotwell as one combined system (see Fig. 59) let one kg of steam be supplied from boilers, then 0·1 kg of steam is bled off to heater, and 0·9 kg of steam passes on to the condenser and as water to the hotwell.

$$\frac{\text{Enthalpy of heating steam and water entering system}}{} = \frac{\text{Enthalpy of feed water leaving system}}{}$$

$$0·1(556 + x \times 2168) + 0·9 \times 167·5 = 1 \times 419·1$$
$$55·6 + 216·8x + 150·8 = 419·1$$
$$216·8x = 212·7$$
$$x = 0·9809 \quad \text{Ans.}$$

17. Tables page 4, steam 16 bar, $\quad h_f = 859, \quad h_{fg} = 1935$
              „   8 bar, $\quad h_f = 721, \quad h_{fg} = 2048$

Enthalpy after throttling $=$ Enthalpy before
$$721 + x \times 2048 = 859 + 0·98 \times 1935$$
$$x \times 2048 = 2034$$
$$x = 0·9931 \quad \text{Ans.}$$

18. Tables page 4, steam 12 bar, $h_f = 798$, $h_{fg} = 1986$
       ,,   1·2 bar, $t_s = 104·8$, $h_g = 2683$

Degree of superheat of throttled steam
$$= 116 - 104·8 = 11·2°C$$

Enthalpy before throttling $=$ Enthalpy after
$$798 + x \times 1986 = 2683 + 2 \times 11·2$$
$$x \times 1986 = 1907·4$$
$$x = 0·9603 \quad \text{Ans.}$$

19. Tables page 4, 15 bar, $h_f = 845$, $h_{fg} = 1947$
            1·1 bar, $t_s = 102·3$, $h_g = 2680$

Dryness fraction by separator,

$$x_1 = \frac{m_2}{m_2 + m_1} = \frac{10}{10·55} = 0·9479$$

Dryness fraction by throttling calorimeter:
Enthalpy before throttling $=$ Enthalpy after
$$845 + x_2 \times 1947 = 2680 \times 2\,(111 - 102·3)$$
$$x_2 \times 1947 = 1852·4$$
$$x_2 = 0·9513$$

Dryness fraction of steam:
$$x = x_1 \times x_2$$
$$= 0·9479 \times 0·9513$$
$$= 0·9018 \quad \text{Ans.}$$

20. Tables page 3, for a saturated steam temperature of 29°C, pressure of steam is 0·04 bar and specific volume 34·8 m³/kg:
$\therefore$ Mass of steam in the volume of 8·7 m³

$$= \frac{8·7}{34·8} = 0·25 \text{ kg} \quad \text{Ans. (i)}$$

$$\text{Total press.} = \frac{\text{partial press.}}{\text{due to air}} + \frac{\text{partial press.}}{\text{due to steam}}$$

Press. due to air $=$ total press. $-$ steam press.
$$= 0·06 - 0·04$$
$$= 0·02 \text{ bar} = 2 \text{ kN/m}^2$$

Taking R for air $= 287$ J/kg K $= 0·287$ kJ/kg K, and the characteristic equation $pV = mRT$:

$$m = \frac{pV}{RT} = \frac{2 \times 8\cdot7}{0\cdot287 \times (29 + 273)}$$

$$= 0\cdot2007 \text{ kg} \quad \text{Ans. (ii)}$$

21. Tables page 3, for a saturation steam temperature of 39°C, pressure of steam is 0·07 bar and $v_g = 20\cdot53$ m³/kg.

For wet steam 0·07 bar, dryness 0·86,

$$v = 0\cdot86 \times 20\cdot53 \text{ m}^3/\text{kg}$$

Let V = volume of condenser, then mass of steam in V m³

$$\text{of space} = \frac{V}{0\cdot86 \times 20\cdot53} = 0\cdot056\,63V \text{ kg} \quad .... \qquad .... \qquad .... \quad \text{(i)}$$

$$\begin{aligned}
\text{Partial press. due to air} \;&=\; \text{total press.} - \text{steam press.} \\
&=\; 0\cdot08 - 0\cdot07 \\
&=\; 0\cdot01 \text{ bar} = 1 \text{ kN/m}^2
\end{aligned}$$

From $pV = mRT$:

$$\text{Mass of air} = \frac{1 \times V}{0\cdot287 \times (39 + 273)}$$

$$= 0\cdot011\,16V \text{ kg} \qquad .... \qquad .... \quad \text{(ii)}$$

Mass ratio of steam to air

$$\begin{aligned}
&=\; 0\cdot056\,63V \;:\; 0\cdot011\,16V \\
&=\; 5\cdot072 \qquad :\; 1 \quad \text{Ans.}
\end{aligned}$$

22. 1st Case: Absolute press. $= 1\cdot9 + 1 = 2\cdot9$ bar

Tables page 4, when temp. of steam is 130°C,

$$p = 2\cdot7 \text{ bar}, \quad v_g = 0\cdot6686 \text{ m}^3/\text{kg}$$

$$\text{Mass of steam} = \frac{4\cdot25}{0\cdot6686} = 6\cdot356 \text{ kg} \quad \text{Ans. (ia)}$$

$$\begin{aligned}
\text{Air pressure} \;&=\; \text{total press.} - \text{steam press.} \\
&=\; 2\cdot9 - 2\cdot7 = 0\cdot2 \text{ bar} = 20 \text{ kN/m}^2
\end{aligned}$$

$$pV = mRT$$

$$\therefore m = \frac{20 \times 4\cdot25}{0\cdot287 \times (130 + 273)}$$

$$= 0\cdot7348 \text{ kg} \quad \text{Ans. (ib)}$$

2nd Case: Absolute press. $= 6{\cdot}25 + 1 = 7{\cdot}25$ bar
Tables page 4, when temp. of steam is 165°C,
$$p = 7 \text{ bar}, \quad v_g = 0{\cdot}2728 \text{ m}^3/\text{kg}$$

$$\text{Mass of steam } = \frac{4{\cdot}25}{0{\cdot}2728} = 15{\cdot}58 \text{ kg} \quad \text{Ans. (iia)}$$

Air pressure $=$ total press. $-$ steam press.
$$= 7{\cdot}25 - 7 = 0{\cdot}25 \text{ bar} = 25 \text{ kN/m}^2$$

$$pV = mRT$$

$$\therefore m = \frac{25 \times 4{\cdot}25}{0{\cdot}287 \times (165 + 273)}$$

$$= 0{\cdot}845 \text{ kg} \quad \text{Ans. (iib)}$$

SOLUTIONS TO TEST EXAMPLES 11

1. Tables page 4, 17 bar, $s_f = 2{\cdot}372$ $s_{fg} = 4{\cdot}028$

$$s = 2{\cdot}372 + 0{\cdot}95 \times 4{\cdot}028$$
$$= 6{\cdot}198 \text{ kJ/kg K} \quad \text{Ans.}$$

2. Tables page 4, 195°C (14 bar) $s_f = 2{\cdot}284$ $s_{fg} = 4{\cdot}185$

$$s = 2{\cdot}284 + 0{\cdot}9 \times 4{\cdot}185$$
$$= 6{\cdot}05 \text{ kJ/kg K} \quad \text{Ans.}$$

3. Tables page 4, 15 bar,

$$T_{sat} = 198{\cdot}3 + 273 = 471{\cdot}3 \text{ K}$$
$$h_{fg} = 1947$$
$$T = 300 + 273 = 573 \text{ K}$$

$$s = c_w \ln\frac{T_{sat}}{273} + \frac{h_{fg}}{T_{sat}} + c_{sup}\ln\frac{T}{T_{sat}}$$

$$= 4{\cdot}24 \ln\frac{471{\cdot}3}{273} + \frac{1947}{471{\cdot}3} + 2{\cdot}43 \ln\frac{573}{471{\cdot}3}$$

$$= 4{\cdot}24 \ln 1{\cdot}726 + 4{\cdot}131 + 2{\cdot}43 \ln 1{\cdot}216$$
$$= 4{\cdot}24 \times 0{\cdot}5457 + 4{\cdot}131 + 2{\cdot}43 \times 0{\cdot}1954$$
$$= 2{\cdot}314 + 4{\cdot}131 + 0{\cdot}4746$$
$$= 6{\cdot}9196 \text{ kJ/kg K} \quad \text{Ans. (i)}$$

Tables page 7, $s = 6{\cdot}919$ kJ/kg K   Ans. (ii)

4. Tables page 4, 5·5 bar, $s_g = 6{\cdot}790$

  ,,   ,,   3, 0·2 bar, $s_f = 0{\cdot}832$, $s_{fg} = 7{\cdot}075$

Entropy after expansion $=$ Entropy before

$$0{\cdot}832 + x \times 7{\cdot}075 = 6{\cdot}79$$
$$x \times 7{\cdot}075 = 5{\cdot}958$$
$$x = 0{\cdot}8422 \quad \text{Ans.}$$

5. Tables page 7, by interpolation:

     15 bar 350°C $s = 7{\cdot}102$

     20 bar 350°C $s = 6{\cdot}957$

For   5 bar increase $s = 0{\cdot}145$ decrease

$$\text{,,} \quad 2 \text{ bar} \quad \text{,,} \quad s = \frac{2}{5} \times 0.145 = 0.058 \text{ decrease}$$

$$\therefore 17 \text{ bar } 350°\text{C}, \quad s = 7.102 - 0.058 = 7.044$$

Page 4, 1.7 bar, $\quad s_f = 1.475 \qquad s_{fg} = 5.707$

Entropy after expansion = Entropy before

$$1.475 + x \times 5.707 = 7.044$$
$$x \times 5.707 = 5.569$$
$$x = 0.9759 \quad \text{Ans.}$$

6. Tables page 4, 22 bar, $\quad h_g = 2801, \quad s_g = 6.305$

$$7 \text{ bar}, \quad h_g = 2764$$
$$1.4 \text{ bar}, \quad s_f = 1.411, \quad s_{fg} = 5.835$$

THROTTLING PROCESS:

$$\text{Enthalpy after} = \text{Enthalpy before}$$
$$\therefore \text{ Enthalpy at 7 bar} = 2801$$

Throttled steam is therefore superheated.

Tables page 7, interpolating:

$$\text{Enthalpy of superheat} = 2801 - 2764 = 37$$
$$\text{At 7 bar,} \quad h = 2846 \text{ for } 200°\text{C}$$
$$h = 2764 \text{ ,, } 165°\text{C (sat. temp.)}$$
$$\text{increase } h = \quad 82 \text{ ,, } \quad 35°\text{C increase}$$

$$\text{difference in temp. for } h = 37, = \frac{37}{82} \times 35 = 15.8°\text{C}$$

$$\therefore \text{ Degree of superheat at 7 bar} = 15.8°\text{C} \quad \text{Ans. (i)}$$

Entropy of steam at 7 bar with 15.8°C of superheat:

$$7 \text{ bar } 200°\text{C} \quad s = 6.888$$
$$\text{,, } 165°\text{C} \quad s = 6.709$$
$$\text{for increase } 35°\text{C} \quad s = 0.179 \text{ increase}$$

$$\text{,, } \quad \text{,, } \quad 15.8°\text{C} \quad s = \frac{15.8}{35} \times 0.179 = 0.0808$$

$\therefore$ Entropy at 7 bar, 15.8°C superheat

$$= 6.709 + 0.0808 = 6.7898$$
$$\text{Increase in entropy} = 6.7898 - 6.305$$
$$= 0.4848 \text{ kJ/kg K} \quad \text{Ans. (ii)}$$

ISENTROPIC EXPANSION from 7 to 1.4 bar:

$$\text{Entropy after} = \text{Entropy before}$$
$$1.411 + x \times 5.835 = 6.7898$$
$$x \times 5.835 = 5.3788$$
$$x = 0.9217 \quad \text{Ans. (iii)}$$

SOLUTIONS TO TEST EXAMPLES 12

1. Mean height of indicator diagram

$$= \frac{\text{area}}{\text{length}} = \frac{1445}{80}\ \text{mm}$$

$$\text{Mean effec. press.} = \frac{1445}{80} \times 0\cdot27$$

$$= 4\cdot877\ \text{bar} = 487\cdot7\ \text{kN/m}^2$$
$$\text{ip of HP} = p_m ALn$$
$$= 487\cdot7 \times 0\cdot7854 \times 0\cdot6^2 \times 1\cdot05 \times 1\cdot65 \times 2$$
$$= 477\cdot8\ \text{kW} \quad \text{Ans. (i)}$$
$$\text{Total ip} = 477\cdot8 \times 4 = 1911\ \text{kW} \quad \text{Ans. (ii)}$$

2.
$$\text{ip} = p_m ALn$$
$$= 405 \times 0\cdot7854 \times 0\cdot275^2 \times 0\cdot4 \times 3 \times 2$$
$$= 57\cdot73\ \text{kW} \quad \text{Ans. (i)}$$

See Chapter 7 for brake power and mechanical efficiency of reciprocating engines.

$$\text{bp} = T\omega$$

brake power is in kW when:
$$T = \text{braking torque in kN m}$$
$$= \text{force [kN]} \times \text{radius [m]}$$
$$\omega = \text{angular velocity in rad/s}$$
$$\therefore\ \text{bp} = 1\cdot9 \times 1\cdot2 \times 3 \times 2\pi$$
$$= 42\cdot98\ \text{kW} \quad \text{Ans. (ii)}$$

$$\text{Mech. effic.} = \frac{\text{brake power}}{\text{indicated power}} = \frac{42\cdot98}{57\cdot73}$$

$$= 0\cdot7446\ \text{or}\ 74\cdot46\% \quad \text{Ans. (iii)}$$

3. Let stroke volume $= 1$
then volume up to cut-off $= 0\cdot34$

$$\text{ratio of expansion} = \frac{\text{volume at end of expansion}}{\text{volume at beginning of expansion}}$$

$$= \frac{1}{0 \cdot 34} = 2 \cdot 941$$

Referring to hypothetical pV diagram, neglecting clearance,

Admission area $= 7 \cdot 5 \times 0 \cdot 34 = 2 \cdot 55$

Expansion area $= p_1 V_1 \ln r$

$\qquad = 7 \cdot 5 \times 0 \cdot 34 \times \ln 2 \cdot 941$

$\qquad = 7 \cdot 5 \times 0 \cdot 34 \times 1 \cdot 0787 = 2 \cdot 751$

Gross area $= 2 \cdot 55 + 2 \cdot 751 = 5 \cdot 301$

Mean press. $=$ mean height $=$ area $\div$ length

Since length is unity, then,

Mean gross pressure $= 5 \cdot 301$ bar   Ans. (i)

Mean effective press. $=$ mean gross press. — back press.

$\qquad = 5 \cdot 301 - 1 \cdot 7$

$\qquad = 3 \cdot 601$ bar Ans. (ii)

4. Let stroke vol. $= 1$, then clearance vol. $= 0 \cdot 05$

$$\text{Ratio of expansion} = \frac{1 + 0 \cdot 05}{0 \cdot 45 + 0 \cdot 05} = 2 \cdot 1$$

Admission area $= 8 \cdot 3 \times 0 \cdot 45 = 3 \cdot 735$

Expansion area $= p_1 V_1 \ln r$

$\qquad = 8 \cdot 3 \times 0 \cdot 5 \times \ln 2 \cdot 1$

$\qquad = 8 \cdot 3 \times 0 \cdot 5 \times 0 \cdot 7419 = 3 \cdot 079$

Gross area $= 3 \cdot 735 + 3 \cdot 079 = 6 \cdot 814$

Hypothetical mean gross pressure

$$= \frac{\text{gross area}}{\text{length}} = \frac{6 \cdot 814}{1} = 6 \cdot 814 \text{ bar}$$

Hypothetical mean effective pressure

$\qquad =$ mean gross press. — back press.

$\qquad = 6 \cdot 814 - 2 \cdot 4 = 4 \cdot 414$ bar   Ans. (i)

Actual $p_m = 4 \cdot 414 \times 0 \cdot 72 = 3 \cdot 178$ bar   Ans. (ii)

5.

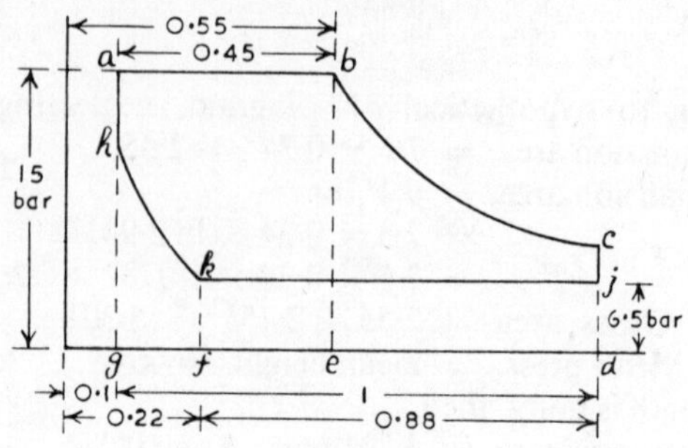

Fig. 104

Let stroke volume $= 1$
clearance volume $= 10$ per cent of $1 = 0.1$
Volume at beginning of expansion
$$= 0.45 + 0.1 = 0.55$$
Volume at end of expansion
$$= 1 + 0.1 = 1.1$$
Ratio of expansion $= 1.1 \div 0.55 = 2.0$
Volume at beginning of compression
$$= 1.1 - 0.88 = 0.22$$
Volume at end of compression $= 0.1$
Ratio of compression $= 0.22 \div 0.1 = 2.2$
Admission area $abeg = 15 \times 0.45 = 6.75$
Expansion area $bcde = p_1 V_1 \ln r$
$$= 15 \times 0.55 \times \ln 2.0$$
$$= 15 \times 0.55 \times 0.6931 = 5.719$$
Gross area $= 6.75 + 5.719 = 12.469$ .... (i)
Compression area $hkfg = p_k V_k \ln r_c$
$$= 6.5 \times 0.22 \times \ln 2.2$$
$$= 6.5 \times 0.22 \times 0.7885 = 1.127$$
Exhaust area $jkfd = 6.5 \times 0.88 = 5.72$
Total back press. area $= 1.127 + 5.72 = 6.847$ .... (ii)
Effective area $abcjkh =$ gross area — back press. area
$$= 12.469 - 6.847 = 5.622$$
Mean height $=$ area $\div$ length $= 5.622 \div 1 = 5.622$
$\therefore$ Mean effective press. $= 5.622$ bar   Ans.

6. $\quad ip = p_m A L n$
ip of HP $= 517 \times 0.7854 \times 0.6^2 \times 1.05 \times 1.8 \times 2$
$$= 552.6 \text{ kW}\quad \text{Ans. (ia)}$$

$$\text{ip of IP} = 207 \times 0.7854 \times 0.95^2 \times 1.05 \times 1.8 \times 2$$
$$= 554.6 \text{ kW} \quad \text{Ans. (ib)}$$

$$\text{ip of LP} = 83 \times 0.7854 \times 1.5^2 \times 1.05 \times 1.8 \times 2$$
$$= 554.6 \text{ kW} \quad \text{Ans. (ic)}$$

p_m of HP referred to LP

$$= \frac{5.17 \times 0.6^2}{1.5^2} = 0.8271 \text{ bar}$$

p_m of IP referred to LP

$$= \frac{2.07 \times 0.95^2}{1.5^2} = 0.8303$$

p_m of LP referred to LP $= 0.83$
p_m all referred to LP
$$= 0.8271 + 0.8303 + 0.83 = 2.4874 \text{ bar} \quad \text{Ans. (ii)}$$

Total ip $= 552.6 + 554.6 + 554.6 = 1661.8 \text{ kW} \quad \text{Ans. (iii)}$

As a check on total ip
$$= 248.74 \times 0.7854 \times 1.5^2 \times 1.05 \times 1.8 \times 2$$
$$= 1662 \text{ kW}$$

7. p_m in LP $= 3.1 \div 4 = 0.775 \text{ bar}$

$$p_m \text{ in HP} = \frac{0.775 \times 1.9^2}{0.68^2} = 6.052 \text{ bar}$$

$$p_m \text{ in IP}_1 = \frac{0.775 \times 1.9^2}{0.96^2} = 3.036 \text{ bar}$$

$$p_m \text{ in IP}_2 = \frac{0.775 \times 1.9^2}{1.34^2} = 1.559 \text{ bar}$$

$\therefore$ Mean effective pressures are,
6.052, 3.036, 1.559 and 0.775 bar   Ans. (i)

$$\text{Total ip} = p_m A L n$$
$$= 310 \times 0.7854 \times 1.9^2 \times 4$$
$$= 3518 \text{ kW} \quad \text{Ans. (ii)}$$

8. $m = a + bP$:

$$18{\cdot}7 = a + b \times 2{\cdot}25 \quad .... \quad .... \quad \text{(i)}$$
$$13{\cdot}3 = a + b \times 1{\cdot}5 \quad .... \quad .... \quad \text{(ii)}$$
$$5{\cdot}4 = b \times 0{\cdot}75 \quad \text{by subtraction}$$
$$b = 7{\cdot}2$$

From (i) $\quad 18{\cdot}7 = a + 7{\cdot}2 \times 2{\cdot}25$
$$a = 18{\cdot}7 - 16{\cdot}2 = 2{\cdot}5$$

$\therefore$ Willans' law is:

$$m = 2{\cdot}5 + 7{\cdot}2P \quad \text{Ans. (i)}$$

When $P = 2000\,\text{kW} = 2\,\text{MW}$:

$$m = 2{\cdot}5 + 7{\cdot}2 \times 2$$
$$= 16{\cdot}9\,\text{Mg/h} \quad \text{Ans. (ii)}$$

$$\text{Specific consumption} = \frac{16{\cdot}9}{2}$$

$$= 8{\cdot}45\,\text{Mg/MWh}$$
$$= 8{\cdot}45\,\text{kg/kWh} \quad \text{Ans. (iii)}$$

9. Specific steam consumption

$$= \frac{28\,500}{3000} = 9{\cdot}5\,\text{kg/kWh} \quad \text{Ans. (i)}$$

Tables page 7, 15 bar 300°C, $\quad h = 3039$
,,   3, 0·14 bar, $\quad h_f = 220$

$$\text{Thermal eff.} = \frac{3600}{\text{spec. steam cons. [kg/kWh]} \times (h_1 - h_{f2})}$$

$$= \frac{3600}{9{\cdot}5 \times (3039 - 220)}$$

$$= 0{\cdot}1345 \text{ or } 13{\cdot}45\% \quad \text{Ans. (ii)}$$

10. Tables page 7, 15 bar 250°C, $\quad h = 2925 \quad s = 6{\cdot}711$
,,   3, 0·16 bar, $\qquad h_f = 232 \quad s_f = 0{\cdot}772$
$$h_{fg} = 2369 \quad s_{fg} = 7{\cdot}213$$

Entropy after expansion $=$ Entropy before
$$0{\cdot}772 + x \times 7{\cdot}213 = 6{\cdot}711$$
$$x \times 7{\cdot}213 = 5{\cdot}939$$
$$\therefore \text{ dryness fraction } x = 0{\cdot}8235 \quad \text{Ans. (i)}$$

At 0·16 bar, $\quad h = h_f + xh_{fg}$
$$= 232 + 0{\cdot}8235 \times 2369 = 2183$$

Rankine efficiency $= \dfrac{h_1 - h_2}{h_1 - h_{f2}}$

$$= \frac{2925 - 2183}{2925 - 232} = \frac{742}{2693}$$

$$= 0{\cdot}2755 \text{ or } 27{\cdot}55\% \quad \text{Ans. (ii)}$$

SOLUTIONS TO TEST EXAMPLES 13

1. $v\,[\text{m/s}] = \sqrt{2 \times \text{spec. enthalpy drop}\,[\text{J/kg}]}$

$$= \sqrt{2 \times 60 \times 10^3} = 346 \cdot 4\ \text{m/s} \quad \text{Ans. (i)}$$

area $[\text{m}^2] \times$ velocity $[\text{m/s}] =$ mass flow $[\text{kg/s}] \times$ spec. vol.
$$[\text{m}^3/\text{kg}]$$

$$\text{area}\,[\text{m}^2] = \frac{0 \cdot 2 \times 1 \cdot 5}{346 \cdot 4}$$

$$= 8 \cdot 66 \times 10^{-4}\,\text{m}^2 = 866\ \text{mm}^2 \quad \text{Ans. (ii)}$$

2. Tables page 4, 8 bar, $h_g = 2769$

                5 bar, $h_f = 640$ $h_{fg} = 2109$

                     $v_g = 0 \cdot 3748$

Enthalpy drop $= 2769 - (640 + 0 \cdot 97 \times 2109)$

$$= 83\ \text{kJ/kg}$$

Velocity $= \sqrt{2 \times 83 \times 10^3} = 407 \cdot 4\ \text{m/s} \quad \text{Ans. (i)}$

Spec. vol. of steam at exit $= 0 \cdot 97 \times 0 \cdot 3748\ \text{m}^3/\text{kg}$

$$\text{Mass flow}\,[\text{kg/s}] = \frac{\text{volume flow}\,[\text{m}^3/\text{s}]}{\text{spec. vol.}\,[\text{m}^3/\text{kg}]}$$

$$= \frac{\text{area}\,[\text{m}^2] \times \text{velocity}\,[\text{m/s}]}{\text{spec. vol.}\,[\text{m}^3/\text{kg}]}$$

$$= \frac{14 \cdot 5 \times 10^{-4} \times 407 \cdot 4}{0 \cdot 97 \times 0 \cdot 3748}$$

$$= 1 \cdot 625\ \text{kg/s} \quad \text{Ans. (ii)}$$

3. Tables page 4, 14 bar, $h_g = 2790$ $v_g = 0 \cdot 1408$

               10 bar, $h_f = 763$ $h_{fg} = 2015$

                    $v_g = 0 \cdot 1944$

$$p_1 V_1^{1 \cdot 135} = p_2 V_2^{1 \cdot 135}$$

$$14 \times 0 \cdot 1408^{1 \cdot 135} = 10 \times v_2^{1 \cdot 135}$$

$$v_2 = 0 \cdot 1408 \times \sqrt[1 \cdot 135]{1 \cdot 4} = 0 \cdot 1893\ \text{m}^3/\text{kg}$$

Spec. vol. of dry steam at 10 bar is $0 \cdot 1944\ \text{m}^3/\text{kg}$ therefore,

$$\text{dryness after expansion} = \frac{0 \cdot 1893}{0 \cdot 1944} = 0 \cdot 974 \quad \text{Ans. (i)}$$

$$\text{Spec. enthalpy drop} = 2790 - (763 + 0.974 \times 2015)$$
$$= 64 \text{ kJ/kg} \quad \text{Ans. (ii)}$$

$$\text{Velocity} = \sqrt{2 \times 64 \times 10^3} = 357.8 \text{ m/s} \quad \text{Ans. (iii)}$$
$$\text{Volume flow [m}^3\text{/s]} = \text{area [m}^2\text{]} \times \text{velocity [m/s]}$$

For mass flow of 1 kg/s:

$$\text{Area [m}^2\text{]} = \frac{0.1893}{357.8} = 5.293 \times 10^{-4} \text{ m}^2$$

$$5.293 \times 10^{-4} \times 10^6 = 529.3 \text{ mm}^2 \quad \text{Ans. (iv)}$$

4. Referring to Fig. 90:
$$v_{a1} = v_1 \sin \alpha_1 = 840 \times \sin 18° = 259.6 \text{ m/s}$$
$$v_{w1} = v_1 \cos \alpha_1 = 840 \times \cos 18° = 798.9 \text{ m/s}$$
$$x = v_{w1} - u = 798.9 - 360 = 438.9 \text{ m/s}$$

$$\tan \beta_1 = \frac{v_{a1}}{x} = \frac{259.6}{438.9} = 0.5914$$

$\therefore$ Entrance and exit angles $= 30° 36'$ Ans. (i)
Neglecting friction, $v_{r2} = v_{r1}$ and, since $\beta_2 = \beta_1$ then:
$$u + v_{w2} = x \text{ and } v_{a2} = v_{a1}$$
$$v_{w2} = 438.9 - 360 = 78.9 \text{ m/s}$$

$$\tan \phi = \frac{v_{w2}}{v_{a2}} = \frac{78.9}{259.6} = 0.3040$$

$$\therefore \quad \phi = 16° 55', \alpha_2 = 90° - 16° 55' = 73° 5'$$

$$v_2 = \frac{v_{a2}}{\cos \phi} = \frac{259.6}{\cos 16° 55'} = 271.3 \text{ m/s}$$

Velocity of steam at exit $= 271.3$ m/s
at 16° 55′ to axis $\Big\}$ Ans. (ii)
or 73° 5′ to blade movement

5. Referring to Fig. 90:
Angle between u and $v_{r1} = 180° - \beta_1$
$$= 180° - 33° = 147°$$
Angle between v_1 and $v_{r1} = \beta_1 - \alpha_1$
$$= 33° - 20° = 13°$$

By sine rule:

$$\frac{u}{\sin 13°} = \frac{v_1}{\sin 147°}$$

$$u = \frac{450 \times \sin 13°}{\sin 147°} = 185 \cdot 9 \text{ m/s Ans.(i)}$$

$$\text{Rotational speed [rev/s]} = \frac{\text{linear velocity [m/s]}}{\text{circumference [m]}}$$

$$= \frac{185 \cdot 9}{\pi \times 0 \cdot 66} = 89 \cdot 64 \text{ rev/s   Ans. (ii)}$$

6. Referring to Fig. 90:

$$v_{a1} = v_1 \sin \alpha_1 = 750 \times \sin 22° = 280 \cdot 9 \text{ m/s}$$
$$v_{w1} = v_1 \cos \alpha_1 = 750 \times \cos 22° = 695 \cdot 5 \text{ m/s}$$
$$x = v_{w1} - u = 695 \cdot 5 - 225 = 470 \cdot 5 \text{ m/s}$$

$$\tan \beta_1 = \frac{v_{a1}}{x} = \frac{280 \cdot 9}{470 \cdot 5} = 0 \cdot 5971$$

Entrance angle of blades $= 30° \, 51'$ Ans. (i)

$$v_{r1} = \frac{v_{a1}}{\sin \beta_1} = \frac{280 \cdot 9}{\sin 30° \, 51'} = 548 \text{ m/s}$$

$$v_{r2} = 0 \cdot 85 \times 548 = 465 \cdot 8 \text{ m/s}$$
$$v_{w2} = v_{r2} \cos \beta_2 - u$$
$$= 465 \cdot 8 \times \cos 31° - 225 = 174 \cdot 3 \text{ m/s}$$
$$v_{a2} = v_{r2} \sin \beta_2 = 465 \cdot 8 \times \sin 31° = 239 \cdot 9 \text{ m/s}$$

$$\tan \alpha_2 = \frac{v_{a2}}{v_{w2}} = \frac{239 \cdot 9}{174 \cdot 3} = 1 \cdot 376 \qquad \therefore \alpha_2 = 54°$$

$$v_2 = \frac{v_{w2}}{\cos \alpha_2} = \frac{174 \cdot 3}{\cos 54°} = 296 \cdot 6 \text{ m/s}$$

Velocity at exit from blades $= 296 \cdot 6$ m/s
at 54° to blade motion
or 36° to turbine axis $\Bigg\}$ Ans. (ii)

7. Tables page 7, 20 bar 300°C, $h = 3025$
      ,,   4, 3 bar, $h_f = 561$, $h_{fg} = 2164$
Enthalpy drop through nozzles
$$= 3025 - (561 + 0.98 \times 2164) = 344 \text{ kJ/kg}$$
$$\text{Velocity} = \sqrt{2 \times 344 \times 10^3} = 829.4 \text{ m/s} \text{ Ans. (i)}$$

Referring to Fig. 90, $v_1 = 829.4$
$$v_{a1} = v_1 \sin \alpha_1 = 829.4 \times \sin 20° = 283.7 \text{ m/s}$$
$$v_{w1} = v_1 \cos \alpha_1 = 829.4 \times \cos 20° = 779.4 \text{ m/s}$$
$$x = v_{w1} - u = 779.4 - 380 = 399.4 \text{ m/s}$$

$$\tan \beta_1 = \frac{v_{a1}}{x} = \frac{283.7}{399.4} = 0.7104$$

Entrance angle of blades $= 35° 23'$ Ans. (ii)

8. Ref. Fig. 90:

$$v_{w1} = v_1 \cos \alpha_1 = 900 \times \cos 18° = 855.9 \text{ m/s}$$
$$x = v_{w1} - u = 855.9 - 330 = 525.9 \text{ m/s}$$
Blades are symmetrical, $\therefore \beta_2 = \beta_1$
Friction factor $= 0.9$
$$\therefore v_{w2} + u = 0.9 \times x$$
$$v_{w2} = 0.9 \times 525.9 - 330 = 143.3 \text{ m/s}$$
$$v_w = v_{w1} + v_{w2} = 855.9 + 143.3 = 999.2 \text{ m/s}$$
Force on blades [N] $= \dot{m}[\text{kg/s}] \times v_w[\text{m/s}]$
$$= 0.095 \times 999.2$$
$$= 94.93 \text{ N} \text{ Ans. (i)}$$

Power [W] $=$ force [N] $\times$ blade velocity [m/s]
$$= 94.93 \times 330$$
$$= 3.132 \times 10^4 \text{ W} = 31.32 \text{ kW}$$
Ans. (ii)

Change of axial velocity $= v_{a1} - v_{a2}$
$$v_{a1} = v_1 \sin \alpha_1 = 900 \times \sin 18° = 278.1 \text{ m/s}$$
Since $\beta_2 = \beta_1$ then $v_{a2} = 0.9 v_{a1}$
$$v_{a1} - v_{a2} = v_{a1} - 0.9 v_{a1} = 0.1 v_{a1}$$
$$= 27.81 \text{ m/s}$$
Axial thrust $= \dot{m}(v_{a1} - v_{a2})$
$$= 0.095 \times 27.81 = 2.642 \text{ N} \text{ Ans. (iii)}$$

9. Velocity at nozzle exit v_1

$$= \sqrt{2 \times 312 \cdot 5 \times 10^3 \times 0 \cdot 9}$$
$$= 750 \text{ m/s}$$

Ref. Fig. 90:

$$v_{a1} = v_1 \sin \alpha_1 = 750 \times \sin 20° = 256 \cdot 5 \text{ m/s}$$
$$v_{w1} = v_1 \cos \alpha_1 = 750 \times \cos 20° = 704 \cdot 9 \text{ m/s}$$

$$x = \frac{v_{a1}}{\tan \beta_1} = \frac{256 \cdot 5}{\tan 35°} = 366 \cdot 4 \text{ m/s}$$

$$\begin{aligned}
\text{Blade velocity} = u = v_{w1} - x \\
= 704 \cdot 9 - 366 \cdot 4 \\
= 338 \cdot 5 \text{ m/s} \quad \text{Ans. (i)}
\end{aligned}$$

Absolute velocity of exit steam is in the direction of the turbine axis, therefore $\alpha_2 = 90°$

$$\tan \beta_2 = \frac{v_2}{u} = \frac{204}{338 \cdot 5} = 0 \cdot 6027$$

Exit angle of blades $= 31° \, 5'$ Ans. (ii)

$$v_{r1} = \frac{v_{a1}}{\sin \beta_1} = \frac{256 \cdot 5}{\sin 35°} = 447 \cdot 2 \text{ m/s}$$

$$v_{r2} = \frac{v_2}{\sin \beta_2} = \frac{204}{\sin 31° \, 5'} = 395 \cdot 2 \text{ m/s}$$

Loss of kinetic energy of steam across the blades
$$= \tfrac{1}{2} \dot{m}(v_{r1}^2 - v_{r2}^2)$$
$$= \tfrac{1}{2} \times 1 \times (447 \cdot 2^2 - 395 \cdot 2^2)$$
$$= 21\,900 \text{ J} = 21 \cdot 9 \text{ kJ/kg steam} \quad \text{Ans. (iii)}$$

$$\begin{aligned}
\text{Axial thrust} &= \dot{m}(v_{a1} - v_{a2}) \\
&= 1 \times (256 \cdot 5 - 204) \\
&= 52 \cdot 5 \text{ N/kg of steam} \quad \text{Ans. (iv)}
\end{aligned}$$

$$\text{Power} = \dot{m} v_w u$$

Since the steam leaves the turbine axially, that is, at 90° to the blade movement, there is no velocity of whirl at exit, $v_{w2} = 0$
$$\therefore v_w = v_{w1}$$

$$\begin{aligned}
\text{Power} &= 1 \times 704 \cdot 9 \times 338 \cdot 5 \\
&= 2 \cdot 386 \times 10^5 \text{ W} \\
&= 238 \cdot 6 \text{ kW/kg of steam} \quad \text{Ans. (v)}
\end{aligned}$$

$$\text{Blade efficiency} = \frac{\text{work done on blades}}{\text{work supplied}}$$

$$= \frac{\dot{m} v_w u \,[\text{J/s}]}{\frac{1}{2}\dot{m} v_1^{\,2}\,[\text{J/s}]} = \frac{2uv_w}{v_1^{\,2}}$$

$$= \frac{2 \times 338 \cdot 5 \times 704 \cdot 9}{750^2}$$

$$= 0 \cdot 8484 \text{ or } 84 \cdot 84\% \quad \text{Ans. (vi)}$$

10.   Referring to Fig. 92:

$$
\begin{aligned}
v_{a1} &= v_1 \sin \alpha_1 = 243 \times \sin 23^\circ = 94 \cdot 95 \text{ m/s} \\
v_{w1} &= v_1 \cos \alpha_1 = 243 \times \cos 23^\circ = 223 \cdot 7 \text{ m/s} \\
x &= v_{w1} - u = 223 \cdot 7 - 159 = 64 \cdot 7 \text{ m/s}
\end{aligned}
$$

$$\tan \beta_1 = \frac{v_{a1}}{x} = \frac{94 \cdot 95}{64 \cdot 7} = 1 \cdot 468$$

Blade inlet angle $= 55^\circ 44'$ Ans. (i)

Since the combined vector diagram of inlet and exit velocities is symmetrical, $v_{w2} = x$,

$$v_w = v_{w1} + v_{w2} = 223 \cdot 7 + 64 \cdot 7 = 288 \cdot 4 \text{ m/s}$$

$$
\begin{aligned}
\text{Force on blades} &= \dot{m} v_w \\
&= 0 \cdot 9 \times 288 \cdot 4 = 259 \cdot 5 \text{ N} \quad \text{Ans. (ii)} \\
\text{Power [W]} &= \text{force [N]} \times \text{velocity [m/s]} \\
&= 259 \cdot 5 \times 159 \\
&= 4 \cdot 127 \times 10^4 \text{ W} = 41 \cdot 27 \text{ kW} \quad \text{Ans. (iii)}
\end{aligned}
$$

11.   Referring to Fig. 94:

$$
\begin{aligned}
T_1 &= 21 + 273 = 294 \text{ K} \\
T_3 &= 680 + 273 = 953 \text{ K}
\end{aligned}
$$

$$\gamma = \frac{c_P}{c_V} = \frac{1 \cdot 005}{0 \cdot 718} = 1 \cdot 4$$

$$\frac{p_2}{p_1} = \frac{p_3}{p_4} = r_p = \frac{5 \cdot 7}{1} = 5 \cdot 7$$

$$r_p^{(\gamma-1)/y} = 5.7^{0.4/1.4} = 1.644$$

$$\frac{T_2}{T_1} = \left\{\frac{p_2}{p_1}\right\}^{\frac{\gamma-1}{\gamma}}$$

$$T_2 = 294 \times 1.644 = 483.3 \text{ K}$$

Temperature at end of compression
$$= 483.3 - 273 = 210.3°\text{C} \quad \text{Ans. (i)}$$

$$\frac{T_4}{T_3} = \left\{\frac{p_4}{p_3}\right\}^{\frac{\gamma-1}{\gamma}}$$

$$T_4 = \frac{953}{1.644} = 579.7 \text{ K}$$

Alternatively, $\dfrac{T_4}{T_3} = \dfrac{T_1}{T_2}$ because pressure ratios are equal

$$\therefore T_4 = \frac{953 \times 294}{483.3} = 579.7 \text{ K}$$

Temperature at end of expansion
$$= 579.7 - 273 = 306.7°\text{C} \quad \text{Ans. (ii)}$$

Heat energy supplied per kg
$$= m \times c_P \times (T_3 - T_2)$$
$$= 1 \times 1.005 \times (953 - 483.3)$$
$$= 472.1 \text{ kJ/kg} \quad \text{Ans. (iii)}$$

Increase in internal energy per kg from inlet to exhaust
$$= m \times c_V \times (T_4 - T_1)$$
$$= 1 \times 0.718 \times (579.7 - 294)$$
$$= 205.1 \text{ kJ/kg} \quad \text{Ans. (iv)}$$

$$\text{Ideal thermal effic.} = 1 - \frac{1}{r_p^{(\gamma-1)/\gamma}}$$

$$= 1 - \frac{1}{1.644} = 0.3918 \quad \text{Ans. (v)}$$

Alternatively,

$$\text{Thermal effic.} = 1 - \frac{T_4 - T_1}{T_3 - T_2}$$

$$= 1 - \frac{579.7 - 294}{953 - 483.3} = 0.3919$$

SOLUTIONS TO TEST EXAMPLES 14

1. Tables page 4, steam 8 bar, $h_f = 721$, $h_{fg} = 2048$
Enthalpy of steam $\quad h_1 = h_f + xh_{fg}$
$$= 721 + 0.98 \times 2048 = 2728$$
Tables page 2, enthalpy of water at 43°C by interpolation,

$$h \text{ at } 44°C = 184.2$$
$$h \text{ at } 42°C = 175.8$$
$$\text{difference for } 2°C = \quad 8.4$$
$$\text{,, \quad ,, } 1°C = \quad 4.2$$
$$\therefore \ h \text{ at } 43°C = 175.8 + 4.2 = 180$$

$$\text{Fuel consumption} = \frac{1.2 \times 10^3}{24} = 50 \text{ kg/h}$$

$$\text{Boiler efficiency} = \frac{\text{Heat energy transferred to water and steam}}{\text{Heat energy supplied by fuel}}$$

$$= \frac{\text{mass of steam/hour} \times (h_1 - h_w)}{\text{mass of fuel/hour} \times \text{cal. value}}$$

$$0.7 = \frac{\text{steam/hour} \times (2728 - 180)}{50 \times 42 \times 10^3}$$

$$\text{steam/hour} = \frac{0.7 \times 50 \times 42 \times 10^3}{2548}$$

$$= 576.9 \text{ kg/h} \quad \text{Ans.}$$

2. Tables page 2, water 100°C, $\ h = 419.1$
$\quad$,, $\quad$,, 7, steam 25 bar 300°C by interpolation:

$$h \text{ at } 20 \text{ bar } 300°C = 3025$$
$$h \text{ at } 30 \text{ bar } 300°C = 2995$$
$$\text{for increase of } 10 \text{ bar, decrease} = 30$$
$$\text{,, \quad ,, \quad ,, } 5 \text{ bar, \quad ,, } = 15$$
$$\therefore h \text{ at } 25 \text{ bar } 300°C = 3025 - 15 = 3010$$

$$\text{Efficiency} = \frac{\text{Heat energy transferred to water and steam}}{\text{Heat energy supplied by fuel}}$$

$$= \frac{\text{steam/h} \times (h_1 - h_w)}{\text{fuel/h} \times \text{cal. value}}$$

$$= \frac{11\,350 \times (3010 - 419)}{875 \times 42{\cdot}3 \times 10^3}$$

$$= 0{\cdot}7949 \text{ or } 79{\cdot}49\% \quad \text{Ans.}$$

3. Tables page 2, water 48°C, $h_w = 200{\cdot}9$
Let h = specific enthalpy of steam leaving boiler
Mass of steam produced = mass of feed water
$$= 0{\cdot}258 \times 3600 \text{ kg/h}$$
Heat energy transferred to water and steam
$$= 0{\cdot}258 \times 3600 \times (h - 200{\cdot}9) \text{ kJ/h}$$
Heat energy supplied by the fuel
$$= 81 \times 43{\cdot}5 \times 10^3 \text{ kJ/h}$$

$$\text{Efficiency} = \frac{\text{Heat energy transferred to steam}}{\text{Heat energy supplied}}$$

$$0{\cdot}68 = \frac{0{\cdot}258 \times 3600 \times (h - 200{\cdot}9)}{81 \times 43{\cdot}5 \times 10^3}$$

$$h - 200{\cdot}9 = \frac{0{\cdot}68 \times 81 \times 43{\cdot}5 \times 10^3}{0{\cdot}258 \times 3600}$$

$$h = 2580 + 200{\cdot}9 = 2780{\cdot}9 \text{ kJ/kg}$$

Since steam is dry saturated we look at tables for a pressure where h_g is near to 2780·9
$$\text{Page 4,} \quad h_g = 2781 \text{ for a pressure of 11 bar}$$
$$\text{,, } 5, \quad h_g = 2779 \text{ for a pressure of 65 bar}$$
The former is the obvious pressure.
$$\text{Pressure} = 11 \text{ bar} \quad \text{Ans.}$$

4. Tables page 2, h_{fg} at 100°C $= 2256{\cdot}7$
Per kg of fuel burned:
Heat energy transferred to water and steam
$$= 15 \times 2256{\cdot}7 \text{ kJ}$$
Heat energy supplied $= 1 \times 41{\cdot}9 \times 10^3 \text{ kJ}$

$$\text{Efficiency} = \frac{\text{Heat energy given to steam}}{\text{Heat energy supplied}}$$

$$= \frac{15 \times 2256 \cdot 7}{41 \cdot 9 \times 10^3}$$

$$= 0 \cdot 8081 \text{ or } 80 \cdot 81\% \quad \text{Ans.}$$

5. Tables page 4, water 130°C, $h = 546$
   „   „   7, steam 30 bar 375°C, by interpolation,
   h at 30 bar 400°C $= 3231$
   h at 30 bar 350°C $= 3117$
   difference for 50°C $= $ 114
   „   „   25°C $= $ 57
$\therefore$ h at 30 bar 375°C $= 3117 + 57 = 3174$

Heat energy transferred to steam per hour
$= 30\,000 \times (3174 - 546) = 30\,000 \times 2628 \text{ kJ/h}$

$$\text{Hourly fuel consumption} = \frac{53 \times 10^3}{24} = 2209 \text{ kg/h}$$

Heat energy supplied by fuel per hour
$$= 2209 \times 42 \times 10^3 \text{ kJ/h}$$

$$\text{Efficiency} = \frac{30\,000 \times 2628}{2209 \times 42 \times 10^3}$$

$$= 0 \cdot 85 \text{ or } 85\% \quad \text{Ans. (i)}$$

Heat energy supplied to plant by fuel [kJ/s $=$ kW]

$$= \frac{2209 \times 42 \times 10^3}{3600}$$

Energy converted into engine power
$$= 0 \cdot 13 \times \text{energy supplied}$$

$$= \frac{0 \cdot 13 \times 2209 \times 42 \times 10^3}{3600} = 3349 \text{ kW} \quad \text{Ans. (ii)}$$

Tables page 2, h_{f_g} at 100°C $= 2256{\cdot}7$
Evaporative capacity from and at 100°C

$$= \frac{30000 \times 2628}{2256{\cdot}7} = 34920 \text{ kg/h} \quad \text{Ans. (iii)}$$

Equivalent evaporation, per kg fuel, from and at 100°C

$$= \frac{34920}{2209} = 15{\cdot}81 \text{ kg steam/kg fuel} \quad \text{Ans. (iv)}$$

6. Solids in initially + solids put in = solids in finally

$$\begin{array}{ccccc}
\text{water in} & \text{initial} & \text{amount} & \text{feed} & \text{water in} & \text{final} \\
\text{boiler} \times \text{p.p.m.} & + & \text{of feed} \times \text{p.p.m.} & = & \text{boiler} \times \text{p.p.m.}
\end{array}$$

$$3{\cdot}5 \times 40 + 0{\cdot}875 \times 24 \times \text{feed p.p.m.} = 3{\cdot}5 \times 2500$$
$$21 \times \text{feed p.p.m.} = 8610$$
$$\text{feed p.p.m.} = 410 \quad \text{Ans.}$$

7. c.v. $= 33{\cdot}7\,C + 144\left(H_2 - \dfrac{O_2}{8}\right) + 9{\cdot}3\,S$

$$= 33{\cdot}7 \times 0{\cdot}81 + 144\left(0{\cdot}05 - \frac{0{\cdot}056}{8}\right) + 9{\cdot}3 \times 0{\cdot}01$$

$$= 27{\cdot}3 + 6{\cdot}193 + 0{\cdot}093$$
$$= 33{\cdot}586 \text{ MJ/kg} \quad \text{Ans.}$$

8. $\quad$ c.v. $= 33{\cdot}7\,C + 144\left(H_2 - \dfrac{O_2}{8}\right)$

$$= 33{\cdot}7 \times 0{\cdot}852 + 144\left(0{\cdot}12 - \frac{0{\cdot}016}{8}\right)$$

$$= 28{\cdot}71 + 16{\cdot}99$$
$$= 45{\cdot}7 \text{ MJ/kg} \quad \text{Ans. (i)}$$

$$\text{Air reqd.} = \frac{100}{23}\left\{2\tfrac{2}{3}\,C + 8\left(H_2 - \frac{O_2}{8}\right)\right\}$$

$$= \frac{100}{23} \{2\tfrac{2}{3} \times 0{\cdot}852 + 8 \times 0{\cdot}118\}$$

$$= \frac{100}{23} \times 3{\cdot}216 = 13{\cdot}98 \text{ kg air/kg fuel} \quad \text{Ans. (ii)}$$

9. Since there is no oxygen present, all the hydrogen is available for combustion.

$$\text{Air reqd.} = \frac{100}{23} \{2\tfrac{2}{3}\,C + 8\,H_2 + S\}$$

$$= \frac{100}{23} \{2\tfrac{2}{3} \times 0{\cdot}855 + 8 \times 0{\cdot}144 + 0{\cdot}001\}$$

$$= \frac{100}{23} \times 3{\cdot}433 = 14{\cdot}93 \text{ kg air/kg fuel}$$

$$pV = mRT$$
$$\text{where } p = 1 \text{ bar} \times 10^2 = 100 \text{ kN/m}^2$$
$$T = 15 + 273 = 288 \text{ K}$$

$$V = \frac{14{\cdot}93 \times 0{\cdot}287 \times 288}{100}$$

$$= 12{\cdot}34 \text{ m}^3 \text{ air/kg fuel} \quad \text{Ans.}$$

10. Theoretical minimum air/kg fuel

$$= \frac{100}{23}\left\{ 2\tfrac{2}{3}\,C + 8\left(H_2 - \frac{O_2}{8}\right) + S \right\}$$

$$= \frac{100}{23}\left\{ 2\tfrac{2}{3} \times 0{\cdot}861 + 8\left(0{\cdot}125 - \frac{0{\cdot}004}{8}\right) + 0{\cdot}01 \right\}$$

$$= \frac{100}{23} \times 3{\cdot}302 = 14{\cdot}35 \text{ kg air/kg fuel}$$

$$\text{Actual air} = 1{\cdot}4 \times 14{\cdot}35 = 20{\cdot}09 \text{ kg/kg fuel}$$
$$\text{Air supply per hour} = 20{\cdot}09 \times 400$$
$$= 8036 \text{ kg/h} \quad \text{Ans. (i)}$$

Heat energy received by air in heater, per hour

$$= \text{mass} \times \text{spec. heat} \times \text{temp. rise}$$
$$= 8036 \times 1{\cdot}005 \times (130 - 18)$$
$$= 9{\cdot}042 \times 10^5 \, \text{kJ/h}$$
$$= 904{\cdot}2 \, \text{MJ/h} \quad \text{Ans. (ii)}$$

11. Per kg of fuel:

$$\text{Available hydrogen} = H_2 - \frac{O_2}{8}$$

$$= 0{\cdot}13 - \frac{0{\cdot}02}{8} = 0{\cdot}1275 \, \text{kg}$$

$$\text{Cal. value} = 33{\cdot}7\,C + 144\left(H_2 - \frac{O_2}{8}\right)$$

$$= 33{\cdot}7 \times 0{\cdot}85 + 144 \times 0{\cdot}1275$$
$$= 28{\cdot}64 + 18{\cdot}36 = 47 \, \text{MJ/kg} \quad \text{Ans. (i)}$$

$$\text{Theor. air} = \frac{100}{23}\left\{2\tfrac{2}{3}\,C + 8\left(H_2 - \frac{O_2}{8}\right)\right\}$$

$$= \frac{100}{23}\{2\tfrac{2}{3} \times 0{\cdot}85 + 8 \times 0{\cdot}1275\}$$

$$= \frac{100}{23} \times 3{\cdot}287 = 14{\cdot}29 \, \text{kg air/kg fuel}$$

Actual air $= 1{\cdot}5 \times 14{\cdot}29 = 21{\cdot}44 \, \text{kg air/kg fuel}$ Ans. (ii)

Products of combustion per kg of fuel burned

$$= 21{\cdot}44 + 1 \, \text{kg fuel} = 22{\cdot}44 \, \text{kg}$$

Heat energy carried away

$$= \text{mass} \times \text{spec. heat} \times \text{temp. rise}$$
$$= 22{\cdot}44 \times 1{\cdot}005 \times (280 - 31) = 5614 \, \text{kJ/kg fuel}$$

as a percentage of the heat energy supplied

$$= \frac{5614}{47 \times 10^3} \times 100 = 11{\cdot}95\% \quad \text{Ans. (iii)}$$

12.   Heat energy released $=$ Heat energy transferred
      by fuel   to water and fittings

$$\text{mass of oil} \times \text{c.v.} = \frac{\text{equiv. mass}}{\text{of water}} \times \frac{\text{spec.}}{\text{heat}} \times \frac{\text{temp.}}{\text{rise}}$$

$$0{\cdot}75 \times 10^{-3} \times \text{cv [kJ/kg]} = (1{\cdot}8 + 0{\cdot}47) \times 4{\cdot}2 \times 3{\cdot}3$$

$$\text{c.v.} = \frac{2{\cdot}27 \times 4{\cdot}2 \times 3{\cdot}3 \times 10^3}{0{\cdot}75}$$

$$= 41{\cdot}94 \times 10^3 \text{ kJ/kg}$$
$$= 41{\cdot}94 \text{ MJ/kg} \quad \text{Ans.}$$

13. $\text{Available hydrogen} = H_2 - \dfrac{O_2}{8}$

$$= 0{\cdot}13 - \frac{0{\cdot}02}{8} = 0{\cdot}1275 \text{ kg}$$

$\text{Cal. value} = 33{\cdot}7 \times 0{\cdot}84 + 144 \times 0{\cdot}1275$
$$= 28{\cdot}31 + 18{\cdot}36 = 46{\cdot}67 \text{ MJ/kg} \quad \text{Ans. (i)}$$

$$\text{Theor. air} = \frac{100}{23} \times \text{oxygen required}$$

$$= \frac{100}{23} \{2\tfrac{2}{3} \times 0{\cdot}84 + 8 \times 0{\cdot}1275\}$$

$$= \frac{100}{23} \times 3{\cdot}26 = 14{\cdot}18 \text{ kg air/kg fuel} \quad \text{Ans. (ii)}$$

Per kg of fuel burned:
Mass of gases in the products from
22 kg air $+$ (1 kg fuel $-$ 0·01 kg incombustibles)
$$= 22{\cdot}99 \text{ kg}$$
Mass of oxygen in 22 kg of air
$$= 0{\cdot}23 \times 22 = 5{\cdot}06 \text{ kg}$$
$\text{Surplus oxygen} = 5{\cdot}06 - 3{\cdot}26 = 1{\cdot}8 \text{ kg}$
Mass of nitrogen in 22 kg of air
$$= 0{\cdot}77 \times 22 = 16{\cdot}94 \text{ kg}$$
$CO_2 \text{ formed} = 3\tfrac{2}{3} \times 0{\cdot}84 = 3{\cdot}08 \text{ kg}$
$H_2O \text{ formed} = 9 \times 0{\cdot}13 = 1{\cdot}17 \text{ kg}$
% composition of gases by mass   Ans. (iii):

$$CO_2 = \frac{3 \cdot 08}{22 \cdot 99} \times 100 = 13 \cdot 4\%$$

$$H_2O = \frac{1 \cdot 17}{22 \cdot 99} \times 100 = 5 \cdot 09\%$$

$$O_2 = \frac{1 \cdot 8}{22 \cdot 99} \times 100 = 7 \cdot 83\%$$

$$N_2 = \frac{16 \cdot 94}{22 \cdot 99} \times 100 = 73 \cdot 68\%$$

14. Per kg of fuel burned:

$$\text{Minimum air} = \frac{100}{23}\left\{2\tfrac{2}{3}\,C + 8\left(H_2 - \frac{O_2}{8}\right)\right\}$$

$$= \frac{100}{23}\left\{2\tfrac{2}{3} \times 0 \cdot 855 + 8\left(0 \cdot 119 - \frac{0 \cdot 016}{8}\right)\right\}$$

$$= \frac{100}{23} \times 3 \cdot 216 = 13 \cdot 98 \text{ kg air/kg fuel}$$

$$CO_2 \text{ formed} = 3\tfrac{2}{3} \times 0 \cdot 855 = 3 \cdot 135 \text{ kg}$$

When air supply is minimum:
Mass of products of combustion

$$= 13 \cdot 98 \text{ kg air} + (1 \text{ kg fuel} - 0 \cdot 01 \text{ kg}$$
$$\text{impurities})$$
$$= 13 \cdot 98 + 0 \cdot 99 = 14 \cdot 97 \text{ kg}$$

$$\% CO_2 = \frac{3 \cdot 135}{14 \cdot 97} \times 100 = 20 \cdot 94\% \quad \text{Ans. (i)}$$

When air supply is 25% excess:
Mass of products of combustion

$$= 1 \cdot 25 \times 13 \cdot 98 \text{ kg air} + 0 \cdot 99 = 18 \cdot 47 \text{ kg}$$

$$\% CO_2 = \frac{3 \cdot 135}{18 \cdot 47} \times 100 = 16 \cdot 97\% \quad \text{Ans. (ii)}$$

When air supply is 50 % excess:
Mass of products of combustion
$$= 1{\cdot}5 \times 13{\cdot}98 + 0{\cdot}99 = 21{\cdot}96 \text{ kg}$$

$$\% \, CO_2 = \frac{3{\cdot}135}{21{\cdot}96} \times 100 = 14{\cdot}28\% \quad \text{Ans. (iii)}$$

When air supply is 75 % excess:
Mass of products of combustion
$$= 1{\cdot}75 \times 13{\cdot}98 + 0{\cdot}99 = 25{\cdot}45 \text{ kg}$$

$$\% \, CO_2 = \frac{3{\cdot}135}{25{\cdot}45} \times 100 = 12{\cdot}31\% \quad \text{Ans. (iv)}$$

SOLUTIONS TO TEST EXAMPLES 15

Refer to Fig. 101 for all solutions

1. Specific enthalpy gain of refrigerant through evaporator
$$= h_1 - h_4 = 320 - 135 = 185 \text{ kJ/kg}$$
(h_4 being equal to h_3 because there is no change of enthalpy in the throttling process through the expansion valve)

$$
\begin{aligned}
\text{Refrig. effect [kJ/h]} &= \text{mass flow [kg/h]} \times (h_1 - h_4) \text{ [kJ/kg]} \\
&= 5 \times 60 \times 185 \\
&= 5{\cdot}55 \times 10^4 \text{ kJ/h or } 55{\cdot}5 \text{ MJ/h} \quad \text{Ans.}
\end{aligned}
$$

2. From tables page 13, Freon - 12,

$$5{\cdot}673 \text{ bar} \qquad h_f = 54{\cdot}87$$
$$1{\cdot}509 \text{ bar} \qquad t = -20°C \qquad h_f = 17{\cdot}82 \qquad h_g = 178{\cdot}73$$
$$h_{fg} = h_g - h_f = 178{\cdot}73 - 17{\cdot}82 = 160{\cdot}91$$

Since the saturation temperature at $1{\cdot}509$ bar is $-20°C$ and the refrigerant at this pressure leaves the evaporator at $-5°C$, it is superheated by $15°$.

$$h \text{ at } 1{\cdot}509 \text{ bar superheated } 15° = 187{\cdot}75$$

Throttling between condenser exit and evaporator inlet:

$$
\begin{aligned}
\text{Enthalpy after } (h_4) &= \text{Enthalpy before } (h_3) \\
17{\cdot}82 + x_4 \times 160{\cdot}91 &= 54{\cdot}87 \\
x_4 \times 160{\cdot}91 &= 37{\cdot}05 \\
x_4 &= 0{\cdot}2303 \quad \text{Ans. (i)}
\end{aligned}
$$

$$
\begin{aligned}
\text{Refrig. effect/kg} &= h_1 - h_4 = h_1 - h_3 \\
&= 187{\cdot}75 - 54{\cdot}87 \\
&= 132{\cdot}88 \text{ kJ/kg} \quad \text{Ans. (ii)}
\end{aligned}
$$

3. From tables page 12, NH_3,

$$8{\cdot}57 \text{ bar} \qquad h_f = 275{\cdot}1 \qquad h_g = 1462{\cdot}6$$
$$1{\cdot}902 \text{ bar} \qquad h_f = 89{\cdot}8 \qquad h_g = 1420{\cdot}0 \qquad v_g = 0{\cdot}6237$$
$$h_{fg} = 1420 - 89{\cdot}8 = 1330{\cdot}2$$

Specific enthalpy drop through condenser
$$= h_2 - h_3 = 1462 \cdot 6 - 275 \cdot 1 = 1187 \cdot 5 \text{ kJ/kg}$$

Heat rejected in condenser (by 2 kg)
$$= 2 \times 1187 \cdot 5 = 2375 \text{ kJ/min} \quad \text{Ans. (i)}$$

Specific enthalpy gain in evaporator
$$= h_1 - h_4 = h_1 - h_3$$
$$= (89 \cdot 8 + 0 \cdot 96 \times 1330 \cdot 2) - 275 \cdot 1$$
$$= 1366 \cdot 8 - 275 \cdot 1 = 1091 \cdot 7 \text{ kJ/kg}$$

Refrigerating effect
$$= 2 \times 1091 \cdot 7 = 2183 \cdot 4 \text{ kJ/min} \quad \text{Ans. (ii)}$$

Specific volume of refrigerant leaving evaporator and entering compressor $= 0 \cdot 96 \times 0 \cdot 6237 \text{ m}^3/\text{kg}$

Volume taken into compressor per minute
$$= 2 \times 0 \cdot 96 \times 0 \cdot 6237 = 1 \cdot 198 \text{ m}^3/\text{min} \quad \text{Ans. (iii)}$$

4. Since there is a change only in the dryness fraction of the refrigerant through the evaporator, the enthalpy of saturated liquid (h_f) and of evaporation (h_{fg}) being the same for h_1 as for h_4 then:

Specific enthalpy gain of the CO_2 through evaporator
$$= h_1 - h_4$$
$$= (h_f + 0 \cdot 92 h_{fg}) - (h_f + 0 \cdot 28 h_{fg})$$
$$= (0 \cdot 92 - 0 \cdot 28) \times 290 \cdot 7$$
$$= 0 \cdot 64 \times 290 \cdot 7 = 186 \cdot 1 \text{ kJ/kg} \quad \text{Ans. (i)}$$

Heat to be extracted from water to make one kg of ice

$$= 4 \cdot 2 \times 14 + 335 + 2 \cdot 04 \times 5$$
$$= 58 \cdot 8 + 335 + 10 \cdot 2 = 404 \text{ kJ/kg}$$

Let m [kg] $=$ mass of ice made per second
when $0 \cdot 5$ kg/s $=$ mass flow of CO_2

Heat transfer:

$$\text{from water} = \text{to } CO_2$$
$$m \times 404 = 0 \cdot 5 \times 186 \cdot 1$$

$$m = \frac{0.5 \times 186.1}{404} = 0.2303 \text{ kg/s}$$

Mass of ice in tonnes per 24 hours

$$= \frac{0.2303 \times 3600 \times 24}{10^3} = 19.9 \text{ tonne/day} \quad \text{Ans. (ii)}$$

5. Quantity of heat extracted per kg of water
$$= 4.2 \times 18 + 335 + 2.04 \times 7$$
$$= 75.6 + 335 + 14.28 = 424.88 \text{ kJ/kg}$$

Heat energy extracted per second (refrigerating effect)

$$= \frac{2.5 \times 10^3}{24 \times 3600} \times 424.88 \text{ kJ/s}$$

Energy supplied per second, kJ/s = kW = 2.25

$$\text{Coeff. of performance} = \frac{\text{heat energy extracted}}{\text{heat energy supplied}}$$

$$= \frac{2.5 \times 10^3 \times 424.88}{24 \times 3600 \times 2.25}$$

$$= 5.464 \quad \text{Ans. (i)}$$

If ice at 0°C was made from water at 0°C, heat extracted would be equal to enthalpy of fusion only = 335 kJ/kg

$\therefore$ Capacity from and at 0°C

$$= \frac{2.5 \times 424.88}{335} = 3.171 \text{ tonne/day} \quad \text{Ans. (ii)}$$

6. From tables page 13, Freon - 12,
   5.673 bar, sat. temp. = 20°C
   $\therefore$ at 50°C refrigerant is superheated by 30°
   h_2 (compressor discharge) = 216.75
   1.826 bar, sat. temp. = —15°C
   $\therefore$ at 0°C refrigerant is superheated by 15°

h_1 (compressor suction and evaporator exit) = 190·15
h_3 (condenser outlet) = h_f at 5·673 bar = 54·87
h_4 (evaporator inlet) = h_3 = 54·87

$$\text{Coeff. of performance} = \frac{\text{refrigerating effect [kJ/kg]}}{\text{work transfer [kJ/kg]}}$$

$$= \frac{h_1 - h_4}{h_2 - h_1}$$

$$= \frac{190\cdot15 - 54\cdot87}{216\cdot75 - 190\cdot15} = \frac{135\cdot28}{26\cdot6}$$

$$= 5\cdot087 \quad \text{Ans.}$$

7. Tables page 12, NH_3:

$$14\cdot7 \text{ bar, sat. temp.} = 38°C$$

$\therefore$ vapour is superheated $(63 - 38) = 25°$

By interpolation,
14·7 bar 50° supht, h = 1620·1
    ,,   ,, no supht, h = 1472·6
        increase for 50° = 147·5
            ,,   ,, 25° = $\frac{1}{2} \times 147\cdot5 = 73\cdot75$
$\therefore$ 14·7 bar 25° supht, h = 1472·6 + 73·75 = 1546·35 = h_2
14·7 bar 50° supht, s = 5·340
    ,,   ,, no supht, s = 4·898
        increase for 50° = 0·442
            ,,   ,, 25° = $\frac{1}{2} \times 0\cdot442 = 0\cdot221$
$\therefore$ 14·7 bar 25° supht, s = 4·898 + 0·221 = 5·119 = s_2
2·077 bar, h_f = 98·8 h_g = 1422·7
            h_{fg} = 1422·7 — 98·8 = 1323·9
            s_f = 0·404 s_g = 5·593
            s_{fg} = 5·593 — 0·404 = 5·189

Isentropic compression in compressor:

$$s_1 = s_2$$
$$0\cdot404 + x_1 \times 5\cdot189 = 5\cdot119$$
$$x_1 = 0\cdot9086$$
$$h_1 = 98\cdot8 + 0\cdot9086 \times 1323\cdot9 = 1301\cdot8$$
$$h_4 = h_3 = h_f \text{ at } 14\cdot7 \text{ bar} = 362\cdot1$$

$$\text{Refrig. effect [kJ/s]} = (h_1 - h_4) \text{ [kJ/kg]} \times \text{mass flow [kg/s]}$$
$$= (1301 \cdot 8 - 362 \cdot 1) \times 0 \cdot 15$$
$$= 140 \cdot 9 \text{ kJ/s} \quad \text{Ans. (i)}$$

$$\text{Work transfer [kJ/s]} = (h_2 - h_1) \text{ [kJ/kg]} \times \text{mass flow [kg/s]}$$
$$= (1546 \cdot 35 - 1301 \cdot 8) \times 0 \cdot 15$$
$$= 36 \cdot 68 \text{ kJ/s} \quad \text{Ans. (ii)}$$

$$\text{Coeff. of performance} = \frac{\text{refrigerating effect}}{\text{work transfer}}$$

$$= \frac{140 \cdot 9}{36 \cdot 68} = 3 \cdot 842 \quad \text{Ans. (iii)}$$

MISCELLANEOUS PROBLEMS
SECOND CLASS

1. The efficiency of an auxiliary boiler is 70% when dry saturated steam at 8 bar is generated from feed water at 43°C. If the calorific value of the fuel is 42·5 MJ/kg and it is burned at the rate of 6 tonne per day, calculate (i) the hourly steam production, (ii) the equivalent evaporation per kg of fuel from and at 100°C.

2. An electric motor was tested by coupling it to a dynamometer and the brake load was 112 N at 40 rev/s. A steady flow of water passed through the brake and was raised in temperature from 15°C to 48°C. If the power absorbed by this brake is given by $Wn/330$ where W is the brake load in newtons and n is the rotational speed in rev/s, and assuming 98% of the heat generated at the brake is carried away by the cooling water, find the quantity of water passing through the dynamometer in litres per minute.

3. Steam at 15·0 bar is throttled in a reducing valve to 8·0 bar. If the reduced pressure steam is dry saturated, calculate the dryness fraction of the steam before throttling.

4. The bore of an I.C. engine is exactly 300 mm at 20°C. The diameters of the piston at 20°C are 298 mm at the crown and 299 mm at mid-depth of the body. Under working conditions the mean temperatures are: piston crown 250°C, piston body 100°C, cylinder 180°C. Take the coefficients of linear expansion of the piston and cylinder materials as $1·2 \times 10^{-5}$ and $1·1 \times 10^{-5}/°C$ respectively and calculate the diametrical clearances at the crown and mid-depth of the piston under working conditions.

5. An indicator card taken off one cylinder of a six-cylinder, two-stroke, single-acting engine, has an area of 2850 mm² and length 75 mm when running at 1·75 rev/s. One millimetre height on the card represents 0·2 bar. If the cylinder bore is 550 mm and stroke 850 mm, calculate the indicated power of the engine assuming equal powers are developed in all cylinders.

6. A vessel of volume 0·65 m³ contains air at 27·6 bar and 18°C. Calculate the final pressure after 3·5 kg of air is added if the final temperature is 20·5°C. Take R for air $= 0·287$ kJ/kg K.

7. One kilogramme of steam at 7·0 bar, 0·95 dry, is expanded according to the law $pV^{1·3} = $ a constant until the pressure is 3·5 bar. Calculate the final dryness fraction of the steam.

8. A turbine plant consisting of H.P. and L.P. units is supplied with steam at 15 bar 300°C. The steam is expanded in the H.P. and leaves at 2·5 bar 0·97 dry. At this point some steam is bled off to the feed heater, the remaining steam passes to the L.P. where it is expanded to 0·14 bar 0·84 dry. If the same quantity of work transfer takes place in each unit, calculate the amount of steam bled off expressed as a percentage of the steam supplied.

9. The mass analysis of a fuel is 87% carbon, 11% hydrogen, and 2% oxygen. Calculate the volume of air in cubic metres at 1·0 bar and 25°C required for perfect combustion per kg of fuel. Take the values: R for air $= 0·287$ kJ/kg K. Mass analysis of air $= 23%$ oxygen, 77% nitrogen. Atomic weights, hydrogen 1, carbon 12, oxygen 16.

10. Gas at pressure 0·95 bar, volume 0·2 m³ and temperature 17°C, is compressed until the pressure is 2·75 bar and volume 0·085 m³, calculate the compression index and the final temperature.

11. A liquid of density 0·8 g/ml, specific heat 2·5 kJ/kg K, and temperature 27°C, is mixed with another liquid of density 0·82 g/ml, specific heat 1·9 kJ/kg K and temperature 55°C, in the ratio of one of the first liquid to three of the second by volume. Find the resulting temperature.

12. A boiler working at 16 bar produces 9000 kg of steam per hour from feed water at 95°C, the dryness fraction of the steam being 0·98. If the boiler efficiency is 87% and the calorific value of the fuel 42 MJ/kg, calculate the daily fuel consumption.

13. The scavenge ports of a two-stroke diesel engine are just covered when the piston is 800 mm from the top of its stroke. The contents of the cylinder at this instant are at a pressure of 1·21 bar and temperature 40°C. The piston diameter is 700 mm and the clearance is equivalent to 70 mm. Find the mass of air taken in per cycle if the scavenge efficiency is 0·95. R for air $= 0·287$ kJ/kg K.

14. The diameter of an air compressor cylinder is 130 mm, the stroke is 180 mm, and the clearance volume is 73 cm^3. The pressure in the cylinder at the beginning of the stroke is 1·0 bar and the pressure during delivery is constant at 4·6 bar. Taking the law of compression as $pV^{1·2} = $ constant, calculate the distance moved by the piston during the delivery period and express this as a fraction of the stroke.

15. In a single-stage impulse turbine the steam leaves the nozzles at a velocity of 500 m/s at 18° to the plane of rotation of the blades, and the linear velocity of the blades is 230 m/s. Neglecting friction across the blades and assuming the steam leaves the blade wheel in an axial direction, calculate (i) the inlet angle of the blades so that the steam enters without shock, (ii) the outlet blade angle.

16. In an opposed piston single-acting two-stroke engine, the strokes of the pistons vary inversely as the masses of the reciprocating parts. The mass of the top piston with its connecting parts is 25% greater than that of the bottom piston, and the combined stroke is 2340 mm. Find the strokes of the top and bottom pistons. The cylinder diameters are 600 mm, there are 6 cylinders, and when running at 1·75 rev/s the indicated mean effective pressure is 7·24 bar. If the mechanical efficiency is 0·86, calculate the ip and bp.

17. In an NH$_3$ refrigerating plant the ammonia leaves the condenser as a saturated liquid at 10·34 bar. The evaporator pressure is 2·265 bar and the refrigerant leaves the evaporator as a vapour 0·95 dry. If the circulation of the refrigerant through the plant is 4 kg/min, calculate (i) the dryness fraction at inlet to the evaporator, (ii) the refrigerating effect per minute, and (iii) the volume of refrigerant taken into the compressor per minute.

18. A glass tube of uniform bore is closed at one end and open at the other, it contains air imprisoned by a column of mercury 40 mm long. When the tube is held vertically with the closed end at the bottom, the length of the air column is 217·8 mm, and when held vertically with the closed end at the top the length of the air column is 242 mm. Calculate the atmospheric pressure and express it in mm Hg, kN/m^2 and bar.

19. One side of a steel plate is exposed to a stream of hot gases and the other side to an air stream. The plate is 13 mm thick

and the surface temperatures of the plate are at 230°C and 150°C. If the coefficient of thermal conductivity of the steel is 44 W/m K, calculate the heat transferred per square metre of plate area per minute.

20. A volume of 0·8 m³ of steam at 17 bar 0·95 dry is passed through a reducing valve and throttled to 6 bar. Calculate the dryness fraction and the volume after throttling.

21. At the beginning of a voyage a boiler contains 6 tonne of water having 120 p.p.m. dissolved solids. The feed rate is 1250 kg/h and after 24 hours the boiler water contains 1080 p.p.m. dissolved solids. Calculate the average dissolved solids in the feed water.

22. 1·5 m³ of wet steam at 2·8 bar are blown into 36 kg of water at 16°C and the resulting temperature of the mixture is 55°C. Calculate the dryness fraction of the steam.

23. In a turbine plant the consumption of fuel is 0·42 kg per shaft kilowatt hour when burning fuel of calorific value 42·5 MJ/kg. Calculate the overall efficiency of the plant.

24. The air in a ship's saloon is maintained at 19°C and is changed twice every hour from the outside atmosphere which is at 7°C. The saloon is 27 m by 15 m by 3 m high. Calculate the kilowatt loading to heat this air, taking the saloon to be at atmospheric pressure = 1·013 bar, R for air = 0·287 kJ/kg K, specific heat = 1·005 kJ/kg K.

25. A six-cylinder, single-acting, four-stroke oil engine, of 200 mm stroke and 225 mm bore runs at 5 rev/s when the mean effective pressure is 8·5 bar. If the mechanical efficiency is 85% calculate the indicated and brake powers.

26. A steel shaft 255 mm diameter has a brass sleeve shrunk on it. Find the temperature to which both should be raised from an initial temperature of 16°C so that the diameter of the sleeve is one millimetre more than the diameter of the shaft. Neglect strains due to shrinkage and take the coefficients of linear expansion for brass and steel as $2·45 \times 10^{-5}$/°C and $1·2 \times 10^{-5}$/°C respectively.

27. The clearance volume in a four-stroke compression-ignition engine is 128 cm³, the bore is 120 mm, stroke 150 mm, and

length of connecting rod 300 mm. The air induction valve closes when the crank is $26°$ past bottom dead centre and the pressure in the cylinder is then $1·0$ bar. Taking the compression index as $1·35$, calculate the pressure in the cylinder at top centre assuming compression continues to the top of the piston stroke.

28. A boiler working at 15 bar generates 7000 kg of steam per hour. The steam leaves the boiler steam drum dry and saturated and then passes through the superheater tubes at constant pressure. The flue gases enter the nests of superheaters at $822°C$ and leaves at $690°C$. The fuel consumption is 750 kg per hour and 24 kg of air are supplied per kg of fuel burned. Find (i) the temperature of the superheated steam, (ii) the mass of injection water to the de-superheater, at $21°C$, required to desuperheat each kg of steam. Take the specific heat of the flue gases as $1·007$ kJ/kg K.

29. A gas initially at 12 bar, $216°C$ and volume 9900 cm^3 is expanded in a cylinder. The volumetric expansion is 6 and the index of expansion is $1·33$. Calculate the final volume, pressure, and temperature.

30. One kilogramme of water is heated in a sealed vessel to $165°C$ under a pressure of 7 bar. An escape valve in the top is now opened and the temperature falls to $100°C$ causing some of the water to flash-evaporate. Calculate the mass of water flashed off as steam, assuming no transfer of heat to or from the vessel.

31. Steam leaves the nozzles and enters the blade wheel of a single-stage impulse turbine at a velocity of 840 m/s and at an angle of $20°$ to the plane of rotation. The blade velocity is 350 m/s and the exit angle of the blades is $25° \, 12'$. Due to friction, the steam loses 20% of its relative velocity across the blades. Calculate (i) the blade inlet angle, (ii) the magnitude and direction of the absolute velocity of the steam at exit.

32. The mass composition of a fuel oil is $84·6\%$ carbon, $11·4\%$ hydrogen, $0·4\%$ sulphur, $2·4\%$ oxygen, and $1·2\%$ impurities. Calculate the calorific value of the fuel and the theoretical mass of air required for complete combustion of one kg of fuel. Take the values:

	CALORIFIC VALUE MJ/kg	ATOMIC WEIGHT
Hydrogen	144	1
Carbon	33·7	12
Sulphur	9·3	32
Oxygen	—	16

Mass analysis of air = 23% oxygen, 77% nitrogen.

33. In a Freon-12 refrigerating plant, the refrigerant leaves the condenser with a specific enthalpy of 50 kJ/kg. The pressure in the evaporator is 1·826 bar and the refrigerant leaves the evaporator at this pressure and at a temperature of 0°C. Calculate (i) the dryness fraction of the freon at inlet to the evaporator, and (ii) the refrigerating effect per minute if the flow rate of the refrigerant is 0·4 kg/s.

34. A glass vessel is filled with mercury and contains 125 ml at 15°C. Calculate the volume of mercury which will overflow when heated to 75°C. Take the coefficient of cubical expansion of mercury as $1·8 \times 10^{-4}/°C$ and coefficient of linear expansion of glass as $8·5 \times 10^{-6}/°C$.

35. The mean area of indicator cards taken from a cylinder of a double-acting two-stroke engine is 346 mm² and the length is 75 mm. The spring used in the indicator deflects one mm under a force of 30 N and the movement of the stylus is six times that of the indicator piston. The diameter of the indicator piston is 7 mm. Calculate (i) the mean effective pressure. If the diameter of the engine cylinder is 600 mm, stroke 900 mm, and rotational speed 2·1 rev/s, calculate (ii) the indicated power per cylinder.

36. At the entrance of a nozzle of circular cross-section, the velocity of the steam is 457 m/s and the specific volume is 0·2765 m³/kg. At exit, the velocity is 1524 m/s and specific volume 7·404 m³/kg. If the mass flow of steam through the nozzle is 0·315 kg/s, calculate the entrance and exit diameters in millimetres.

37. A quantity of air of volume 0·2 m³ at 1·1 bar and 15°C is heated at constant pressure until its temperature is 150°C, and then compressed to 7·15 bar according to the law $pV^{1·32} =$ a constant. Calculate (i) the amount of heat energy transferred to the air at constant pressure, and (ii) the temperature at the end of compression. For air, $R = 0·287$ kJ/kg K, $c_P = 1·005$ kJ/kg K.

38. In a four-stroke diesel engine the maximum lift of the exhaust valve is 50 mm. It opens at 32° before bottom dead centre and closes at 18° after top dead centre. The valve is full open from 18° before bottom dead centre to 6° before top dead centre. If the engine runs at 2 rev/s, calculate (i) the time the valve is open, (ii) the time it is full open, (iii) the mean velocity of opening in m/s, (iv) the mean velocity of closing in m/s.

39. Calculate (i) the mass of 113 litre of nitrogen at 5·86 bar and 61°C, and (ii) the volume of this mass of nitrogen at 1·38 bar and 15°C. R for nitrogen $= 297 \, \text{J/kg K}$.

40. The mass analysis of a hydrocarbon fuel A is 88·5% carbon and 11·5% hydrogen. Another hydrocarbon fuel B requires 6% more air than fuel A for complete combustion. Calculate the mass analysis of fuel B taking the following values: Atomic weights, carbon 12, hydrogen 1, oxygen 16, mass content of oxygen in air $= 23\%$.

41. The stroke of an internal combustion engine is 90 mm and the clearance volume is equivalent to a linear clearance of 15 mm. If the clearance is reduced by 2·5 mm, find the pressure at the end of compression before and after the alteration, taking the initial pressure in each case as 1·0 bar and the index of compression as 1·33.

42. The kinetic energy of the steam jet leaving the nozzles of a single-stage impulse turbine is equivalent to 250 kJ/kg. The entrance and exit angles of the blades are equal at 35°, and the steam leaves the blade wheel in an axial direction. Neglecting friction across the blades and assuming shockless flow, calculate (i) the nozzle angle, and (ii) the blade velocity.

43. A four-cylinder, single-acting, two-stroke engine develops 300 kW indicated power when the mean effective pressure is 6·28 bar and the speed is 4·5 rev/s.

(a) If the stroke is 25% greater than the cylinder diameter, calculate the diameter of the cylinders and the stroke to the nearest millimetre.

(b) When burning fuel of calorific value 42 MJ/kg the fuel consumption is 0·225 kg/kWh (indicated), find the indicated thermal efficiency.

44. The refrigerating effect of a plant using ammonia as the refrigerant is 800 kJ/min. At the exit points of the components the conditions of the refrigerant are:

Evaporator, dry saturated vapour at 1·902 bar

Compressor, vapour at 7·529 bar and 66°C

Condenser, saturated liquid at 7·529 bar

Calculate (i) the mass flow of the refrigerant through the plant, in kg/min, (ii) the heat rejected in the condenser, in kJ/min, and (iii) the output power of the compressor in kW.

45. Steam enters a desuperheater at 30 bar and 400°C and leaves at the same pressure as dry saturated steam. The temperature of the water injected into the desuperheater is 38°C. Calculate (i) the mass of injection water used per kg of steam desuperheated, and (ii) the percentage change in volume from that occupied by one kg of superheated steam to the volume occupied by the dry saturated steam resulting from the mixture of one kg of superheated steam and its injected water.

46. A gas is compressed in a cylinder from 1 bar and 35°C at the beginning of the stroke to 37 bar at the end of the stroke. If the clearance volume is 850 cm³ and the compression index 1·32, calculate the stroke volume and the temperature at the end of compression.

47. In a six-cylinder two-stroke diesel engine, the fuel and starting-air valves are cam-operated The fuel is admitted from 5° before top dead centre (T.D.C.) to 38° after T.D.C. when running ahead, and from 5° before T.D.C. to 34° after T.D.C. when running astern. Starting air is admitted 8° after T.D.C. for both ahead and astern running. Allowing 10° starting air overlap, find (i) the angle between the centre-lines of the ahead and astern fuel cams, (ii) the angle between the centre-lines of the ahead and astern starting air cams.

48. Gas at 500°C flows on one side of a steel plate of 19 mm thickness and water at 195°C under a pressure of 14 bar flows on the other side. Assuming the surface temperatures of the plate are equal to the fluid temperatures, calculate the mass of water completely evaporated per hour per square metre of plate area. Take λ for steel = 44·5 W/m K.

49. A cylindrical air vessel with flat ends, 6 m long by 2 m diameter, contains air at 30 bar and 24°C. Calculate (i) the mass of

air in the vessel, (ii) the volume of atmospheric air at 1·013 bar and at the same temperature to be pumped into the vessel to increase the pressure to 37·5 bar. Take R for air $= 0·287$ kJ/kg K.

50. During a trial run on a single-cylinder, four-stroke, gas engine, of stroke 380 mm and cylinder diameter 200 mm, the following data were obtained: mean effective pressure 5·5 bar, speed 5·3 rev/s, explosions 126/min, load on brake 362 N, effective radius of brake 800 mm, gas consumption 6·8 m³/h, calorific value of gas 17·5 MJ/m³. Calculate (i) indicated power, (ii) brake power, (iii) mechanical efficiency, (iv) indicated thermal efficiency, (v) overall efficiency.

SOLUTIONS TO SECOND CLASS
MISCELLANEOUS PROBLEMS

1. From steam tables

Steam 8 bar, $h_g = 2769$ kJ/kg

Water 43°C, by interpolation,

$$h \text{ at } 44°C = 184 \cdot 2$$
$$h \text{ at } 42°C = 175 \cdot 8$$
$$\text{increase for } 2° = 8 \cdot 4$$
$$\text{,, \quad ,, } 1° = 4 \cdot 2$$
$$\therefore h \text{ at } 43°C = 175 \cdot 8 + 4 \cdot 2 = 180 \text{ kJ/kg}$$

$$\text{Fuel consumption [kg/}h] = \frac{6 \times 10^3}{24} = 250 \text{ kg/h}$$

$$\text{Boiler efficiency} = \frac{\text{heat energy transferred to steam [kJ/h]}}{\text{heat energy supplied by fuel [kJ/h]}}$$

$$0 \cdot 7 = \frac{\text{mass of steam [kg/h]} \times (2769 - 180)}{250 \times 42 \cdot 5 \times 10^3}$$

$$\text{Mass of steam/h} = \frac{0 \cdot 7 \times 250 \times 42 \cdot 5 \times 10^3}{2589}$$

$$= 2873 \text{ kg/h} \quad \text{Ans. (i)}$$

From tables, h_{fg} at 100°C = 2256·7 kJ/kg

Equivalent evaporation from and at 100°C, per kg of fuel,

$$= \frac{2873 \times (2769 - 180)}{250 \times 2256 \cdot 7}$$

$$= 13 \cdot 18 \text{ kg steam/kg fuel} \quad \text{Ans. (ii)}$$

2. Brake power $= \dfrac{Wn}{330} = \dfrac{112 \times 40}{330} = 13 \cdot 58$ kW

Energy at brake $= 13 \cdot 58$ kJ/s

Energy carried away by water

$$= 0 \cdot 98 \times 13 \cdot 58 = 13 \cdot 3 \text{ kJ/s}$$

From steam tables,
$$\text{Water } 48°C, \quad h = 200\cdot9$$
$$\text{,, } \quad 15°C, \quad h = 62\cdot9$$

Enthalpy gain of cooling water
$$= 200\cdot9 - 62\cdot9 = 138 \text{ kJ/kg}$$

Heat energy transferred to water [kJ/s]
$$= \text{mass flow [kg/s]} \times \text{spec. enthalpy gain [kJ/kg]}$$

$$\therefore \text{ mass flow} = \frac{13\cdot3}{138} = 0\cdot096\,38 \text{ kg/s}$$

$$0\cdot096\,38 \times 60 = 5\cdot783 \text{ kg/min}$$
$$= 5\cdot783 \text{ litre/min} \quad \text{Ans.}$$

3. From steam tables,
$$\text{Steam } 15 \text{ bar}, \quad h_f = 845 \quad h_{fg} = 1947$$
$$\text{,, } \quad 8 \text{ bar}, \quad h_g = 2769$$

$$\text{Enthalpy before throttling} = \text{Enthalpy after}$$
$$845 + x \times 1947 = 2769$$
$$x \times 1947 = 1924$$
$$x = 0\cdot9881 \quad \text{Ans.}$$

4. Linear (diametrical) expansion $= \alpha \times d \times (\theta_2 - \theta_1)$
Increase in cylinder diameter
$$= 1\cdot1 \times 10^{-5} \times 300 \times (180 - 20) = 0\cdot5279 \text{ mm}$$
$$\text{Working diameter} = 300 + 0\cdot5279 = 300\cdot5279 \text{ mm}$$

Increase in piston crown diameter
$$= 1\cdot2 \times 10^{-5} \times 298 \times (250 - 20) = 0\cdot8224 \text{ mm}$$
$$\text{Working diameter} = 298 + 0\cdot8224 = 298\cdot8224 \text{ mm}$$

Increase in piston body diameter
$$= 1\cdot2 \times 10^{-5} \times 299 \times (100 - 20) = 0\cdot2871 \text{ mm}$$
$$\text{Working diameter} = 299 + 0\cdot2871 = 299\cdot2871 \text{ mm}$$

Diametrical clearances at working temperatures:
Piston crown $= 300\cdot5279 - 298\cdot8224 = 1\cdot7055 \text{ mm}$
Piston body $= 300\cdot5279 - 299\cdot2871 = 1\cdot2408 \text{ mm}$ } Ans.

5. Mean height of card $= \dfrac{2850}{75} = 38$ mm

$$\text{Mean effective pressure} = 38 \times 0.2 \text{ bar}$$
$$38 \times 0.2 \times 10^2 = 760 \text{ kN/m}^2$$

Indicated power of 6 cylinders
$$= p_m A L n \times 6$$
$$= 760 \times 0.7854 \times 0.55^2 \times 0.85 \times 1.75 \times 6$$
$$= 1612 \quad \text{Ans.}$$

6. $$pV = mRT$$

Initial mass of air,

$$m = \frac{pV}{RT} = \frac{27.6 \times 10^2 \times 0.65}{0.287 \times (18 + 273)} = 21.48 \text{ kg}$$

Final mass $= 21.48 + 3.5 = 24.98$ kg

$$\text{Final pressure, } p = \frac{mRT}{V}$$

$$= \frac{24.98 \times 0.287 \times (20.5 + 273)}{0.65}$$

$$= 3238 \text{ kN/m}^2 = 32.38 \text{ bar} \quad \text{Ans.}$$

7. From steam tables
$$7 \text{ bar, } v_g = 0.2728 \qquad 3.5 \text{ bar, } v_g = 0.5241$$

Volume of one kg of steam at 7 bar 0.95 dry
$$= 0.95 \times 0.2728 = 0.2592 \text{ m}^3$$
$$p_1 V_1^{1.3} = p_2 V_2^{1.3}$$
$$7 \times 0.2592^{1.3} = 3.5 \times V_2^{1.3}$$

$$V_2 = 0.2592 \times \sqrt[1.3]{\frac{7}{3.5}}$$

$$= 0.4417 \text{ m}^3$$

Specific volume of dry sat. steam at 3·5 bar is 0·5241 m³/kg therefore dryness fraction of expanded steam

$$= \frac{0·4417}{0·5241} = 0·8428 \quad \text{Ans.}$$

8. From steam tables,

$$15 \text{ bar } 300°C, \quad h = 3039$$
$$2·5 \text{ bar}, \quad h_f = 535 \qquad h_{fg} = 2182$$
$$0·14 \text{ bar}, \quad h_f = 220 \qquad h_{fg} = 2376$$

Specific enthalpy at 2·5 bar 0·97 dry
$$= 535 + 0·97 \times 2182 = 2651 \text{ kJ/kg}$$

Specific enthalpy at 0·14 bar 0·84 dry
$$= 220 + 0·84 \times 2376 = 2215 \text{ kJ/kg}$$

Specific enthalpy drop through H.P.
$$= 3039 - 2651 \qquad = 388 \text{ kJ/kg}$$

Specific enthalpy drop through L.P.
$$= 2651 - 2215 \qquad = 436 \text{ kJ/kg}$$

Total enthalpy drop through each unit is to be the same, let 1 kg of steam be supplied and x kg bled off, then 1 kg passes through H.P. and $(1 - x)$ kg passes through L.P.

$$1 \times 388 = (1 - x) \times 436$$
$$436x = 436 - 388$$

$$x = \frac{48}{436} = 0·1101$$

Expressed as a percentage
$$\text{Amount bled off} = 11·01\% \quad \text{Ans.}$$

9. Mass of air required per kg of fuel

$$= \frac{100}{23} \times \text{oxygen required}$$

$$= \frac{100}{23} \left\{ 2\tfrac{2}{3}C + 8 \left(H_2 - \frac{O_2}{8}\right) \right\}$$

$$= \frac{100}{23} \{2\tfrac{2}{3}C + 8H_2 - O_2\}$$

$$= \frac{100}{23} \{2\tfrac{2}{3} \times 0.87 + 8 \times 0.11 - 0.02\}$$

$$= \frac{100}{23} \times 3.18 = 13.82 \text{ kg air/kg fuel}$$

$$pV = mRT$$

$$V = \frac{13.82 \times 0.287 \times (25 + 273)}{1 \times 10^2}$$

$$= 11.82 \text{ m}^3 \text{ air/kg fuel} \quad \text{Ans.}$$

10.
$$p_1 V_1{}^n = p_2 V_2{}^n$$
$$0.95 \times 0.2^n = 2.75 \times 0.085^n$$

$$n \times \log \left\{ \frac{0.2}{0.085} \right\} = \log \left\{ \frac{2.75}{0.95} \right\}$$

$$n \times 0.3716 = 0.4616$$

$$n = \frac{0.4616}{0.3716} = 1.243 \quad \text{Ans. (i)}$$

$$\frac{p_1 V_1}{T_1} = \frac{p_2 V_2}{T_2}$$

$$\frac{0.95 \times 0.2}{(17 + 273)} = \frac{2.75 \times 0.085}{T_2}$$

$$T_2 = \frac{290 \times 2.75 \times 0.085}{0.95 \times 0.2} = 356.8 \text{ K}$$

Final temp. $= 356.8 - 273 = 83.8°C \quad$ Ans. (ii)

11. Ratio of volumes is 1 to 3 therefore,

$$\text{let volume of first liquid} = 1 \text{ ml}$$
$$\text{let \quad ,, \quad ,, second \quad ,, \quad} = 3 \text{ ml}$$

$$\text{Mass of first liquid} = 1 \times 0{\cdot}8 = 0{\cdot}8 \text{ g}$$
$$\text{,, \quad ,, second ,,} = 3 \times 0{\cdot}82 = 2{\cdot}46 \text{ g}$$

Heat transferred = mass $\times$ spec. heat $\times$ temp. change

Let θ = final temperature:

Heat gained by first liquid = Heat lost by second
$$0{\cdot}8 \times 2{\cdot}5 \times (\theta - 27) = 2{\cdot}46 \times 1{\cdot}9 \times (55 - \theta)$$
$$2\theta - 54 = 257{\cdot}1 - 4{\cdot}674\theta$$
$$6{\cdot}674\theta = 311{\cdot}1$$
$$\theta = 46{\cdot}61°C \quad \text{Ans.}$$

12. From steam tables,

$$\text{Steam 16 bar,} \quad h_f = 859 \quad h_{fg} = 1935$$
$$\text{Water 95°C,} \quad h = 398$$

$$\text{Boiler steam } h_1 = 859 + 0{\cdot}98 \times 1935 = 2755$$
$$\text{Feed water } h_w = 398$$

Heat energy transferred to water to make steam
$$= h_1 - h_w = 2755 - 398 = 2357 \text{ kJ/kg}$$
$$= 2357 \times 9000 \text{ kJ per hour}$$

Heat energy released by m kg of fuel per hour
$$= 42 \times 10^3 \times m \text{ kJ per hour}$$

$$\text{Boiler efficiency} = \frac{\text{heat energy transferred to steam}}{\text{heat energy supplied by fuel}}$$

$$0{\cdot}87 = \frac{2357 \times 9000}{42 \times 10^3 \times m}$$

$$m = \frac{2357 \times 9000}{0{\cdot}87 \times 42 \times 10^3} = 580{\cdot}6 \text{ kg/h}$$

Tonnes of fuel used per day
$$= 580{\cdot}6 \times 24 \times 10^{-3} = 13{\cdot}93 \text{ tonne/day} \quad \text{Ans.}$$

13. When scavenge ports are just closed, distance from cylinder cover to piston $= 800 + 70 = 870$ mm

Volume enclosed $= 0{\cdot}7854 \times 0{\cdot}7^2 \times 0{\cdot}87$ m³

Volume of air taken in (scavenge effic. being 0·95)

$$= 0{\cdot}95 \times 0{\cdot}7854 \times 0{\cdot}7^2 \times 0{\cdot}87$$
$$= 0{\cdot}3181 \text{ m}^3$$

$$pV = mRT$$

$$m = \frac{pV}{RT} = \frac{1{\cdot}21 \times 10^2 \times 0{\cdot}3181}{0{\cdot}287 \times (40 + 273)}$$

$$= 0{\cdot}4285 \text{ kg} \quad \text{Ans.}$$

14. Linear clearance [mm] $= \dfrac{\text{volumetric clearance [mm}^3\text{]}}{\text{area of cylinder [mm}^2\text{]}}$

$$= \frac{73 \times 10^3}{0{\cdot}7854 \times 130^2} = 5{\cdot}5 \text{ mm}$$

Representing volumes by linear dimensions:
$$V_1 = 180 + 5{\cdot}5 = 185{\cdot}5 \text{ mm}$$
$$p_1 V_1^{1{\cdot}2} = p_2 V_2^{1{\cdot}2}$$
$$1 \times 185{\cdot}5^{1{\cdot}2} = 4{\cdot}6 \times V_2^{1{\cdot}2}$$

$$V_2 = \frac{185{\cdot}5}{\sqrt[1{\cdot}2]{4{\cdot}6}} = 52 \text{ mm}$$

Movement of piston during delivery period
$$= 52 - 5{\cdot}5 = 46{\cdot}5 \text{ mm} \quad \text{Ans. (i)}$$

as a fraction of the stroke
$$= \frac{46{\cdot}5}{180} = 0{\cdot}2583 \quad \text{Ans. (ii)}$$

15.

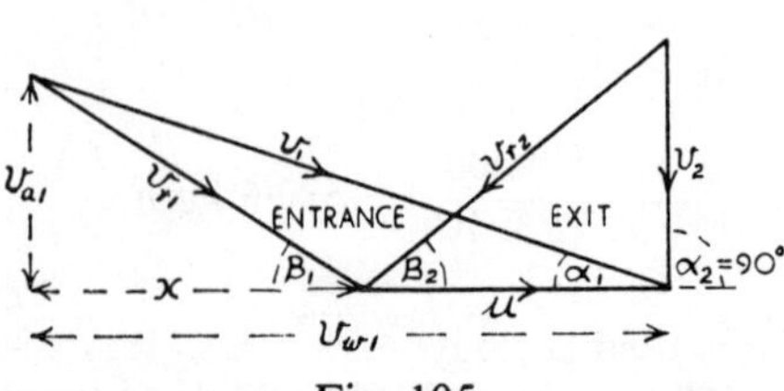

Fig. 105

$$v_{a1} = v_1 \sin \alpha_1 = 500 \times \sin 18^\circ = 154 \cdot 5 \text{ m/s}$$
$$v_{w1} = v_1 \cos \alpha_1 = 500 \times \cos 18^\circ = 475 \cdot 5 \text{ m/s}$$
$$x = v_{w1} - u = 475 \cdot 5 - 230 = 245 \cdot 5$$

$$\tan \beta_1 = \frac{v_{a1}}{x} = \frac{154 \cdot 5}{245 \cdot 5} = 0 \cdot 6292$$

$$\therefore \text{ Inlet blade angle} = 32^\circ\ 11' \quad \text{Ans. (i)}$$

Neglecting friction across the blades,

$$v_{r2} = v_{r1} = \frac{v_{a1}}{\sin \beta_1} = \frac{154 \cdot 5}{\sin 32^\circ\ 11'} = 290 \text{ m/s}$$

$$\cos \beta_2 = \frac{u}{v_{r2}} = \frac{230}{290} = 0 \cdot 7930$$

$$\therefore \text{ Outlet blade angle} = 37^\circ\ 32' \quad \text{Ans. (ii)}$$

16. Ratio of strokes, top : bottom $= 1 : 1 \cdot 25$
Combined stroke is represented by $1 + 1 \cdot 25 = 2 \cdot 25$

$$\text{Stroke of top piston} = \frac{1}{2 \cdot 25} \times 2340 = 1040 \text{ mm} \quad \text{Ans. (i)}$$

$$\text{Stroke of bottom piston} = \frac{1 \cdot 25}{2 \cdot 25} \times 2340 = 1300 \text{ mm} \quad \text{Ans. (ii)}$$

For 6 cylinders:
$$\begin{aligned}
\text{ip} &= p_m A L n \times 6 \\
&= 7 \cdot 24 \times 10^2 \times 0 \cdot 7854 \times 0 \cdot 6^2 \times 2 \cdot 34 \times 1 \cdot 75 \times 6 \\
&= 5030 \text{ kW} \quad \text{Ans. (ii)} \\
\text{bp} &= 5030 \times 0 \cdot 86 = 4326 \text{ kW} \quad \text{Ans. (iii)}
\end{aligned}$$

17. From tables for NH_3

$10 \cdot 34$ bar, $\quad h_f = 303 \cdot 7$
$2 \cdot 265$ bar, $\quad h_f = 107 \cdot 9 \quad\quad h_g = 1425 \cdot 3 \quad\quad v_g = 0 \cdot 5296$
$$h_{fg} = 1425 \cdot 3 - 107 \cdot 9 = 1317 \cdot 4$$

Ref. Fig. 101:

Throttling process through expansion valve:

$$\text{Enthalpy after } (h_4) = \text{Enthalpy before } (h_3)$$
$$107 \cdot 9 + x_4 \times 1317 \cdot 4 = 303 \cdot 7$$
$$x_4 \times 1317 \cdot 4 = 195 \cdot 8$$
$$x_4 = 0 \cdot 1486 \quad \text{Ans. (i)}$$

Enthalpy gain of refrigerant through evaporator
$$= h_1 - h_4$$
$$= (107 \cdot 9 + 0 \cdot 95 \times 1317 \cdot 4) - (107 \cdot 9 + 0 \cdot 1486 \times 1317 \cdot 4)$$
$$= (0 \cdot 95 - 0 \cdot 1486) \times 1317 \cdot 4$$
$$= 1055 \text{ kJ/kg}$$

Refrigerating effect per minute
$$= 4 \times 1055 = 4220 \text{ kJ/min} \quad \text{Ans. (ii)}$$

Specific volume of refrigerant entering compressor
$$= 0 \cdot 95 \times 0 \cdot 5296 \text{ m}^3/\text{kg}$$

Total volume entering compressor per minute
$$= 4 \times 0 \cdot 95 \times 0 \cdot 5296 = 2 \cdot 013 \text{ m}^3/\text{min} \quad \text{Ans. (iii)}$$

18. Let a = cross sect. area of tube in mm^2
H = atmospheric pressure in mm of mercury

When tube is vertical with closed end at the bottom:

Pressure of trapped air $= (H + 40) \text{ mm Hg}$
Volume   ,,   ,,   ,, $= 217 \cdot 8 \times a \text{ mm}^3$

When tube is vertical with closed end at the top:

Pressure of trapped air $= (H - 40) \text{ mm Hg}$
Volume   ,,   ,,   ,, $= 242 \times a \text{ mm}^3$

Temperature is constant, therefore
$$p_1 V_1 = p_2 V_2$$
$$(H + 40) \times 217 \cdot 8 \times a = (H - 40) \times 242 \times a$$
$$217 \cdot 8H + 8712 = 242H - 9680$$
$$24 \cdot 2H = 18\ 392$$
$$H = 760 \text{ mm}$$

Atmospheric pressure $= 760 \text{ mm Hg} \quad \text{Ans. (i)}$

One mm head of mercury $= 133\cdot3$ N/m² (see Chap. 1)

$\therefore$ Atmos. press. $= 760 \times 133\cdot3 = 1\cdot013 \times 10^5$ N/m²

$$= 101\cdot3 \text{ kN/m}^2 \quad \text{Ans. (ii)}$$

$$= 1\cdot013 \text{ bar} \quad \text{Ans. (iii)}$$

19.

$$Q[\text{J}] = \frac{\lambda\,[\text{W/m K}] \times A\,[\text{m}^2] \times t\,[\text{s}] \times (T_1 - T_2)\,[\text{K}]}{S\,[\text{m}]}$$

$$= \frac{44 \times 1 \times 60 \times (230 - 150)}{0\cdot013}$$

$$= 1\cdot625 \times 10^7 \text{ joules/m}^2 \text{ min}$$
$$= 16\cdot25 \text{ MJ/m}^2 \text{ min} \quad \text{Ans.}$$

20. From steam tables,

17 bar,	$h_f = 872$	$h_{fg} = 1923$	$v_g = 0\cdot1167$
6 bar,	$h_f = 670$	$h_{fg} = 2087$	$v_g = 0\cdot3156$

Throttling process:

$$\text{Enthalpy after} = \text{Enthalpy before}$$
$$670 + x \times 2087 = 872 + 0\cdot95 \times 1923$$
$$x \times 2087 = 872 + 1827 - 670$$
$$x \times 2087 = 2029$$
$$x = 0\cdot9721 \quad \text{Ans. (i)}$$

Spec. vol. of steam before throttling
$$= 0\cdot95 \times 0\cdot1167 \text{ m}^3/\text{kg}$$

Mass of steam passed through
$$= \frac{0\cdot8}{0\cdot95 \times 0\cdot1167} \text{ kg}$$

Spec. vol. of steam after throttling
$$= 0\cdot9721 \times 0\cdot3156 \text{ m}^3/\text{kg}$$

Final volume of steam $= \dfrac{0\cdot8 \times 0\cdot9721 \times 0\cdot3156}{0\cdot95 \times 0\cdot1167}$

$$= 2\cdot214 \text{ m}^3 \quad \text{Ans. (ii)}$$

21. Total feed in 24 hours

$$= 1250 \times 10^{-3} \times 24 = 30 \text{ tonne}$$

$$\begin{matrix} \text{Solids in boiler} \\ \text{initially} \end{matrix} + \begin{matrix} \text{solids} \\ \text{put in} \end{matrix} = \begin{matrix} \text{solids in} \\ \text{finally} \end{matrix}$$

$$\begin{matrix} \text{water in} \\ \text{boiler} \end{matrix} \times \begin{matrix} \text{initial} \\ \text{p.p.m.} \end{matrix} + \begin{matrix} \text{amount} \\ \text{of feed} \end{matrix} \times \begin{matrix} \text{feed} \\ \text{p.p.m.} \end{matrix} = \begin{matrix} \text{water in} \\ \text{boiler} \end{matrix} \times \begin{matrix} \text{final} \\ \text{p.p.m.} \end{matrix}$$

$$6 \times 120 + 30 \times \text{feed p.p.m.} = 6 \times 1080$$
$$30 \times \text{feed p.p.m.} = 6480 - 720$$

$$\text{feed p.p.m.} = \frac{5760}{30}$$

$$= 192 \text{ p.p.m.} \quad \text{Ans.}$$

22. From steam tables

Steam 2·8 bar, $\quad h_f = 551 \qquad h_{fg} = 2171 \qquad v_g = 0·6462$
Water 16°C, $\qquad h = 67·1$
Water 55°C, $\qquad h = 230·2$

Let x = dryness fraction
$\quad m$ = mass of steam [kg]

$$\begin{matrix} \text{Total enthalpy of steam} \\ \text{and water before mixing} \end{matrix} = \begin{matrix} \text{Total enthalpy of} \\ \text{water after mixing} \end{matrix}$$

$$m(551 + x \times 2171) + (36 \times 67·1) = (m + 36) \times 230·2$$
$$551m + 2171mx + 2415 = 230·2m + 8287$$
$$320·8m + 2171mx = 5872$$

$$320·8 + 2171x = \frac{5872}{m} \quad \dots \qquad \dots \quad \text{(i)}$$

Also, spec. vol. of steam $= x \times 0·6462 \text{ m}^3/\text{kg}$
$\therefore$ mass of 1·5 m³ of wet steam

$$m = \frac{1·5}{x \times 0·6462} \text{ kg} \dots \text{ (ii)}$$

Substituting for m from (ii) into (i):

$$320{\cdot}8 + 2171x = \frac{5872 \times 0{\cdot}6462x}{1{\cdot}5}$$

$$320{\cdot}8 + 2171x = 2528x$$
$$320{\cdot}8 = 357x$$
$$x = 0{\cdot}8984 \quad \text{Ans.}$$

23. Heat energy supplied by fuel in one hour to produce one kW of power at the shaft
$$= 42{\cdot}5 \times 10^3 \times 0{\cdot}42 \text{ kJ}$$

Energy produced at shaft
$$= 1\,\text{kWh} = 1\,[\text{kW}] \times 3600\,[\text{s}] = 3600 \text{ kJ}$$

$$\text{Overall efficiency} = \frac{\text{energy at shaft}}{\text{energy supplied}}$$

$$= \frac{3600}{42{\cdot}5 \times 10^3 \times 0{\cdot}42}$$

$$= 0{\cdot}2017 \text{ or } 20{\cdot}17\% \quad \text{Ans.}$$

24. Volume of air to be heated every hour
$$= 27 \times 15 \times 3 \times 2 = 2430 \text{ m}^3$$

Mass of air to be heated every hour:
$$pV = mRT$$

$$m = \frac{pV}{RT}$$

$$= \frac{1{\cdot}013 \times 10^2 \times 2430}{0{\cdot}287 \times (19 + 273)} = 2937 \text{ kg/h}$$

Heat energy supplied $=$ mass $\times$ spec. heat $\times$ temp. rise

Energy supplied per second

$$= \frac{2937}{3600} \times 1 \cdot 005 \times (19 - 7)$$

$$= 9 \cdot 837 \text{ kJ/s}$$
Kilowatt loading $= 9 \cdot 837 \text{ kW}$ Ans.

25. For a 4-stroke single-acting engine
$$n = \text{rev/s} \div 2 = 2 \cdot 5$$

For six cylinders:
$$\begin{aligned}
\text{ip} &= p_m ALn \times 6 \\
&= 8 \cdot 5 \times 10^2 \times 0 \cdot 7854 \times 0 \cdot 225^2 \times 0 \cdot 2 \times 2 \cdot 5 \times 6 \\
&= 101 \cdot 4 \text{ kW}   \text{ Ans. (i)}
\end{aligned}$$

$$\begin{aligned}
\text{bp} &= \text{ip} \times \text{mech. efficiency} \\
&= 101 \cdot 4 \times 0 \cdot 85 \\
&= 86 \cdot 18 \text{ kW}   \text{ Ans. (ii)}
\end{aligned}$$

26. Increase in diameter $= \alpha \times d \times (\theta_2 - \theta_1)$
Increase in sleeve diameter
$$= 2 \cdot 45 \times 10^{-5} \times 255 \times (\theta_2 - \theta_1)$$
Increase in shaft diameter
$$= 1 \cdot 2 \times 10^{-5} \times 255 \times (\theta_2 - \theta_1)$$
Difference in increased diameters
$$1 = (2 \cdot 45 - 1 \cdot 2) \times 10^{-5} \times 255 \times (\theta_2 - \theta_1)$$

$$\theta_2 - \theta_1 = \frac{1}{1 \cdot 25 \times 10^{-5} \times 255} = 313 \cdot 7$$

Final temp. $= 16 + 313 \cdot 7 = 329 \cdot 7°\text{C}$ Ans.

27.

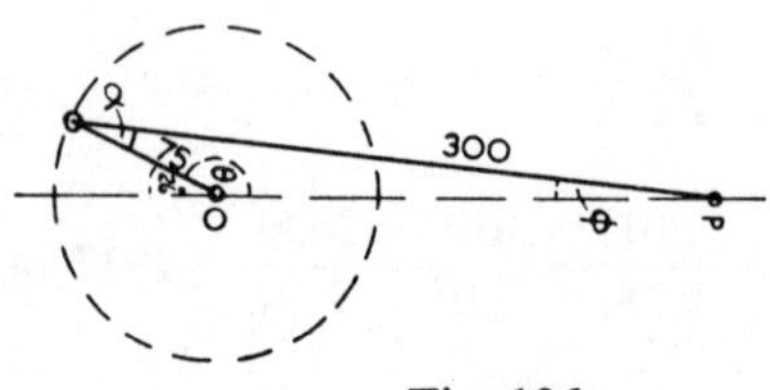

Fig. 106

$$\theta = 180° - 26° = 154°$$
$$\text{crank length} = \tfrac{1}{2}\ \text{stroke} = 75\ \text{mm}$$

$$\text{Sine rule:}\ \frac{300}{\sin 154°} = \frac{75}{\sin \varphi}$$

$$\sin \varphi = \frac{75 \times 0.4384}{300} = 0.1096$$

$$\varphi = 6°\ 18'$$
$$\alpha = 180° - 6°\ 18' - 154° = 19°\ 42'$$

$$\frac{OP}{\sin 19°42'} = \frac{75}{\sin 6°\ 18'}$$

$$OP = \frac{75 \times 0.3371}{0.1096} = 230.7\ \text{mm}$$

Distance from shaft centre to crosshead when crank is on bottom centre = con. rod length — crank length
$$= 300 - 75 = 225\ \text{mm}$$

Distance from shaft centre to crosshead when crank is 26° past bottom centre = OP = 230.7 mm

Therefore crosshead (and piston) has moved up from bottom of stroke a distance of
$$230.7 - 225 = 5.7\ \text{mm}$$

Part of stroke remaining (compression period)
$$= 150 - 5.7 = 144.3\ \text{mm}$$

Piston swept volume during compression
$$= 144.3 \times 0.7854 \times 120^2$$
$$= 1.632 \times 10^6\ \text{mm}^3 = 1632\ \text{cm}^3$$

Volume of air in cylinder at beginning of compression
$$V_1 = 1632 + 128 = 1760\ \text{cm}^3$$

Volume of air in cylinder at end of compression
$$V_2 = \text{clearance space} = 128\ \text{cm}^3$$
$$p_1 V_1^{1.35} = p_1 V_2^{1.35}$$
$$1 \times 1760^{1.35} = p_2 \times 128^{1.35}$$

$$p_2 = \left\{ \frac{1760}{128} \right\}^{1.35} = 34.42\ \text{bar}\quad \text{Ans.}$$

28. From steam tables, 15 bar, $h_g = 2792$

Mass of flue gases per kg of fuel
$$= 1 \text{ kg fuel} + 24 \text{ kg air} = 25 \text{ kg}$$

Heat energy transferred from flue gases per hour
$$= \text{mass} \times \text{spec. heat} \times \text{temp. change}$$
$$= 25 \times 750 \times 1 \cdot 007 \times (822 - 690)$$
$$= 2 \cdot 492 \times 10^6 \text{ kJ/h}$$

Let $h =$ spec. enthalpy of the superheated steam,

Heat energy transferred to steam per hour to superheat it
$$= 7000 \times (h - 2792)$$

Heat gained by steam = Heat lost by gases
$$7000 \, (h - 2792) = 2 \cdot 492 \times 10^6$$

$$h - 2792 = \frac{2 \cdot 492 \times 10^6}{7000}$$

$$h = 356 + 2792 = 3148 \text{ kJ/kg}$$

From superheated steam tables,
$$h \text{ for 15 bar 350°C reads 3148}$$
$$\therefore \text{ Temperature of steam} = 350°C \quad \text{Ans. (i)}$$

From steam tables, water 21°C, $h = 88 \cdot 0$

Enthalpy of 1 kg. sup. steam and m kg water entering desuperheater = Enthalpy of $(1 + m)$ kg of dry sat. steam leaving desuperheater

$$(1 \times 3148) + m \times 88 = (1 + m) \times 2792$$
$$3148 + 88 \, m = 2792 + 2792 \, m$$
$$356 = 2704 \, m$$
$$m = 0 \cdot 1316 \text{ kg} \quad \text{Ans. (ii)}$$

29. Ratio of expansion $= \dfrac{\text{final volume}}{\text{initial volume}}$

$$\therefore \text{ final volume} = 6 \times 9900 = 59\,400 \text{ cm}^3$$
$$\text{or, } 59 \cdot 4 \text{ litre, or, } 0 \cdot 0594 \text{ m}^3 \quad \text{Ans. (i)}$$

$$p_1 V_1^{1 \cdot 33} = p_2 V_2^{1 \cdot 33}$$

where V_1 and V_2 may be represented by 1 and 6 respectively.

$$12 \times 1^{1 \cdot 33} = p_2 \times 6^{1 \cdot 33}$$

$$p_2 = \frac{12}{6^{1 \cdot 33}} = 1 \cdot 108 \text{ bar} \quad \text{Ans. (ii)}$$

$$\frac{p_1 V_1}{T_1} = \frac{p_2 V_2}{T_2}$$

$$\frac{12 \times 1}{(216 + 273)} = \frac{1 \cdot 108 \times 6}{T_2}$$

$$T_2 = \frac{489 \times 1 \cdot 108 \times 6}{12} = 270 \cdot 9 \text{ K}$$

$$270 \cdot 9 - 273 = -2 \cdot 1 °\text{C} \quad \text{Ans. (iii)}$$

30. From steam tables,
$$\text{Water } 165°\text{C (7 bar), } h_f = 697$$

Steam 100°C (atmospheric pressure),
$$h_f = 419 \cdot 1 \qquad h_{fg} = 2256 \cdot 7 \qquad h_g = 2675 \cdot 8$$

Heat energy released by one kg of water in falling in temperature from 165°C to 100°C
$$= 697 - 419 \cdot 1 = 277 \cdot 9 \text{ kJ}$$

This is transferred to m kg of water at 100°C causing it to evaporate, heat energy received by m kg of water at 100°C to evaporate it into steam at 100°C
$$= m \times 2256 \cdot 7 \text{ kJ}$$
$$\therefore \ m \times 2256 \cdot 7 = 277 \cdot 9$$
$$m = 0 \cdot 1231 \text{ kg} \quad \text{Ans.}$$

Alternatively:
$$\text{Enthalpy before release} = \text{Enthalpy after}$$

Enthalpy of 1 kg of
 water at 165°C
$$= \frac{\text{Enthalpy of } m \text{ kg}}{\text{of steam at 100°C}} + \frac{\text{Enthalpy of } (1 - m) \text{ kg}}{\text{of water at 100°C}}$$

$$697 = m \times 2675 \cdot 8 + (1 - m) \times 419 \cdot 1$$
$$697 = 2675 \cdot 8 \, m + 419 \cdot 1 - 419 \cdot 1 \, m$$
$$277 \cdot 9 = 2256 \cdot 7 \, m$$
$$m = 0 \cdot 1231 \text{ kg (as above)}$$

31. Referring to Fig. 90,

$$v_{a1} = v_1 \sin \alpha_1 = 840 \times \sin 20° = 287·4 \text{ m/s}$$
$$v_{w1} = v_1 \cos \alpha_1 = 840 \times \cos 20° = 789·4 \text{ m/s}$$
$$x = v_{w1} - u = 789·4 - 350 = 439·4 \text{ m/s}$$

$$\tan \beta_1 = \frac{V_{a1}}{x} = \frac{287·4}{439·4} = 0·6539$$

$$\therefore \text{ Blade inlet angle} = 33° 11' \quad \text{Ans. (i)}$$

$$v_{r1} = \frac{v_{a1}}{\sin \beta_1} = \frac{287·4}{\sin 33° 11'} = 525 \text{ m/s}$$

$$v_{r2} = 0·8 \times 525 = 420 \text{ m/s}$$
$$v_{a2} = v_{r2} \times \sin \beta_2 = 420 \times \sin 25° 12' = 178·8 \text{ m/s}$$
$$v_{w2} + u = v_{r2} \times \cos \beta_2 = 420 \times \cos 25° 12' = 380 \text{ m/s}$$
$$v_{w2} = 380 - 350 = 30 \text{ m/s}$$

$$\tan \varphi = \frac{v_{w2}}{v_{a2}} = \frac{30}{178·8} = 0·1678$$

$$\varphi = 9° 32'$$

$$v_2 = \frac{v_{a2}}{\cos \varphi} = \frac{178·8}{\cos 9° 32'} = 181·3 \text{ m/s}$$

Absolute velocity of steam at exit:

$$\left. \begin{array}{l} \text{Magnitude} = 181·3 \text{ m/s} \\ \text{Direction to axis} = 9° 32' \\ \text{or, Direction to plane of wheel} = 80° 28' \end{array} \right\} \quad \text{Ans. (ii)}$$

32. Available hydrogen $= H_2 - \dfrac{O_2}{8}$

$$= 0·114 - \frac{0·024}{8} = 0·111 \text{ kg}$$

$$\text{c.v.} = 33·7 \text{ C} + 144 \left[H_2 - \frac{O_2}{8} \right] + 9·3 \text{ S}$$

$$= 33·7 \times 0·846 + 144 \times 0·111 + 9·3 \times 0·004$$
$$= 28·51 + 15·99 + 0·0372$$
$$= 44·5372 \text{ MJ/kg} \quad \text{Ans. (i)}$$

$$\text{Air reqd.} = \frac{100}{23} \times \text{oxygen required}$$

$$= \frac{100}{23}\left\{ 2\tfrac{2}{3}\,C + 8\left[H_2 - \frac{O_2}{8}\right] + S\right\}$$

$$= \frac{100}{23}\{2\tfrac{2}{3} \times 0\cdot846 + 8 \times 0\cdot111 + 0\cdot004\}$$

$$= \frac{100}{23} \times 3\cdot148$$

$$= 13\cdot69 \text{ kg air/kg fuel}\quad\text{Ans. (ii)}$$

33. From tables, Freon-12, 1·826 bar,
$$h_f = 22\cdot33 \qquad h_g = 180\cdot97$$
$$h_{fg} = 180\cdot97 - 22\cdot33 = 158\cdot64$$

Sat. temp. at 1·826 bar $= -15°C$ therefore freon at evaporator outlet at 0°C is superheated by 15°.

$$1\cdot826 \text{ bar } 15° \text{ superheat,} \qquad h = 190\cdot15$$

Referring to Fig. 101,
Throttling effect through expansion valve between condenser outlet and evaporator inlet:

$$\text{Enthalpy after throttling } (h_4) = \text{Enthalpy before } (h_3)$$
$$22\cdot33 + x_4 \times 158\cdot64 = 50$$
$$x_4 \times 158\cdot64 = 27\cdot67$$
$$x_4 = 0\cdot1744 \quad\text{Ans. (i)}$$

$$\text{Refrig. effect/kg} = h_1 - h_4 \text{ (note } h_4 = h_3)$$
$$= 190\cdot15 - 50$$
$$= 140\cdot15 \text{ kJ/kg}$$

Refrigerating effect per minute
$$= 0\cdot4 \times 60 \times 140\cdot15$$
$$= 3364 \text{ kJ/min}\quad\text{Ans. (ii)}$$

34. Coeff. of cubical exp. $= 3 \times$ coeff. of linear exp.

For glass, $\quad \beta = 3 \times 8.5 \times 10^{-6} = 2.55 \times 10^{-5}/°C$

Increase in volume $= \beta \times$ orig. vol. $\times$ temp. rise

For mercury, increase $= 1.8 \times 10^{-4} \times 125 \times 60$

For vessel, increase $= 2.55 \times 10^{-5} \times 125 \times 60$

$$
\begin{aligned}
\text{Overflow} &= \text{difference in volumetric expansion} \\
&= (18 - 2.55) \times 10^{-5} \times 125 \times 60 \\
&= 1.159 \text{ ml} \quad \text{Ans.}
\end{aligned}
$$

35. For one mm movement of the stylus (one mm on height of card) indicator piston deflection $= \frac{1}{6}$ mm and this would be under a force of $30 \div 6 = 5$ N in the indicator cylinder.

$$
\begin{aligned}
\text{force [N]} &= \text{pressure [N/m}^2\text{]} \times \text{area [m}^2\text{]} \\
5 &= \text{pressure [N/m}^2\text{]} \times 0.7854 \times 7^2 \times 10^{-6}
\end{aligned}
$$

$\therefore$ pressure scale on card per mm of height

$$
= \frac{5}{0.7854 \times 7^2 \times 10^{-6}}
$$

$$
= 1.299 \times 10^5 \text{ N/m}^2 = 1.299 \text{ bar}
$$

$$
\text{Mean height of diagram [mm]} = \frac{\text{area [mm}^2\text{]}}{\text{length [mm]}}
$$

$$
= \frac{346}{75} \text{ mm}
$$

$$
\therefore \text{Mean effective press.} = \frac{346}{75} \times 1.299
$$

$$
= 5.994 \text{ bar} \quad \text{Ans. (i)}
$$

$$
\text{ip} = p_m ALn
$$

where $n = 2 \times$ rev/s for a double-acting 2-stroke

$$
\begin{aligned}
\text{ip} &= 5.994 \times 10^2 \times 0.7854 \times 0.6^2 \times 0.9 \times 2.1 \times 2 \\
&= 640.4 \text{ kW} \quad \text{Ans. (ii)}
\end{aligned}
$$

36. Cross-sect. area [m^2] $= \dfrac{\text{volume flow [m}^3\text{/s]}}{\text{velocity [m/s]}}$

$$
= \frac{\text{mass flow [kg/s]} \times \text{spec. vol. [m}^3\text{/kg]}}{\text{velocity [m/s]}}
$$

$$\text{Diameter} = \sqrt{\frac{\text{area}}{0\cdot7854}}$$

$$\text{Entrance diameter} = \sqrt{\frac{0\cdot315 \times 0\cdot2765}{457 \times 0\cdot7854}}$$

$$= 0\cdot015\ 58\ \text{m} = 15\cdot58\ \text{mm}\quad \text{Ans. (i)}$$

$$\text{Exit diameter} = \sqrt{\frac{0\cdot315 \times 7\cdot404}{1524 \times 0\cdot7854}}$$

$$= 0\cdot044\ 14\ \text{m} = 44\cdot14\ \text{mm}\quad \text{Ans. (ii)}$$

37. $pV = mRT$

$$m = \frac{pV}{RT} = \frac{1\cdot1 \times 10^2 \times 0\cdot2}{0\cdot287 \times (15 + 273)} = 0\cdot2662\ \text{kg}$$

Heat energy supplied [kJ]
  $= \text{mass [kg]} \times \text{spec. heat [kJ/kg K]} \times \text{temp. rise [K]}$
  $= 0\cdot2662 \times 1\cdot005 \times (150 - 15)$
  $= 36\cdot1\ \text{kJ}\quad \text{Ans. (i)}$

$$p_1 V_1^{1\cdot32} = p_2 V_2^{1\cdot32}$$
$$1\cdot1 \times 0\cdot2^{1\cdot32} = 7\cdot15 \times V_2^{1\cdot32}$$

$$V_2 = 0\cdot2 \times \sqrt[1\cdot32]{\frac{1\cdot1}{7\cdot15}} = \frac{0\cdot2}{\sqrt[1\cdot32]{6\cdot5}}$$

$$= 0\cdot048\ 44\ \text{m}^3$$

$$\frac{p_1 V_1}{T_1} = \frac{p_2 V_2}{T_2}$$

$$\frac{1\cdot1 \times 0\cdot2}{(15 + 273)} = \frac{7\cdot15 \times 0\cdot048\ 44}{T_2}$$

$$T_2 = \frac{288 \times 7\cdot15 \times 0\cdot048\ 44}{1\cdot1 \times 0\cdot2} = 453\cdot4\ \text{K}$$

$$453\cdot4 - 273 = 180\cdot4°\text{C}\quad \text{Ans. (ii)}$$

38. 2 rev/s $= 720°$ of crank movement per second
Total open period of valve

$$= 32 + 180 + 18 = 230° \text{ of crank movement}$$

$$\text{Time of open period} = \frac{230}{720} = 0.3195 \text{ s} \quad \text{Ans. (i)}$$

$$\text{Full open period} = 18 + 180 - 6 = 192° \text{ of crank}$$
$$\text{movement}$$

$$\text{Time of full opening} = \frac{192}{720} = 0.2667 \text{ s} \quad \text{Ans. (ii)}$$

Angular displacement to open

$$= 32 - 18 = 14° \text{ of crank movement}$$

$$\text{Time to open} = \frac{14}{720} = 0.019\ 45 \text{ s}$$

$$\text{Mean velocity of opening [m/s]} = \frac{\text{distance} \ [m]}{\text{time} \ [s]}$$

$$= \frac{50 \times 10^{-3}}{0.019\ 45} = 2.571 \text{ m/s} \quad \text{Ans. (iii)}$$

Angular displacement to close
$$= 6 + 18 = 24°$$

$$\text{Time to close} = \frac{24}{720} = 0.033\ 33 \text{ s}$$

Mean velocity of closing

$$= \frac{50 \times 10^{-3}}{0.033\ 33} = 1.5 \text{ m/s} \quad \text{Ans. (iv)}$$

39. $\qquad pV = mRT \qquad \therefore \ m = \frac{pV}{RT}$

$$\text{where } p = 5.86 \times 10^2 = 586 \text{ kN/m}^2$$
$$V = 113 \times 10^{-3} = 0.113 \text{ m}^3$$

$$R = 297 \times 10^{-3} = 0.297 \text{ kJ/kg K}$$
$$T = 61 + 273 = 334 \text{ K}$$

$$m = \frac{586 \times 0.113}{0.297 \times 334} = 0.6676 \text{ kg} \quad \text{Ans. (i)}$$

$$V = \frac{mRT}{p} = \frac{0.6676 \times 0.297 \times (15 + 273)}{1.38 \times 10^2}$$

$$= 0.4138 \text{ m}^3 \text{ or } 413.8 \text{ litre} \quad \text{Ans. (ii)}$$

40. Theoretical air reqd. per kg of fuel A

$$= \frac{100}{23}\{2\tfrac{2}{3} \times 0.885 + 8 \times 0.115\}$$

$$= \frac{100}{23} \times 3.28 \text{ kg}$$

Theoretical air reqd. per kg of fuel B

$$= \frac{100}{23}\{2\tfrac{2}{3}\,C + 8\,H_2\}$$

and this is 6 % more than for fuel A, therefore,

$$\frac{100}{23}\{2\tfrac{2}{3}\,C + 8\,H_2\} = 1.06 \times \frac{100}{23} \times 3.28$$

Cancelling $\dfrac{100}{23}$ and multiplying throughout by $\dfrac{3}{8}$:

$$C + 3\,H_2 = 1.304 \quad \text{....} \quad \text{....} \quad \text{....} \quad \text{(i)}$$

Also, fractional analysis of fuel B :
$$C + H_2 = 1$$
$$\therefore \ C = 1 - H_2 \quad \text{....} \quad \text{....} \quad \text{....} \quad \text{(ii)}$$

Substituting value of C from (ii) into (i) :
$$C + 3H_2 = 1.304$$
$$1 - H_2 + 3\,H_2 = 1.304$$

$$2\,H_2 = 0.304$$
$$H_2 = 0.152$$
$$\text{and, } C = 1 - 0.152 = 0.848$$

$\therefore$ Mass analysis of fuel B =
$$84.8\% \text{ carbon, } 15.2\% \text{ hydrogen} \quad \text{Ans.}$$

41. Before alteration:
$$V_1 = 90 + 15 = 105 \qquad V_2 = 15$$
$$p_1 V_1^{1.33} = p_2 V_2^{1.33}$$
$$1 \times 105^{1.33} = p_2 \times 15^{1.33}$$

$$p_2 = \left\{\frac{105}{15}\right\}^{1.33} = 13.3 \text{ bar} \quad \text{Ans. (i)}$$

After alteration:
$$V_1 = 90 + 12.5 = 102.5 \qquad V_2 = 12.5$$
$$1 \times 102.5^{1.33} = p_2 \times 12.5^{1.33}$$

$$p_2 = \left\{\frac{102.5}{12.5}\right\}^{1.33} = 16.43 \text{ bar} \quad \text{Ans. (ii)}$$

42.

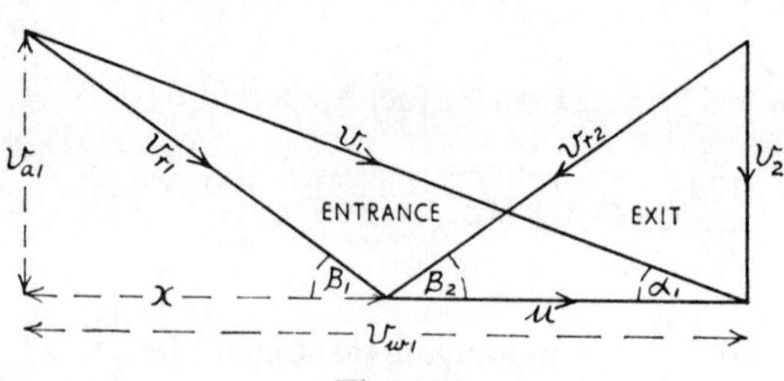

Fig. 107

Kinetic energy $= \frac{1}{2}mv^2$
Kinetic energy being in joules when m is the mass in kg and v is the velocity in m/s, hence,

$$250 \times 10^3 = \frac{1}{2} \times 1 \times v^2$$
$$v = \sqrt{2 \times 250 \times 10^3} = 707.1 \text{ m/s}$$

With no friction, $v_{r2} = v_{r1}$

and, since $\beta_2 = \beta_1$
then $x = u = \tfrac{1}{2}v_{w1}$

$$\frac{v_{a1}}{x} = \tan\beta_1 \qquad \therefore \; \frac{v_1 \sin\alpha_1}{\tfrac{1}{2}\,v_1 \cos\alpha_1} = \tan 35°$$

v_1 cancels, $\sin\alpha_1 \div \cos\alpha_1 = \tan\alpha_1$ therefore
$$2\tan\alpha_1 = \tan 35°$$

$$\tan\alpha_1 = \frac{0\cdot7002}{2} = 0\cdot3501$$

$$\therefore \; \text{Nozzle angle } \alpha_1 = 19° \, 18' \quad \text{Ans. (i)}$$

Blade velocity $u = \tfrac{1}{2}v_{w1} = \tfrac{1}{2}v_1 \cos\alpha_1$
$$= \tfrac{1}{2} \times 707\cdot1 \times \cos 19° \, 18'$$
$$= 333\cdot7\,\text{m/s} \quad \text{Ans. (ii)}$$

43.   Let d = diameter of cylinder
then $1\cdot25d$ = stroke

For 4 cylinders:
$$\text{ip} = pALn \times 4$$
$$300 = 6\cdot28 \times 10^2 \times 0\cdot7854d^2 \times 1\cdot25d \times 4\cdot5 \times 4$$

$$d = \sqrt[3]{\frac{300}{628 \times 0\cdot7854 \times 1\cdot25 \times 4\cdot5 \times 4}}$$

$$= 0\cdot3\,\text{m} = 300\,\text{mm}$$

$$\left.\begin{array}{l}\text{Cylinder diameter} = 300\,\text{mm} \\ \text{Stroke} = 1\cdot25 \times 300 = 375\,\text{mm}\end{array}\right\}\text{Ans. (a)}$$

Indicated thermal efficiency

$$= \frac{3\cdot6\,[\text{MJ/kWh}]}{\text{kg fuel/ind. kWh} \times \text{c.v. [MJ/kg]}}$$

$$= \frac{3\cdot6}{0\cdot225 \times 42}$$

$$= 0\cdot381 \text{ or } 38\cdot1\% \quad \text{Ans. (b)}$$

44. From NH_3 tables,

$$1\cdot902 \text{ bar}, \quad h_g = 1420$$
$$7\cdot529 \text{ bar}, \quad h_f = 256, \quad \text{sat. temp.} = 16°C$$

$\therefore$ at 66°C vapour is superheated 50°

$$7\cdot529 \text{ bar } 50° \text{ superheat}, \quad h = 1591\cdot7$$

Referring to Fig. 101:

Enthalpy gain per kg through evaporator

$$= h_1 - h_4 \qquad (\text{note } h_4 = h_3)$$
$$= 1420 - 256 = 1164 \text{ kJ/kg}$$

Refrigerating effect [kJ/min]

$$= \text{mass flow [kg/min]} \times \text{enthalpy gain [kJ/kg]}$$

$$\therefore m = \frac{800}{1164} = 0\cdot6873 \text{ kg/min} \quad \text{Ans. (i)}$$

Enthalpy drop per kg through condenser

$$= h_2 - h_3$$
$$= 1591\cdot7 - 256 = 1335\cdot7 \text{ kJ/kg}$$

Heat rejected in condenser

$$= 0\cdot6873 \times 1335\cdot7 = 918\cdot1 \text{ kJ/min} \quad \text{Ans. (ii)}$$

Enthalpy gain per kg in compressor

$$= h_2 - h_1$$
$$= 1591\cdot7 - 1420 = 171\cdot7 \text{ kJ/kg}$$

Energy given to refrigerant in compressor

$$= 0\cdot6873 \times 171\cdot7 \text{ kJ/min}$$

Power [kW $=$ kJ/s]

$$= \frac{0\cdot6873 \times 171\cdot7}{60}$$

$$= 1\cdot967 \text{ kW} \quad \text{Ans. (iii)}$$

45. From saturated steam tables:

$$\text{Steam 30 bar}, \quad h_g = 2803 \quad v_g = 0\cdot066\,65$$
$$\text{Water 38°C}, \quad h = 159\cdot1$$

From superheated steam tables,

$$\text{Steam 30 bar 400°C,} \quad h = 3231 \quad v = 0.0993$$

Let m[kg] = mass of injection water per kg of superheated

steam.

Mixing in desuperheater:

$$\text{Enthalpy before mixing} = \text{Enthalpy after}$$
$$1 \times 3231 + m \times 159.1 = (1 + m) \times 2803$$
$$3231 + 159.1\,m = 2803 + 2803\,m$$
$$428 = 2643.9\,m$$
$$m = 0.1618 \text{ kg} \quad \text{Ans. (i)}$$

Volume of one kg superheated steam $= 0.0993 \text{ m}^3$
Volume of $(1 + m)$ kg of dry saturated steam

$$= 1.1618 \times 0.066\,65 = 0.077\,43 \text{ m}^3$$

Percentage change in volume

$$= \frac{0.0993 - 0.077\,43}{0.0993} \times 100$$

$$= 22.03\% \quad \text{Ans. (ii)}$$

46. Let V_1 = volume of gas in cylinder at beginning of stroke, this is stroke volume + clearance volume.

V_2 = volume of gas in cylinder at end of stroke, this is the clearance volume.

$$p_1 V_1^{1.32} = p_2 V_2^{1.32}$$
$$1 \times V_1^{1.32} = 37 \times 850^{1.32}$$

$$V_1 = 850 \times {}^{1.32}\sqrt{37} = 13\,100 \text{ cm}^3$$
$$\text{Stroke volume} = 13\,100 - 850 = 12\,250 \text{ cm}^3$$
$$= 12.25 \text{ litre or } 0.012\,25 \text{ m}^3 \quad \text{Ans. (i)}$$

$$\frac{p_1 V_1}{T_1} = \frac{p_2 V_2}{T_2}$$

$$\frac{1 \times 13\,100}{(35 + 273)} = \frac{37 \times 850}{T_2}$$

$$T_2 = \frac{308 \times 37 \times 850}{13\,100} = 739.4 \text{ K}$$

$$739.4 - 273 = 466.4°C \quad \text{Ans. (ii)}$$

47. Being a two-stroke engine, the rotational speed of the cam shaft is the same as the crank shaft, therefore the positions of opening and closing of the valves relative to the crank are the same as for the cam shaft.

Open period of ahead fuel cam
$$= 5 + 38 = 43°$$

Centre-line of ahead fuel cam
$$= \tfrac{1}{2} \text{ of } 43° = 21 \cdot 5° \text{ from } 5° \text{ before T.D.C.}$$
$$21 \cdot 5 - 5 = 16 \cdot 5° \text{ from T.D.C.}$$

Open period of astern fuel cam
$$= 5 + 34 = 39°$$

Centre-line of astern fuel cam
$$= \tfrac{1}{2} \text{ of } 39° = 19 \cdot 5° \text{ from } 5° \text{ before T.D.C.}$$
$$19 \cdot 5 - 5 = 14 \cdot 5° \text{ from T.D.C. (reverse direction)}$$
$\therefore$ Angle between centre-lines of ahead and astern fuel cams
$$= 16 \cdot 5 + 14 \cdot 5 = 31° \quad \text{Ans. (i)}$$

Angle between cranks of a 2-stroke, 6-cylinder engine

$$= \frac{360°}{\text{no. of cylinders}} = 60°$$

Open period of starting air cams
$$= 60° + 10° \text{ overlap} = 70°$$

Centre-line of ahead starting air cam
$$= \tfrac{1}{2} \text{ of } 70 + 8 = 43° \text{ from T.D.C.}$$

Centre-line of astern starting air cam
$$= 43° \text{ from T.D.C. reverse direction}$$
$\therefore$ Angle between centre-lines of ahead and astern starting air cams
$$= 2 \times 43 = 86° \quad \text{Ans. (ii)}$$

48. Quantity of heat conducted:

$$Q[\text{J}] = \frac{\lambda[\text{W/m K}] \times A\,[\text{m}^2] \times t\,[\text{s}] \times (T_1 - T_2)\,[\text{K}]}{S\,[\text{m}]}$$

$$= \frac{44\cdot5 \times 1 \times 3600 \times (500 - 195)}{0\cdot019}$$

$$= 2\cdot571 \times 10^9\,J \;=\; 2\cdot571 \times 10^6\,kJ$$

From steam tables, 14 bar 195°C,
enthalpy of evaporation $\quad h_{fg} \;=\; 1960\,kJ/kg$
$\therefore$ Mass [kg] of water evaporated

$$= \frac{2\cdot571 \times 10^6}{1960} \;=\; 1312\,kg \quad Ans.$$

49. Volume of vessel $= 0\cdot7854 \times 2^2 \times 6 = 18\cdot85\,m^3$
Mass of air at 30 bar and 24°C (297 K):

$$m = \frac{pV}{RT}$$

$$= \frac{30 \times 10^2 \times 18\cdot85}{0\cdot287 \times 297} = 663\cdot2\,kg \quad Ans.\ (i)$$

Mass of air at 37·5 bar and 297 K:

$$m = \frac{37\cdot5 \times 10^2 \times 18\cdot85}{0\cdot287 \times 297} = 829\,kg$$

Mass of air to be pumped in
$$= 829 - 663\cdot2 \;=\; 165\cdot8\,kg$$

Volume of 165·8 kg of air at 1·013 bar, 297 K:

$$V = \frac{mRT}{p}$$

$$= \frac{165\cdot8 \times 0\cdot287 \times 297}{1\cdot013 \times 10^2} = 139\cdot5\,m^3 \quad Ans.\ (ii)$$

Alternative solution to part (ii):
Extra mass to increase pressure by 7·5 bar at the same

temperature $= 663\cdot2 \times \dfrac{7\cdot5}{30} = 165\cdot8\,kg$

Volume at 1·013 bar, 297 K

$$= 18\cdot85 \times \frac{7\cdot5}{1\cdot013} = 139\cdot6\,m^3$$

50. Power strokes per second $= 126 \div 60 = 2\cdot1$

$$\begin{aligned}
\text{ip} &= pALn \\
&= 5\cdot5 \times 10^2 \times 0\cdot7854 \times 0\cdot2^2 \times 0\cdot38 \times 2\cdot1 \\
&= 13\cdot79\,\text{kW} \quad \text{Ans. (i)} \\
\text{bp [kW]} &= T\,[\text{kNm}] \times \omega\,[\text{rad/s}] \\
&= 0\cdot362 \times 0\cdot8 \times 2\pi \times 5\cdot3 \\
&= 9\cdot645\,\text{kW} \quad \text{Ans. (ii)}
\end{aligned}$$

$$\text{Mech. eff.} = \frac{\text{brake power}}{\text{indicated power}}$$

$$= \frac{9\cdot645}{13\cdot79} = 0\cdot6993 \text{ or } 69\cdot93\% \quad \text{Ans. (iii)}$$

Indicated thermal efficiency

$$= \frac{\text{heat turned into work in cylinder [kJ/h]}}{\text{heat energy supplied [kJ/h]}}$$

$$= \frac{13\cdot79 \times 3600}{6\cdot8 \times 17\cdot5 \times 10^3}$$

$$= 0\cdot4173 \text{ or } 41\cdot73\% \quad \text{Ans. (iv)}$$

$$\begin{aligned}
\text{Overall eff.} &= \text{ind. thermal eff.} \times \text{mech. eff.} \\
&= 0\cdot4173 \times 0\cdot6993 \\
&= 0\cdot2918 \text{ or } 29\cdot18\% \quad \text{Ans. (v)}
\end{aligned}$$

FIRST CLASS
MISCELLANEOUS PROBLEMS

1. During a Morse test on a four-cylinder internal combustion engine, the speed was kept constant at 20 rev/s. With all cylinders working it developed a brake power of 20 kW and the specific fuel consumption was 0·33 kg/kWh (brake). With each cylinder cut out in turn the average torque developed by the three remaining cylinders was 108 Nm. Calculate (i) the indicated power of the engine, (ii) the specific fuel consumption per indicated kWh, (iii) the indicated thermal efficiency. Take the calorific value of the fuel as 43 MJ/kg.

2. A vessel of volume 15 m³ contains air and dry saturated steam at a total pressure of 0·09 bar and temperature 39°C. Taking R for air $= 0·287$ kJ/kgK calculate the masses of steam and air in the vessel, and give the mass ratio of air to steam.

3. The stroke of the piston in a diesel engine is 1200 mm and the clearance is equal to 9% of the stroke volume. The index of compression is 1·35 and the final compression pressure is 32 bar. If the depth of the clearance is reduced by 5 mm, find the alteration in the final compression pressure.

4. The mass analysis of the fuel burned in a boiler is 87% carbon, 11% hydrogen, and 2% oxygen, and the fuel is burned at the rate of 1·8 tonne/h. Calculate the mass flow rate [kg/s] of each of the constituents of the flue gases if combustion is complete and no excess air is supplied. Express the composition of the flue gases as percentages by mass.
Atomic weights: hydrogen 1, carbon 12, nitrogen 14, oxygen 16. Mass composition of air: 23% oxygen, 77% nitrogen.

5. Dry saturated steam enters a convergent-divergent nozzle at 9 bar, and the pressures at the throat and exit are 5 bar and 0·14 bar respectively. The specific enthalpy drop of the steam from entrance to throat is 107 kJ/kg, and from entrance to exit it is 633 kJ/kg. Assuming 8% of the enthalpy drop is lost to friction in the divergent part of the nozzle, calculate the areas in mm² of the nozzle at the throat and exit to pass 23 kg of steam per minute.

6. In a Freon-12 refrigerating plant the compressor takes the refrigerant in at 0·8071 bar and discharges it at 12·19 bar and 65°C. At condenser outlet the Freon is saturated liquid at 12·19 bar. If compression is isentropic and the flow of the refrigerant is 15 kg/min, calculate the refrigerating effect and the coefficient of performance.

7. A surface feed heater is supplied with steam direct from the main steam line at 14 bar, 0·98 dry, the drain from the heater being led direct to the hotwell. The condensate from the condenser is discharged into the hotwell at 44°C. If the temperature of the feed water is 100°C, calculate the percentage of the main steam bled off to the heater.

8. The initial condition of a gas at the beginning of compression is 1 bar, 9600 cm³ and 20°C, and the final condition at the end of compression is 720 cm³ and 512°C. Calculate (i) the index of compression, (ii) the work done during compression, (iii) change in internal energy, (iv) the transfer of heat during compression. For this gas, take $R = 0·264$ kJ/kg K and $c_v = 0·657$ kJ/kg K. The area under the compression curve $pV^n = (p_1V_1 - p_2V_2)/n - 1$.

9. Steam at 15 bar 0·88 dry is mixed with steam at the same pressure superheated to 250°C in the ratio by mass of 3 of wet steam to 2 of superheated steam. The mixture is then passed through a reducing valve and throttled to 8 bar. Assuming no heat losses, calculate the dryness fraction of the steam (i) after mixing and before throttling, (ii) after throttling.

10. The walls of a cold room consists of an outer layer of wood of thickness 30 mm and thermal conductivity 0·18 W/m K, and a cork lining of thickness 70 mm and thermal conductivity 0·05 W/m K. If the rate of heat transfer from and to each exposed surface is 10 W/m² K and the heat flow through the wall is 24 W/m², calculate, (i) the temperature differences across the thicknesses of the wood and cork, (ii) the total temperature difference between the outside atmosphere and inside of room, and (iii) the temperature of the room when the external ambient temperature is 20°C.

11. In an engine working on the ideal diesel cycle, the compression ratio is 13·5 and fuel is admitted for 0·1 of the stroke. The temperature of the air at beginning of compression is 38°C.

Taking the index of compression and expansion $= 1\cdot4$, $c_{\mathrm{p}} = 1\cdot005$ kJ/kg K, $c_{\mathrm{v}} = 0\cdot718$ kJ/kg K, calculate the temperatures (i) at the end of compression, (ii) at the end of combustion, (iii) at the end of expansion, and calculate also (iv) the ideal thermal efficiency.

12. In a test on a single-cylinder, two-stroke, diesel engine, the mean effective pressure was $8\cdot9$ bar, running speed $2\cdot3$ rev/s, brake load 8 kN, brake radius $1\cdot25$ m, specific fuel consumption $0\cdot251$ kg/kW h (brake). The diameter of the cylinder is 360 mm, stroke 780 mm, calorific value of the fuel $41\cdot5$ MJ/kg. Calculate (i) the indicated power, (ii) brake power, (iii) indicated thermal efficiency, and (iv) the total heat energy loss per second.

13. Steam is supplied to an engine at $17\cdot5$ bar 300°C and expanded isentropically to $0\cdot07$ bar. Calculate (i) the dryness fraction of the steam at the end of expansion, (ii) the Rankine efficiency. Show by a sketch how the Rankine cycle appears on a temperature-entropy diagram.

14. In a single-stage impulse turbine the steam enters the nozzle at 7 bar 300°C and is discharged at $1\cdot2$ bar dry saturated, directed at 20° to the plane of rotation. The blade velocity is 40% of the steam jet velocity and the relative velocity of the steam at exit is 80% of the relative velocity at entrance. The outlet angle of the blades is 35° and the steam flow $0\cdot5$ kg/s. Calculate (i) the axial thrust, (ii) the power developed.

15. The output of an electric generator is 7500 kW and the electrical efficiency is 95% (neglecting all other losses). The volume flow of cooling air blown through the windings is $16\cdot5$ m³/s at 21°C at a gauge pressure equal to 153 mm water, and this maintains the temperature of the windings at 55°C. Calculate the difference in temperature between the windings and the exit temperature of the cooling air, taking R for air $= 0\cdot287$ kJ/kg K, $c_{\mathrm{p}} = 1\cdot005$ kJ/kg K, and atmospheric pressure $= 1\cdot013$ bar.

16. The pressure, volume and temperature of a gas mixture sample in a closed vessel is $1\cdot01$ bar, 500 cm³ and 20°C, and is composed of 14% carbon dioxide and 86% nitrogen by volume. Taking R for $CO_2 = 0\cdot189$ kJ/kg K, calculate (i) the partial pressure of each gas, (ii) the mass of carbon dioxide in the sample.

17. Sketch a bomb calorimeter and describe an experiment to determine the calorific value of a fuel, and state whether the

result so obtained is the higher or lower value. In such an experiment, the sample of fuel burned was 1·5 grammes, quantity of water in calorimeter 1·65 litres, water equivalent of bomb and fittings 450 grammes, and rise in temperature recorded 7·15°C, calculate the calorific value. Specific heat of water = 4·2 kJ/kg K.

18. Dry saturated steam at 16 bar is throttled to 9 bar and then expanded isentropically in an engine to 1·0 bar, calculate (i) the change in entropy, (ii) the final condition of the steam.

19. In a compression ignition engine working on the ideal dual-combustion cycle, the volumetric compression ratio is 12·5:1. The cycle consists of (a) adiabatic compression from 1·013 bar, 35°C, (b) heat received at constant volume to a maximum pressure of 40 bar, (c) heat received at constant pressure to a maximum temperature of 1425°C, (d) adiabatic expansion to the initial volume, (e) heat rejected at constant volume. Make a sketch of the pV diagram and calculate the mean effective pressure. Take $\gamma = 1·4$ for air and products of combustion, area under an

$$\text{adiabatic curve} = \frac{p_1 V_1 - p_2 V_2}{\gamma - 1}$$

20. The steam consumption of an engine is 18·9 Mg/h when the power developed is 1800 kW, and 25·5 Mg/h when the power is 3000 kW. State Willans' law for this engine and calculate the specific steam consumption in kg/kWh when the power is 2500 kW.

21. The mass analysis of a fuel burned in a boiler is 85·5% carbon, 13·5% hydrogen, and 1% oxygen, and the air supply is 25% in excess of the minimum required for complete combustion. Calculate (i) the percentage mass analysis of the wet flue gases, (ii) the percentage volumetric analysis of the dry flue gases. Take mass composition of air = 23% oxygen, 77% nitrogen. Atomic weights: hydrogen 1, carbon 12, nitrogen 14, oxygen 16.

22. Two copper pipes of a telemotor system are each 76 m long and 13 mm internal diameter when filled with oil at 16°C. Calculate the quantity of oil in litres that will be released into the replenishing tank from the pipes when the temperature rises to 30°C. Neglect the amount of oil in the cylinders.

Coefficient of linear expansion of copper $= 1·7 \times 10^{-5}/°C$
Coefficient of volumetric expansion of oil $= 7·7 \times 10^{-4}/°C$

23. A gas of initial conditions 32 bar, 0·5 m³ and 520°C, is expanded behind a piston until the volume is 6 m³ and temperature 30°C. Calculate (i) the index of expansion, (ii) change in internal energy, (iii) heat flow during expansion. Take R for the gas $= 0·287$ kJ/kg K and $c_V = 0·718$ kJ/kg K. Area under the

curve $pV^n = \dfrac{p_1 V_1 - p_2 V_2}{n-1}.$

24. A two-stroke engine develops an indicated power of 2700 kW, the specific fuel consumption is 0·225 kg/kW h (indicated), and 18 kg of air is supplied per kg of fuel burned. The exhaust gases pass through an exhaust gas boiler, entering at 327°C and leaving at 215°C. Steam is generated in the boiler at 15 bar, 0·96 dry, from feed water at 50°C. If the efficiency of the boiler is 0·8, calculate the mass of steam generated per hour, taking the specific heat of the gases as 1·05 kJ/kg K.

25. Steam is expanded in the first stage of a turbine from 40 bar 450°C to 2·5 bar, it is then passed through a reheater and its temperature is raised to 300°C at constant pressure of 2·5 bar. The steam now passes through the final stage and expanded to 0·06 bar. Assume the reheat cycle to be ideal and calculate the Rankine efficiency.

26. The mean dimensions of an engine-room are 19 m long by 12 m broad and its permeability is 75%. A CO_2 fire extinguishing system is fitted which is capable of producing an average saturation of 25% of CO_2 gas in the engine-room to a height of 7 m. Calculate (i) the mass of CO_2 required at 1·013 bar and 16°C taking R for $CO_2 = 0·189$ kJ/kg K, and (ii) the total volume of the storage bottles if the specific volume of CO_2 liquid is $1·24 \times 10^{-3}$ m³/kg.

27. In a refrigerating plant using Freon as the refrigerating agent, the refrigerant leaves the condenser as saturated liquid at 15°C, leaves the evaporator and enters the compressor at 1·509 bar and —5°C, and is delivered from the compressor into the condenser at 4·914 bar 45°C. Calculate the coefficient of performance.

28. A sample of steam at 20 bar was drawn from the steam drum of a boiler and passed through a throttling calorimeter, the temperature of the steam in the calorimeter was 150°C and the

pressure 1·25 bar. Calculate (i) the dryness fraction of the steam. If the boiler steam is superheated at constant pressure to 350°C, calculate (ii) the heat energy supplied per kg of steam in the superheaters, and (iii) the percentage increase in volume due to drying and superheating.

29. A closed vessel of 0·7 m³ internal volume contains a gas at 0·58 bar 18°C of gas constant $R = 0·27$ kJ/kgK, calculate the mass of gas in the vessel. If now 0·35 kg of another gas at 18°C and gas constant $R = 0·29$ kJ/kgK is also admitted into the vessel, calculate the final pressure of the mixture.

30. The ratio of compression of an engine working on the constant volume cycle is 8·6:1. At the beginning of compression the temperature is 32°C and at the end of heat supply at constant volume the temperature is 1650°C. Taking the index of compression and expansion as 1·4, calculate (i) the temperature at the end of compression, (ii) temperature at the end of expansion, (iii) the theoretical thermal efficiency.

31. The power absorbed by a single-acting, single-stage, reciprocating air compressor is 13·58 kW when the mean piston speed is 2·8 m/s and rotational speed 3·5 rev/s. The air is compressed from 1·0 bar and delivered at 10 bar, the index of the law of compression being 1·32. Neglecting clearance, calculate (i) the stroke of the compressor piston, (ii) the cylinder diameter, and (iii) the mean effective pressure. Note: area under a pV^n

curve is given by $\dfrac{p_1V_1 - p_2V_2}{n - 1}$.

32. In an engine working on the ideal diesel cycle, the temperature of the air at the beginning of compression is 37°C, compression takes place according to the law $pV^{1·4} = $ a constant, and the volumetric compression ratio is 13:1. At the end of compression the air receives heat energy at constant pressure and is then expanded to the original volume. If 1 kg of fuel is burned per 35 kg of air compressed, the calorific value of the fuel being 42 MJ/kg, calculate (i) the temperature at the end of compression, (ii) the temperature at the end of heat reception, (iii) the volumetric expansion ratio. Take $c_p = 1·02$ kJ/kgK.

33. At a certain stage of a reaction turbine, the steam leaves the guide blades at a velocity of 135 m/s, the exit angle being 20°.

The linear velocity of the moving blades is 87 m/s. Assuming the channel section of fixed and moving blades to be identical, and assuming ideal conditions, calculate (i) the entrance angle of the moving blades, (ii) the stage power per kg/s steam flow.

34. Steam is generated in a boiler at 25 bar 300°C from feed water at 130°C at the rate of 20 000 kg/h, and the engine develops 2300 kW. The daily fuel consumption is 33 tonne and the calorific value of the fuel is 43 MJ/kg. Calculate (i) the boiler efficiency, (ii) the equivalent evaporation per kg of fuel from and at 100°C, (iii) the overall efficiency of the plant.

35. In an open cycle gas turbine plant, a heat exchanger is included to heat the air before entering the combustion chamber by the exhaust gases from the engine. The gases enter the heat exchanger at 300°C and 140 m/s and leave at 240°C and 10 m/s. The air enters the exchanger at 200°C and the air/fuel ratio is 84. Calculate the temperature of the air at the exchanger exit, taking the specific heats as 1·1 kJ/kg K for the gases, and 1·005 kJ/kg K for air.

36. A CO_2 refrigerating machine produces 250 kg of ice per hour at -10°C from water at 15°C. The refrigerant enters the evaporator 0·2 dry and leaves 0·95 dry. The compressor is single-acting, runs at 4·15 rev/s, and the stroke/bore ratio is 2:1. Calculate (i) the mass flow of the refrigerant through the circuit, and (ii) the diameter and stroke of the compressor piston. Take the following values:

Specific heat of ice $= 2·04$ kJ/kg K
Latent heat of fusion $= 335$ kJ/kg
Specific heat of water $= 4·2$ kJ/kg K
CO_2 vapour at evaporator pressure,
$$h_{fg} = 290·2 \text{ kJ/kg}, \quad v_g = 0·02168 \text{ m}^3/\text{kg}$$

37. Air is taken into a single-stage air compressor at 1 bar and delivered at 5 bar. The piston swept volume is 1440 cm³ and the clearance volume is 40 cm³. Taking the index of compression and expansion as 1·3, calculate (i) the fraction of the stroke when the delivery valves open, (ii) the fraction of the stroke when the suction valves open, (iii) the mean indicated pressure.

Note, the area under a pV^n curve is $\dfrac{p_1 V_1 - p_2 V_2}{n - 1}$

38. Steam is supplied to a turbine at 30 bar 350°C and the condenser pressure is 0·045 bar. The power developed is 5 MW when the steam consumption is 22·5 Mg/h. Calculate (i) the ideal efficiency of the Rankine cycle, (ii) the actual efficiency of the engine, (iii) the efficiency ratio.

39. A sample of steam at 10 bar is tested by a combined separating and throttling calorimeter, the data obtained were:
Mass of water collected in separator = 0·113 kg
Mass of condensed water after throttling = 3·03 kg
Pressure of steam in throttling calorimeter = 1·2 bar
Temperature of steam in throttling calorimeter = 109·8°C
Take specific heat of superheated steam at calorimeter pressure as 2·02 kJ/kg K and calculate the dryness fraction of the sample.

40. The piston swept volume of an engine working on the ideal dual combustion cycle is 0·1068 m³ and the clearance volume is 8900 cm³. At the beginning of compression the pressure is 1 bar and temperature 42°C. The maximum pressure in the cycle is 45 bar and maximum temperature 1500°C. Taking $\gamma = 1\cdot4$ and $c_v = 0\cdot715$ kJ/kg K for air and products of combustion, calculate the proportion of heat received at constant volume to that received at constant pressure.

41. One kilogramme of water is heated in a closed vessel to 175·4°C, the pressure being 9 bar. An escape valve in the top of the vessel is now opened, the pressure falls to atmospheric, the temperature to 100°C, and the steam formed by flash-evaporation escapes. Calculate the mass of water remaining in the vessel. If the experiment is to be repeated without adding more water to the vessel, find to what temperature the water must be raised so that an equal mass flashes off when the escape valve is again opened and the temperature falls to 100°C.

42. At one stage of a steam turbine the inlet velocity of the steam to the rotating blades is 590 m/s and the relative velocity at outlet is 620 m/s. The blade speed is 320 m/s and the inlet and outlet angles are 37° and 26° respectively. Calculate the force on the blades and the power developed at this stage for a steam flow of 0·075 kg/s.

43. A single-acting air compressor takes in air at 1 bar and delivers it at 4 bar, the cylinder is 300 mm diameter, stroke 450

mm, and it runs at 5 rev/s. Initially the index of compression was 1·15 and after running for some time the index of compression was found to be 1·35. Neglecting clearance, calculate the power absorbed in each case and the percentage increase in power.

$$\text{Area under } pV^n \text{ curve} = \frac{p_1 V_1 - p_2 V_2}{n - 1}$$

44. A steam pipe 140 mm outside diameter and 25 m long is lagged with insulating material of thermal conductivity 0·13 W/m K. Steam passes along the pipe at the rate of 1200 kg/h, entering at 18 bar dry saturated and leaving at the same pressure 0·985 dry. The outside surface temperature of the lagging is 35°C and the inside surface may be taken as equal to the steam temperature. Calculate the thickness of the lagging taking the rate of heat transfer through the insulating material, in J/s per unit length of pipe, as:

$$\frac{2\pi\lambda(T_1 - T_2)}{\ln r_2/r_1}$$

45. A vessel contains steam only, the pressure being 0·5 bar and dryness fraction 0·9. After admitting a further 0·26 kg of steam the pressure becomes 0·9 bar and dryness fraction 0·95. Calculate the volume of the vessel.

46. Steam is supplied to a turbine at 20 bar 400°C and exhausts at 0·04 bar and 0·85 dry. At the stage in the turbine where the pressure is 1·4 bar, 13·4% of the steam is withdrawn and passed to the feed heater and this heats the feed water to the saturation temperature of the tapped off steam. Compare the thermal efficiencies with and without feed heating.

47. Air is compressed in a cylinder according to the law $pV^n = $ a constant. The initial condition of the air is 0·125 m³, 1·01 bar and 19°C, and the final condition is 36 bar and 508°C. Taking $c_p = 1·005$ kJ/kg K and $c_v = 0·718$ kJ/kg K, calculate (i) the index of compression, (ii) the mass of air compressed, (iii) the work done during compression, (iv) the change of internal energy, (v) the transfer of heat to or from the air.

$$\text{Area under the compression curve} = \frac{p_1 V_1 - p_2 V_2}{n - 1}$$

48. The following data were taken during a test on a four-cylinder, four-stroke, compression ignition engine of cylinder diameter 320 mm   and stroke 480 mm while running at 4 rev/s. Mean effective pressure 7·45 bar, brake load 6 kN on a radius of 960 mm, fuel consumption 49·5 kg/h, calorific value of fuel 44·5 MJ/kg, mass flow of engine cooling water 77 kg/min, water inlet and outlet temperatures 14°C and 47°C. Calculate the indicated and brake thermal efficiencies, percentage of heat carried away in the cooling water, and draw up a heat balance.

Specific heat of cooling water = 4·2 kJ/kg K.

49. In a simple gas turbine working on the ideal cycle, the pressure ratio of both the compressor and turbine is 4·3:1. The temperatures at inlet to the compressor and at inlet to the turbine are 16°C and 600°C respectively. Calculate (i) the temperature at outlet from the compressor, (ii) the temperature at outlet from the turbine, (iii) the heat supplied per kilogramme of working fluid, (iv) the thermal efficiency. Take $\gamma = 1\cdot4$ and $c_p = 1\cdot005$ kJ/kg K.

50. The temperature inside a condenser is 32·9°C and the vacuum gauge reads 714 mm Hg when the barometer stands at 762 mm Hg. If the rate of air extraction is 3·82 m³/min, calculate (i) the mass air leakage per minute, (ii) the mass of vapour associated with this air. R for air = 0·287 kJ/kg K.

SOLUTIONS TO FIRST CLASS
MISCELLANEOUS PROBLEMS

1. Brake power $[kW] = T[kNm] \times \omega[rad/s]$

With one cylinder cut out, brake power developed by three remaining cylinders $= 0.108 \times 2\pi \times 20$

$$= 13.57\,kW$$

$$\text{ip of each cylinder} = 20 - 13.57 = 6.43\,kW$$

$$\text{ip of four cylinders} = 4 \times 6.43 = 25.72\,kW \quad \text{Ans. (i)}$$

$$\text{Fuel consumption} = 0.33 \times 20\,kg/h$$

$$= \frac{0.33 \times 20}{25.72}\,kg/kWh\,\text{(indicated)}$$

$$= 0.2566\,kg/ind.\,kWh \quad \text{Ans. (ii)}$$

Indicated thermal efficiency

$$= \frac{\text{heat energy converted into work in cylinders}}{\text{heat energy supplied}}$$

$$= \frac{3.6\,[MJ/kWh]}{0.2566\,[kg/ind.\,kWh] \times 43\,[MJ/kg]}$$

$$= 0.3262 \text{ or } 32.62\% \quad \text{Ans. (iii)}$$

2. From steam tables, for a saturation temperature of 39°C, pressure is 0.07 bar, and spec. volume 20.53 m³/kg

$\therefore$ mass of steam in volume of 15 m³

$$= \frac{15}{20.53} = 0.7307\,kg \quad \text{Ans. (i)}$$

Partial pressure due to air

$$= \text{total pressure} - \text{steam pressure}$$

$$= 0.09 - 0.07 = 0.02\,bar = 2\,kN/m^2$$

$$pV = mRT$$

mass of air $m = \dfrac{pV}{RT}$

$$= \frac{2 \times 15}{0.287 \times (39 + 273)} = 0.335\,kg \quad \text{Ans. (ii)}$$

Ratio of air to steam
$$= 0{\cdot}335 : 0{\cdot}7307$$

$$= \frac{0{\cdot}335}{0{\cdot}7307} : \frac{0{\cdot}7307}{0{\cdot}7307}$$

$$= 0{\cdot}4584 : 1 \quad \text{Ans. (iii)}$$

3. Representing volumes by linear dimensions:

Before alteration,

V_2 = clearance = $0{\cdot}09 \times 1200 = 108$ mm

V_1 = stroke + clearance = $1200 + 108 = 1308$ mm

$$p_1 V_1^{1.35} = p_2 V_2^{1.35}$$

$$p_1 \times 1308^{1.35} = 32 \times 108^{1.35}$$

$$p_1 = 32 \times \left\{ \frac{108}{1308} \right\}^{1.35} \qquad \dots \qquad \dots \quad \text{(i)}$$

After alteration,

$$V_2 = 108 - 5 = 103 \text{ mm}$$

$$V_1 = 1200 + 103 = 1303 \text{ mm}$$

$$p_1 V_1^{1.35} = p_2 V_2^{1.35}$$

$$p_1 \times 1303^{1.35} = p_2 \times 103^{1.35}$$

$$p_2 = p_1 \times \left\{ \frac{1303}{103} \right\}^{1.35} \qquad \dots \qquad \dots \quad \text{(ii)}$$

Substituting value of p_1 from (i) into (ii),

$$p_2 = 32 \times \left\{ \frac{108}{1308} \right\}^{1.35} \times \left\{ \frac{1303}{103} \right\}^{1.35}$$

$$= 32 \times \left\{ \frac{108 \times 1303}{1308 \times 103} \right\}^{1.35}$$

$$= 33{\cdot}94 \text{ bar}$$

Increase in final compression pressure

$$= 33{\cdot}94 - 32 = 1{\cdot}94 \text{ bar} \quad \text{Ans.}$$

4. Minimum oxygen required for complete combustion per kg of fuel

$$= 2\tfrac{2}{3} C + 8 \left(H_2 - \frac{O_2}{8} \right)$$

$$= 2\tfrac{2}{3}\,C + 8\,H_2 - O_2$$
$$= 2\tfrac{2}{3} \times 0.87 + 8 \times 0.11 - 0.02$$
$$= 3.18 \text{ kg}$$

$$\text{Minimum air} = \frac{100}{23} \times 3.18 = 13.82 \text{ kg}$$

Mass of nitrogen in 13·82 kg of air
$$= 0.77 \times 13.82 = 10.64 \text{ kg}$$
$$(\text{or, } 13.82 \text{ kg air} - 3.18 \text{ kg } O_2 = 10.64 \text{ kg } N_2)$$
$$CO_2 \text{ formed} = 3\tfrac{2}{3} \times 0.87 = 3.19 \text{ kg}$$
$$H_2O \text{ formed} = 9 \times 0.11 = 0.99 \text{ kg}$$

Total gases/kg fuel
$$= 10.64 + 3.19 + 0.99 = 14.82 \text{ kg}$$
$$\text{also, } 13.82 \text{ kg air} + 1 \text{ kg fuel} = 14.82 \text{ kg}$$
At 1·8 tonne of fuel per hour, fuel rate

$$= \frac{1.8 \times 10^3}{3600} = 0.5 \text{ kg/s}$$

Mass flow rate of each of the gases,   Ans. (i):

$$\text{Nitrogen} = 0.5 \times 10.64 = 5.32 \text{ kg/s}$$
$$CO_2 = 0.5 \times 3.19 = 1.595 \text{ kg/s}$$
$$H_2O = 0.5 \times 0.99 = 0.495 \text{ kg/s}$$
As a percentage analysis,   Ans. (ii):

$$\text{Nitrogen} = \frac{10.64}{14.82} \times 100 = 71.79\%$$

$$CO_2 = \frac{3.19}{14.82} \times 100 = 21.53\%$$

$$H_2O = \frac{0.99}{14.82} \times 100 = 6.68\%$$

5. From steam tables,

$$9 \text{ bar,}\quad h_g = 2774$$
$$5 \text{ bar,}\quad h_f = 640 \qquad h_{fg} = 2109 \qquad v_g = 0.3748$$
$$0.14 \text{ bar,}\quad h_f = 220 \qquad h_{fg} = 2376 \qquad v_g = 10.69$$

Enthalpy drop from 9 bar to 5 bar =
$$2774 - (640 + x \times 2109) = 107$$
$$2774 - 640 - 107 = x \times 2109$$
$$2027 = x \times 2109$$
$$\text{dryness at throat,} \quad x = 0{\cdot}9614$$

Specific volume of steam at throat
$$= 0{\cdot}9614 \times 0{\cdot}3748 = 0{\cdot}3603 \text{ m}^3/\text{kg}$$

Velocity [m/s] $= \sqrt{2 \times \text{spec. enthalpy drop [J/kg]}}$
Velocity through throat
$$= \sqrt{2 \times 107 \times 10^3} = 462{\cdot}6 \text{ m/s}$$

area [m²] × velocity [m/s] = mass flow [kg/s] × spec. vol. [m³/kg]

$\therefore$ Area at throat [mm²]

$$= \frac{23 \times 0{\cdot}3603}{60 \times 462{\cdot}6} \times 10^6 = 298{\cdot}5 \text{ mm}^2 \quad \text{Ans. (i)}$$

Effective enthalpy drop from entrance to exit
$$= 0{\cdot}92 \times 633 = 582{\cdot}4 \text{ kJ/kg}$$
Enthalpy drop from 9 bar to 0·14 bar =
$$2774 - (220 + x \times 2376) = 582{\cdot}4$$
$$2774 - 220 - 582{\cdot}4 = x \times 2376$$
$$1971{\cdot}6 = x \times 2376$$
$$\text{dryness at exit,} \quad x = 0{\cdot}8299$$

Specific volume of steam at exit
$$= 0{\cdot}8299 \times 10{\cdot}69 = 8{\cdot}869 \text{ m}^3/\text{kg}$$

Velocity at exit

$$= \sqrt{2 \times 582{\cdot}4 \times 10^3} = 1079 \text{ m/s}$$

Area at exit [mm²]

$$= \frac{\text{mass flow [kg/s]} \times \text{spec. vol. [m}^3/\text{kg]}}{\text{velocity [m/s]}} \times 10^6$$

$$= \frac{23 \times 8{\cdot}869}{60 \times 1079} \times 10^6 = 3150 \text{ mm}^2 \quad \text{Ans. (ii)}$$

6. From Freon-12 tables,

$$0{\cdot}8071 \text{ bar}, \quad h_f = 4{\cdot}42 \quad\quad h_g = 171{\cdot}9$$
$$\therefore h_{fg} = 171{\cdot}9 - 4{\cdot}42 = 167{\cdot}48$$
$$s_f = 0{\cdot}0187 \quad\quad s_g = 0{\cdot}7219$$
$$\therefore s_{fg} = 0{\cdot}7219 - 0{\cdot}0187 = 0{\cdot}7032$$
$$12{\cdot}19 \text{ bar}, \quad h_f = 84{\cdot}94 \text{ sat. temp.} = 50°C$$

$$\therefore \text{ at } 65°C \text{ freon is superheated by } 15°$$

$$h = 218{\cdot}64 \quad\quad s = 0{\cdot}7166$$

Ref. Fig. 101,

Isentropic compression: $s_1 = s_2$
$$0{\cdot}0187 + x_1 \times 0{\cdot}7032 = 0{\cdot}7166$$
$$x_1 \times 0{\cdot}7032 = 0{\cdot}6979$$
$$x_1 = 0{\cdot}9924$$

$$h_1 = h \text{ leaving evaporator} = h \text{ entering compressor}$$
$$= 4{\cdot}42 + 0{\cdot}9924 \times 167{\cdot}48$$
$$= 4{\cdot}42 + 166{\cdot}3 = 170{\cdot}72$$

$$h_4 = h \text{ entering evaporator} = h \text{ leaving condenser } (h_3)$$
$$\text{Refrigerating effect/kg} = h_1 - h_4$$
$$= 170{\cdot}72 - 84{\cdot}94 = 85{\cdot}78 \text{ kJ/kg}$$

Refrigerating effect for flow of 15 kg/min
$$= 15 \times 85{\cdot}78 = 1286 \text{ kJ/min} \quad \text{Ans. (i)}$$

Work transfer in compresser/kg $= h_2 - h_1$
$$= 218{\cdot}64 - 170{\cdot}72 = 47{\cdot}92 \text{ kJ/kg}$$

$$\text{Coeff. of performance} = \frac{\text{refrigerating effect}}{\text{work transfer}}$$

$$= \frac{85{\cdot}78}{47{\cdot}92} = 1{\cdot}79 \quad \text{Ans. (ii)}$$

7. From steam tables:

$$\text{Steam 14 bar}, \quad h_f = 830 \quad h_{fg} = 1960$$
$$\text{Water 44°C}, \quad h = 184{\cdot}2$$
$$\text{Water 100°C}, \quad h = 419{\cdot}1$$

Considering system of heater and hotwell combined as shown in Fig. 59, let x kg of steam be bled off to heater per kg of main steam:

Enthalpy entering system = Enthalpy leaving system

$$\begin{array}{c} h \text{ of } x \text{ kg of} \\ \text{main steam} \end{array} + \begin{array}{c} h \text{ of } (1-x) \text{ kg of} \\ \text{condensate water} \end{array} = \begin{array}{c} h \text{ of } 1 \text{ kg of feed} \\ \text{leaving heater} \end{array}$$

$$x(830 + 0.98 \times 1960) + (1-x) \times 184.2 = 419.1$$
$$2751x + 184.2 - 184.2x = 419.1$$
$$2566.8x = 234.9$$
$$x = 0.091 \ 52 \text{ kg}$$

Bled steam to heater as a percentage of main
steam supply $= 9.152\%$ Ans.

8.
$$T_1 = 20°C + 273 = 293 \text{ K}$$
$$T_2 = 512°C + 273 = 785 \text{ K}$$

$$\frac{T_1}{T_2} = \left\{ \frac{V_2}{T_1} \right\}^{n-1}$$

$$\frac{293}{785} = \left\{ \frac{720}{9600} \right\}^{n-1}$$

$$\log \frac{785}{293} = (n-1) \times \log \frac{9600}{720}$$

$$0.4280 = (n-1) \times 1.1250$$
$$0.428 = 1.125n - 1.125$$
$$1.553 = 1.125n$$
$$n = 1.38 \quad \text{Ans. (i)}$$

From $pV = mRT$

Mass of gas, $m = \dfrac{p_1 V_1}{RT_1}$

$$= \frac{(1 \times 10^2) \times (9600 \times 10^{-6})}{0.264 \times 293} = 0.012 \ 41 \text{ kg}$$

Work done $= \dfrac{p_1 V_1 - p_2 V_2}{n-1}$

$$= \frac{mR(T_1 - T_2)}{n-1}$$

$$= \frac{0 \cdot 012\ 41 \times 0 \cdot 264 \times (293 - 785)}{1 \cdot 38 - 1}$$

$$= \frac{0 \cdot 012\ 41 \times 0 \cdot 264 \times (-492)}{0 \cdot 38}$$

$$= -4 \cdot 242 \text{ kJ}$$

The minus sign indicating work done ON the gas.

Work done on gas $= 4 \cdot 242$ kJ   Ans. (ii)

Change in internal energy,
$$E_2 - E_1 = m \times c_V \times (T_2 - T_1)$$
$$= 0 \cdot 012\ 41 \times 0 \cdot 657 \times 492$$
$$= 4 \cdot 013 \text{ kJ (increase)}   \text{Ans. (iii)}$$

$$\frac{\text{Heat supplied}}{\text{to the gas}} = \frac{\text{increase in}}{\text{internal energy}} + \frac{\text{work done}}{\text{by the gas}}$$

$$= 4 \cdot 013 + (-4 \cdot 242)$$
$$= -0 \cdot 229 \text{ kJ}$$

The minus sign indicating heat rejected

$\therefore$ Transfer of heat from the gas $= 0 \cdot 229$ kJ   Ans. (iv)

9. From steam tables:
$$8 \text{ bar,}\quad h_f = 721 \quad h_{fg} = 2048$$
$$15 \text{ bar,}\quad h_f = 845 \quad h_{fg} = 1947$$
$$15 \text{ bar } 250°\text{C,}\quad h = 2925$$

Mixing process:

Enthalpy before mixing $=$ Enthalpy after
$$3(845 + 0 \cdot 88 \times 1947) + 2 \times 2925 = 5(845 + x \times 1947)$$
$$7677 + 5850 = 4225 + 9735x$$
$$9302 = 9735x$$
$$x = 0 \cdot 9557 \quad \text{Ans. (i)}$$

Throttling process:

Enthalpy before throttling $=$ Enthalpy after
$$845 + 0 \cdot 9557 \times 1947 = 721 + x \times 2048$$
$$1985 = 2048x$$
$$x = 0 \cdot 9694 \quad \text{Ans. (ii)}$$

10 Temperature difference across thickness of wood

$$= \frac{QS_{\mathrm{w}}}{\lambda_{\mathrm{w}}At}$$

$$= \frac{24 \times 0.03}{0.18 \times 1 \times 1} = 4\,\mathrm{K} \quad \text{Ans (ia)}$$

Temperature difference across thickness of cork

$$= \frac{QS_{\mathrm{c}}}{\lambda_{\mathrm{c}}At}$$

$$= \frac{24 \times 0.07}{0.05 \times 1 \times 1} = 33.6\,\mathrm{K} \quad \text{Ans (ib)}$$

Temperature difference between inside and outside atmospheres and their respective exposed surfaces

$$= \frac{Q}{hAt}$$

$$= \frac{24}{10 \times 1 \times 1} = 2.4\,\mathrm{K}$$

Total temperature difference between outside atmosphere and inside atmosphere
$$= 2.4 + 4 + 33.6 + 2.4 = 42.4\,\mathrm{K} \text{ or } 42.4°\mathrm{C} \quad \text{Ans (ii)}$$

Room temperature $= 20 - 42.4 = -22.4°\mathrm{C}$ Ans (iii)

11 Referring to Fig 41, representing volumes in terms of ratios

$$V_1 = 13.5 \qquad V_2 = 1$$
$$\text{stroke volume} = V_1 - V_2 = 12.5$$
$$\text{Fuel admission period} = 0.1 \text{ stroke}$$
$$= 0.1 \times 12.5 = 1.25$$
$$V_3 = 1.25 + 1 = 2.25$$
$$V_4 = V_1 = 13.5$$
$$T_1 = 38°\mathrm{C} + 273 = 311\,\mathrm{K}$$

$$\frac{T_2}{T_1} = \left\{\frac{V_1}{V_2}\right\}^{\gamma-1}$$

$$T_2 = 311 \times 13.5^{0.4} = 880.8 \text{ K}$$

Temperature at end of compression
$$= 880.8 - 273 = 607.8°C \quad \text{Ans. (i)}$$

$$\frac{T_3}{T_2} = \frac{V_3}{V_2}$$

$$T_3 = \frac{880.8 \times 2.25}{1} = 1982 \text{ K}$$

Temperature at end of combustion
$$= 1982 - 273 = 1709°C \quad \text{Ans. (ii)}$$

$$\frac{T_4}{T_3} = \left\{\frac{V_3}{V_4}\right\}^{\gamma-1}$$

$$T_4 = 1982 \times \left\{\frac{2.25}{13.5}\right\}^{0.4} = 968.1 \text{ K}$$

Temperature at end of expansion
$$= 968.1 - 273 = 695.1°C \quad \text{Ans. (iii)}$$

$$\text{Ideal thermal efficiency} = \frac{\text{heat energy converted into work}}{\text{heat energy supplied}}$$

$$= \frac{\text{heat supplied} - \text{heat rejected}}{\text{heat supplied}}$$

$$= 1 - \frac{\text{heat rejected}}{\text{heat supplied}}$$

$$= 1 - \frac{c_V (T_4 - T_1)}{c_P (T_3 - T_2)}$$

$$= 1 - \frac{0.718 (968.1 - 311)}{1.005 (1982 - 880.8)}$$

$$= 1 - 0.4265$$

$$= 0.5735 \text{ or } 57.35\% \quad \text{Ans. (iv)}$$

12. $\text{ip} = p_m ALn$
$= 8{\cdot}9 \times 10^2 \times 0{\cdot}7854 \times 0{\cdot}36^2 \times 0{\cdot}78 \times 2{\cdot}3$
$= 162{\cdot}5 \text{ kW}$ Ans. (i)

$\text{bp} = T\omega$
$= 8 \times 1{\cdot}25 \times 2\pi \times 2{\cdot}3$
$= 144{\cdot}5 \text{ kW}$ Ans. (ii)

Heat energy supplied per second

$$= \frac{0{\cdot}251 \times 144{\cdot}5 \times 41{\cdot}5 \times 10^3}{3600} = 418{\cdot}1 \text{ kJ/s}$$

Heat energy converted into work in cylinder per second
$= 162{\cdot}5 \text{ kJ/s}$

Indicated thermal efficiency

$$= \frac{162{\cdot}5}{418{\cdot}1} = 0{\cdot}3887 \text{ or } 38{\cdot}87\% \quad \text{Ans. (iii)}$$

Total heat energy loss per second
$= \text{heat supplied} - \text{heat converted into work}$
$= 418{\cdot}1 - 162{\cdot}5 = 255{\cdot}6 \text{ kJ/s}$ Ans. (iv)

13. From steam tables:
$0{\cdot}07 \text{ bar,} \quad h_f = 163 \quad h_{fg} = 2409$
$s_f = 0{\cdot}559 \quad s_{fg} = 7{\cdot}715$

17·5 bar 300°C, taking mean values between 15 bar 300°C and 20 bar 300°C

$15 \text{ bar } 300°\text{C,} \quad h = 3039 \qquad s = 6{\cdot}919$
$20 \text{ bar } 300°\text{C,} \quad h = 3025 \qquad s = 6{\cdot}768$
$17{\cdot}5 \text{ bar } 300°\text{C,} \quad h = \tfrac{1}{2}(3039 + 3025) = 3032$
$s = \tfrac{1}{2}(6{\cdot}919 + 6{\cdot}768) = 6{\cdot}8435$

Entropy after expansion $=$ Entropy before
$0{\cdot}559 + x \times 7{\cdot}715 = 6{\cdot}8435$
$x \times 7{\cdot}715 = 6{\cdot}2845$
dryness at exhaust, $x = 0{\cdot}8145$ Ans. (i)

Enthalpy of exhaust
$= 163 + 0{\cdot}8145 \times 2409 = 2125 \text{ kJ/kg}$

$$\text{Rankine efficiency} = \frac{h_1 - h_2}{h_1 - h_{f2}}$$

$$= \frac{3032 - 2125}{3032 - 163} = \frac{907}{2869}$$

$$= 0 \cdot 3161 \text{ or } 31 \cdot 61\% \text{ Ans. (ii)}$$

See Fig. 80 Chapter 12 for Rankine cycle.

14. From steam tables:
$$1 \cdot 2 \text{ bar, } \quad h_g = 2683$$
$$7 \text{ bar } 300°C, \quad h = 3060$$
Enthalpy drop through nozzle
$$= 3060 - 2683 = 377 \text{ kJ/kg}$$
Velocity of steam at nozzle exit [m/s]

$$= \sqrt{2 \times \text{spec. enthalpy drop [J/kg]}}$$

$$= \sqrt{2 \times 377 \times 10^3} = 868 \cdot 4 \text{ m/s}$$

Ref. Fig. 90:
$$v_{w1} = v_1 \cos \alpha_1 = 868 \cdot 4 \times \cos 20° = 816 \text{ m/s}$$
$$u = 40\% \text{ of } v_1 = 0 \cdot 4 \quad \times \quad 868 \cdot 4 = 347 \cdot 4 \text{ m/s}$$
$$x = v_{w1} \quad - \quad u = 816 \quad - \quad 347 \cdot 4 = 468 \cdot 6 \text{ m/s}$$
$$v_{a1} = v_1 \sin \alpha_1 = 868 \cdot 4 \times \sin 20° = 297 \text{ m/s}$$
$$v_{r1} = \sqrt{297^2 + 468 \cdot 6^2} = 554 \cdot 8 \text{ m/s}$$
$$v_{r2} = 0 \cdot 8 \times 554 \cdot 8 = 443 \cdot 8 \text{ m/s}$$
$$v_{a2} = v_{r2} \times \sin \beta_2 = 443 \cdot 8 \times \sin 35° = 254 \cdot 6 \text{ m/s}$$

Axial force on blades [N]
$$= \text{mass flow [kg/s]} \times \text{change of axial velocity [m/s]}$$
$$= 0 \cdot 5 \times (297 - 254 \cdot 6)$$
$$= 21 \cdot 2 \text{ N} \quad \text{Ans. (i)}$$

$$v_{w2} = v_{r2} \cos \beta_2 - u$$
$$= 443 \cdot 8 \times \cos 35° - 347 \cdot 4 = 16 \cdot 2 \text{ m/s}$$

Effective change of velocity, $v_w = v_{w1} + v_{w2}$
$$= 816 + 16 \cdot 2 = 832 \cdot 2 \text{ m/s}$$

Tangential force on blades $= \dot{m} v_w$
$$= 0 \cdot 5 \times 832 \cdot 2 = 416 \cdot 1 \text{ N}$$

$$\text{Power [W]} = \text{force [N]} \times \text{blade velocity [m/s]}$$
$$= 416 \cdot 1 \times 347 \cdot 4$$
$$= 1 \cdot 445 \times 10^5 \text{ W} = 144 \cdot 5 \text{ kW} \quad \text{Ans. (ii)}$$

15. $$\text{Input} = \frac{\text{output}}{\text{efficiency}} = \frac{7500}{0 \cdot 95} \text{ kW}$$

Electrical losses $= 5\%$ of input

$$= 0 \cdot 05 \times \frac{7500}{0 \cdot 95} = 394 \cdot 8 \text{ kW}$$

$\therefore$ heat generated in windings $= 394 \cdot 8$ kJ/s
$$1 \text{ mm water} = 9 \cdot 81 \text{ N/m}^2 \text{ of pressure}$$
Gauge pressure of cooling air
$$= 153 \times 9 \cdot 81 = 1501 \text{ N/m}^2 = 1 \cdot 501 \text{ kN/m}^2$$
Atmospheric press. $= 1 \cdot 013 \text{ bar} = 101 \cdot 3 \text{ kN/m}^2$
Abs. press. of air $= 1 \cdot 501 + 101 \cdot 3 = 102 \cdot 801 \text{ kN/m}^2$
From $pV = mRT$, mass of cooling air per second

$$m = \frac{pV}{RT} = \frac{102 \cdot 8 \times 16 \cdot 5}{0 \cdot 287 \times (21 + 273)} = 20 \cdot 1 \text{ kg/s}$$

Heat energy transferred to air [kJ/s]
$$= \text{mass} \times c_P \times \text{temperature rise}$$
$\therefore 20 \cdot 1 \times 1 \cdot 005 \times \text{temp. rise} = 394 \cdot 8$

$$\text{temp. rise} = \frac{394 \cdot 8}{20 \cdot 1 \times 1 \cdot 005} = 19 \cdot 54 °C$$

Exit temp. of air $= 21 + 19 \cdot 54 = 40 \cdot 54 °C$
Difference in temperature between exit air and windings
$$= 55 - 40 \cdot 54 = 14 \cdot 46 °C \quad \text{Ans.}$$

16. The ratio of partial pressures is the same as the ratio of partial volumes, therefore,
partial press. of $CO_2 = 0 \cdot 14 \times 1 \cdot 01 = 0 \cdot 1414$ bar $\Big\}$ Ans. (i)
,,   ,,   ,, $N_2 = 0 \cdot 86 \times 1 \cdot 01 = 0 \cdot 8686$ bar

For mass of CO_2, $pV = mRT$

where $p = 0.1414 \times 10^2 \, kN/m^2$

$V = 500 \times 10^{-6} \, m^3$

$$m = \frac{pV}{RT} = \frac{0.1414 \times 10^2 \times 500 \times 10^{-6}}{0.189 \times (20 + 273)}$$

$$= 1.276 \times 10^{-4} \, kg \text{ or } 0.1276 \, g \quad \text{Ans. (ii)}$$

17. See Chapter 14 for sketch of bomb calorimeter and description of experiment, the result is the *higher* calorific value.

Mass of 1·65 litres of water $= 1.65 \, kg$

$$\begin{array}{c} \text{Heat energy released} \\ \text{by fuel} \end{array} = \begin{array}{c} \text{Heat energy transferred} \\ \text{to water and bomb} \end{array}$$

mass of fuel $\times$ cal. value

$$= \left\{ \begin{array}{c} \text{mass of} \\ \text{water} \end{array} + \begin{array}{c} \text{water equiv.} \\ \text{of bomb} \end{array} \right\} \times \text{spec. heat} \times \text{temp. rise}$$

$$1.5 \times 10^{-3} \times \text{c.v.} = (1.65 + 0.45) \times 4.2 \times 7.15$$

$$\text{c.v.} = \frac{2.1 \times 4.2 \times 7.15}{1.5 \times 10^{-3}}$$

$$= 4.204 \times 10^4 \, kJ/kg$$
$$= 42.04 \, MJ/kg \quad \text{Ans.}$$

18. From steam tables:

$$16 \, bar, \quad h_g = 2794 \quad s_g = 6.422$$
$$9 \, bar, \quad h_g = 2774$$

Throttling is a constant enthalpy process, therefore h at 9 bar after throttling is equal to h_g at 16 bar before throttling, this is 2794 kJ/kg, hence at 9 bar the steam is superheated, and the enthalpy of superheat is $2794 - 2774 = 20$ kJ/kg.

From superheat tables, at 9 bar:

$$\text{when } h = 2835 \qquad s = 6.753$$
$$\text{,, } \quad h = 2774 \qquad s = 6.623$$
$$\text{For difference of } h = 61 \qquad s = 0.13$$
$$\text{,,} \qquad \text{,,} \qquad \text{,, } h = 20 \qquad s = \frac{20}{61} \times 0.13 = 0.042\,62$$

$\therefore$ entropy at 9 bar $= 6{\cdot}623 + 0{\cdot}042\,62 = 6{\cdot}665\,62$ kJ/kg
Change in entropy from 16 bar dry sat.
$\qquad = 6{\cdot}665\,62 - 6{\cdot}422 = 0{\cdot}243\,62$ increase   Ans. (i)

From steam tables, $1{\cdot}0$ bar:
$$s_f = 1{\cdot}303 \qquad s_{fg} = 6{\cdot}056 \qquad s_g = 7{\cdot}359$$
$\therefore$ steam is wet after isentropic expansion,
$\qquad$ Entropy after expansion $=$ Entropy before
$$1{\cdot}303 + x \times 6{\cdot}056 = 6{\cdot}665\,62$$
$$\text{Final dryness, } x = 0{\cdot}8855 \quad \text{Ans. (ii)}$$

19.

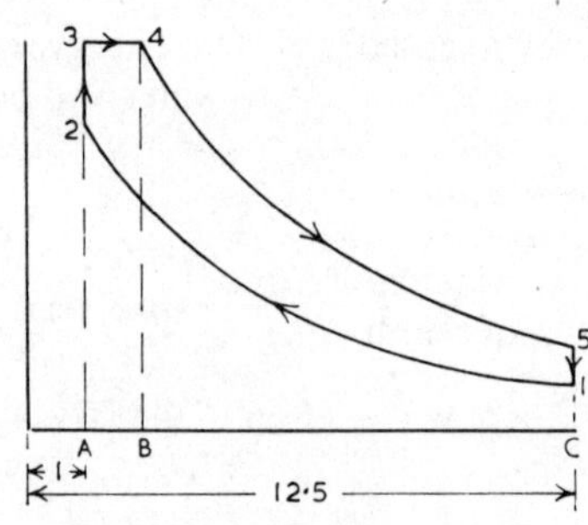

Fig. 108

Representing volumes by ratios:
$$V_1 \text{ and } V_5 = 12{\cdot}5 \qquad V_2 \text{ and } V_3 = 1$$
$$p_1 V_1^{\gamma} = p_2 V_2^{\gamma}$$
$$1{\cdot}013 \times 12{\cdot}5^{1{\cdot}4} = p_2 \times 1^{1{\cdot}4} \qquad p_2 = 34{\cdot}77 \text{ bar}$$

$$\frac{p_1 V_1}{T_1} = \frac{p_2 V_2}{T_2}$$

$$T_2 = \frac{(35 + 273) \times 34{\cdot}77 \times 1}{1{\cdot}013 \times 12{\cdot}5} = 845{\cdot}9 \text{ K}$$

$$\frac{T_3}{T_2} = \frac{p_3}{p_2}$$

$$T_3 = \frac{845{\cdot}9 \times 40}{34{\cdot}77} = 973{\cdot}4 \text{ K}$$

$$\frac{V_4}{V_3} = \frac{T_4}{T_3}$$

$$V_4 = \frac{1 \times (1425 + 273)}{973 \cdot 4} = 1 \cdot 745$$

$$p_4 V_4^{\gamma} = p_5 V_5^{\gamma}$$
$$40 \times 1 \cdot 745^{1 \cdot 4} = p_5 \times 12 \cdot 5^{1 \cdot 4}$$

$$p_5 = 40 \times \left\{ \frac{1 \cdot 745}{12 \cdot 5} \right\}^{1 \cdot 4} = 2 \cdot 541 \text{ bar}$$

Areas representing positive work done:
Area during combustion $= A\,3\,4\,B$
$$= 40 \times (1 \cdot 745 - 1) = 29 \cdot 8$$
Area during expansion $= B\,4\,5\,C$

$$= \frac{p_4 V_4 - p_5 V_5}{\gamma - 1}$$

$$= \frac{40 \times 1 \cdot 745 - 2 \cdot 541 \times 12 \cdot 5}{1 \cdot 4 - 1} = 95 \cdot 07$$

Gross area $= A\,3\,4\,5\,C$
$$= 29 \cdot 8 + 95 \cdot 07 = 124 \cdot 87$$
Area representing negative work done:
Area during compression $= C\,1\,2\,A$

$$= \frac{p_1 V_1 - p_2 V_2}{\gamma - 1}$$

$$= \frac{1 \cdot 013 \times 12 \cdot 5 - 34 \cdot 77 \times 1}{1 \cdot 4 - 1} = -55 \cdot 27$$

Net area representing useful work
$$= 124 \cdot 87 - 55 \cdot 27 = 69 \cdot 6$$
Mean effective pressure $=$ mean height of net area

$$= \frac{\text{area}}{\text{length}} = \frac{69 \cdot 6}{12 \cdot 5 - 1}$$

$$= 6 \cdot 052 \text{ bar} \quad \text{Ans.}$$

20. Let Willans' law be expressed by:
$$m = a + bP$$
where m = mass of steam consumed [Mg/h]
P = power developed [kW]
a and b = constants

Inserting the two pairs of given quantities as a simultaneous equation:

$$25 \cdot 5 = a + b \times 3000 \ \text{....} \quad \text{....} \quad \text{....} \quad \text{....} \quad \text{(i)}$$
$$\underline{18 \cdot 9 = a + b \times 1800 \ \text{....} \quad \text{....} \quad \text{....} \quad \text{....} \quad \text{(ii)}}$$
$$6 \cdot 6 = \quad b \times 1200 \ \text{ by subtraction}$$
$$b = 0 \cdot 0055$$

From (i) $25 \cdot 5 = a + 0 \cdot 0055 \times 3000$
$$a = 9$$

Willans' law: $m = 9 + 0 \cdot 0055P$ Ans. (i)

When P = 2500 kW:
$$m = 9 + 0 \cdot 0055 \times 2500$$
$$= 22 \cdot 75 \ \text{Mg/h}$$

Specific steam consumption

$$= \frac{22 \cdot 75 \times 10^3}{2500} = 9 \cdot 1 \ \text{kg/kWh} \quad \text{Ans. (ii)}$$

21. Working on the basis of 1 kg fuel:

$$\text{Theoretical air} = \frac{100}{23}\left\{2\tfrac{2}{3}\,C + 8\left(H_2 - \frac{O_2}{8}\right)\right\}$$

$$= \frac{100}{23}\{2\tfrac{2}{3}\,C + 8\,H_2 - O_2\}$$

$$= \frac{100}{23}\{2\tfrac{2}{3} \times 0 \cdot 855 + 8 \times 0 \cdot 135 - 0 \cdot 1\}$$

$$= \frac{100}{23} \times 3 \cdot 35 = 14 \cdot 56 \ \text{kg}$$

Excess air $= 0 \cdot 25 \times 14 \cdot 56 = 3 \cdot 64$ kg
Actual air $= 14 \cdot 56 + 3 \cdot 64 = 18 \cdot 2$ kg
Mass products of combustion per kg fuel:
$$CO_2 = 3\tfrac{2}{3}\,C = 3\tfrac{2}{3} \times 0 \cdot 855 = 3 \cdot 135 \ \text{kg}$$

$$H_2O = 9\,H_2 = 9 \times 0\cdot135 = 1\cdot215\,\text{kg}$$
$$O_2 = 23\% \text{ of excess air} = 0\cdot23 \times 3\cdot64 = 0\cdot8372\,\text{kg}$$
$$N_2 = 77\% \text{ of all air} = 0\cdot77 \times 18\cdot2 = 14\cdot014\,\text{kg}$$

Total products $= 1\,\text{kg fuel} + 18\cdot2\,\text{kg air} = 19\cdot2\,\text{kg}$

$\%$ mass analysis,   Ans. (i):

$$CO_2 = \frac{3\cdot135}{19\cdot2} \times 100 = 16\cdot32\%$$

$$H_2O = \frac{1\cdot215}{19\cdot2} \times 100 = 6\cdot33\%$$

$$O_2 = \frac{0\cdot8372}{19\cdot2} \times 100 = 4\cdot36\%$$

$$N_2 = \frac{14\cdot014}{19\cdot2} \times 100 = 72\cdot99\%$$

Dry flue gases $=$ total gases $- H_2O$
$$= 19\cdot2 - 1\cdot215 = 17\cdot985\,\text{kg}$$

$\%$ volumetric analysis of dry flue gases,   Ans. (ii):

DRY FLUE GAS a	RATIO OF MASSES b	MOLECULAR WEIGHT c	RATIO OF VOLUMES $d = b \div c$	VOLUMES AS PERCENTAGES $\dfrac{d}{\text{total vol.}} \times 100$
CO_2	16·32	44	0·371	11·91 %
O_2	4·36	32	0·1362	4·38 %
N_2	72·99	28	2·606	83·71 %

Total $= 3\cdot1132$

22. Internal volume of the two pipes, and volume of oil [mm³]
$$= 2 \times 0\cdot7854 \times 13^2 \times 76 \times 10^3$$
$$= 2\cdot018 \times 10^7\,\text{mm}^3$$

Coeff. of cubical expansion of copper
$$\beta_C = 3 \times \text{coeff. of linear expansion}$$
$$= 3 \times 1\cdot7 \times 10^{-5} = 5\cdot1 \times 10^{-5}/^{\circ}\text{C}$$

Increase in temperature $= 30 - 16 = 14^{\circ}\text{C}$

Increase in internal volume of pipes
$$= \beta_C \times V \times \text{temp. rise}$$
Increase in volume of oil
$$= \beta_O \times V \times \text{temp. rise}$$
difference in volumetric expansion
$$= (\beta_O - \beta_C) \times V \times \text{temp. rise}$$
$$= (77 - 5{\cdot}1) \times 10^{-5} \times 2{\cdot}018 \times 10^7 \times 14$$
$$= 2{\cdot}031 \times 10^5 \text{ mm}^3$$
$$\text{Overspill} = 0{\cdot}2031 \text{ litre}\quad\text{Ans.}$$

23.
$$T_1 = 520 + 273 = 793 \text{ K}$$
$$T_2 = 30 + 273 = 303 \text{ K}$$

$$\frac{T_1}{T_2} = \left\{\frac{V_2}{V_1}\right\}^{n-1}$$

$$\log\left\{\frac{793}{303}\right\} = (n-1) \times \log\left\{\frac{6}{0{\cdot}5}\right\}$$

$$0{\cdot}4179 = (n-1) \times 1{\cdot}0792$$
$$0{\cdot}4179 = 1{\cdot}0792n - 1{\cdot}0792$$
$$1{\cdot}4971 = 1{\cdot}0792n$$

$$n = \frac{1{\cdot}4971}{1{\cdot}0792} = 1{\cdot}388\quad\text{Ans. (i)}$$

From $p_1 V_1 = mRT_1$

$$m = \frac{32 \times 10^2 \times 0{\cdot}5}{0{\cdot}287 \times 793} = 7{\cdot}03 \text{ kg}$$

Increase in internal energy,
$$E_2 - E_1 = m \times c_V \times (T_2 - T_1)$$
$$= 7{\cdot}03 \times 0{\cdot}718 \times (303 - 793)$$
$$= 7{\cdot}03 \times 0{\cdot}718 \times (-490)$$
$$= -2474 \text{ kJ}$$
The minus sign indicates a *decrease*, therefore,
decrease in internal energy $= 2474$ kJ   Ans. (ii)

$$\text{Work done} = \frac{p_1 V_1 - p_2 V_2}{n-1}$$

$$= \frac{mR(T_1 - T_2)}{n - 1}$$

$$= \frac{7 \cdot 03 \times 0 \cdot 287 \times 490}{0 \cdot 388} = 2549 \text{ kJ}$$

Heat supplied $=$ increase in internal energy $+$ work done
$$= -2474 + 2549$$
$$= 75 \text{ kJ} \quad \text{Ans. (iii)}$$

24. Fuel consumption $= 2700 \times 0 \cdot 225$ kg/h
    Mass of exhaust gases per kg of fuel burned
$= 1$ kg fuel $+ 18$ kg air $= 19$ kg
Mass of exhaust gases per hour
$= 2700 \times 0 \cdot 225 \times 19$ kg
Heat energy transferred to steam
$= 0 \cdot 8$ of heat energy given up by gases
$= 0 \cdot 8 \times$ mass $\times$ spec. heat $\times$ temp. drop
$= 0 \cdot 8 \times 2700 \times 0 \cdot 225 \times 19 \times 1 \cdot 05 \times (327 - 215)$
$= 1 \cdot 086 \times 10^6$ kJ/h
From steam tables,
$$\text{Steam 15 bar,} \quad h_f = 845 \quad h_{fg} = 1947$$
$$\text{Water } 50^\circ\text{C,} \quad h = 209 \cdot 3$$
Heat energy required per kg steam
$= (845 + 0 \cdot 96 \times 1947) - 209 \cdot 3 = 2504 \cdot 7$ kJ/kg
Mass of steam generated
$$= \frac{1 \cdot 086 \times 10^6}{2504 \cdot 7} = 433 \cdot 6 \text{ kg/h} \quad \text{Ans.}$$

25. From steam tables,
$$40 \text{ bar } 450^\circ\text{C,} \quad h = 3330 \quad s = 6 \cdot 935$$
$\quad 2 \cdot 5$ bar 300°C,   taking mean values between 2 and 3 bar,
$$2 \text{ bar } 300^\circ\text{C,} \quad h = 3072 \quad s = 7 \cdot 892$$
$$3 \text{ bar } 300^\circ\text{C,} \quad h = 3070 \quad s = 7 \cdot 702$$
$$\therefore 2 \cdot 5 \text{ bar } 300^\circ\text{C,} \quad h = 3071 \quad s = 7 \cdot 797$$
$$2 \cdot 5 \text{ bar sat.,} \quad h_f = 535 \quad s_f = 1 \cdot 607$$
$$h_{fg} = 2182 \quad s_{fg} = 5 \cdot 446$$
$$0 \cdot 06 \text{ bar,} \quad h_f = 152 \quad s_f = 0 \cdot 521$$
$$h_{fg} = 2415 \quad s_{fg} = 7 \cdot 808$$
Referring to notation on Fig. 85:
$$h_1 = h \text{ at } 40 \text{ bar } 450^\circ\text{C} = 3330$$

Isentropic expansion from 40 bar 450°C to 2·5 bar,

Entropy after expansion = Entropy before

$$1\!\cdot\!607 + x \times 5\!\cdot\!446 = 6\!\cdot\!935$$

dryness at 2·5 bar, $x = 0\!\cdot\!9781$

$$\begin{aligned}
h_2 &= h \text{ at } 2\!\cdot\!5 \text{ bar } 0\!\cdot\!9781 \text{ dry}\\
&= 535 + 0\!\cdot\!9781 \times 2182 = 2669\\
h_3 &= h \text{ at } 2\!\cdot\!5 \text{ bar } 300°C \quad = 3071
\end{aligned}$$

Isentropic expansion from 2·5 bar 300°C to 0·06 bar,

Entropy after expansion = Entropy before

$$0\!\cdot\!521 + x \times 7\!\cdot\!808 = 7\!\cdot\!797$$

dryness at 0·06 bar, $x = 0\!\cdot\!9319$

$$\begin{aligned}
h_4 &= h \text{ at } 0\!\cdot\!06 \text{ bar } 0\!\cdot\!9319 \text{ dry}\\
&= 152 + 0\!\cdot\!9319 \times 2415 = 2403\\
h_{4f} &= h_f \text{ at } 0\!\cdot\!06 \text{ bar} = 152
\end{aligned}$$

$$\text{Rankine effic.} = \frac{\text{heat energy given up by steam through engine}}{\text{heat energy supplied in boiler and reheater}}$$

$$= \frac{(h_1 - h_2) + (h_3 - h_4)}{(h_1 - h_{4f}) + (h_3 - h_2)}$$

$$= \frac{(3330 - 2669) + (3071 - 2403)}{(3330 - 152) + (3071 - 2669)}$$

$$= \frac{661 + 668}{3178 + 402} = \frac{1329}{3580}$$

$$= 0\!\cdot\!3712 \text{ or } 37\!\cdot\!12\% \quad \text{Ans.}$$

26. Permeable space $= 0\!\cdot\!75$ of engine-room

$$= 0\!\cdot\!75 \times 19 \times 12 \times 7 \text{ m}^3$$

Volume of CO_2 gas required

$$= 25\% \text{ of permeable space}$$
$$= 0\!\cdot\!25 \times 0\!\cdot\!75 \times 19 \times 12 \times 7 = 299\!\cdot\!3 \text{ m}^3$$

From $pV = mRT$

$$m = \frac{1\!\cdot\!013 \times 10^2 \times 299\!\cdot\!3}{0\!\cdot\!189 \times (16 + 273)} = 555 \text{ kg} \quad \text{Ans. (i)}$$

Vol. of bottles $= 555 \times 1\!\cdot\!24 \times 10^{-3}$

$$= 0\!\cdot\!6882 \text{ m}^3 \quad \text{Ans. (ii).}$$

27   Reference to Freon-12 tables, and Fig 101,

Compressor suction and evaporator exit:
1·509 bar,   sat. temp. = —20°C,
$\therefore$ at —5°C refrigerant is superheated by 15°
$h_1 = h$ at 1·509 bar, supht. 15° = 187·75
Compressor discharge:
4·914 bar,   sat. temp. = 15°C
$\therefore$ at 45°C refrigerant is superheated by 30°
$h_2 = h$ at 4·914 bar, supht. 30° = 214·35
Condenser outlet: $h_3 = h_f$ at 15°C = 50·1
Evaporator inlet : $h_4 = h_3 = 50·1$

Coeff. of performance

$$= \frac{\text{refrigerating effect in evaporator [kJ/kg]}}{\text{work transfer in compressor [kJ/kg]}}$$

$$= \frac{h_1 - h_4}{h_2 - h_1}$$

$$= \frac{187·75 - 50·1}{214·35 - 187·75} = 5·175 \quad \text{Ans.}$$

28. From steam tables,
20 bar, $h_f = 909$ $h_{fg} = 1890$ $v_g = 0·099\,57$
1·25 bar 150°C, taking mean value between 1·0 and 1·5 bar 150°C,

$$h = \tfrac{1}{2}(2777 + 2773) = 2775$$

$$\text{Enthalpy before throttling} = \text{Enthalpy after}$$
$$909 + x \times 1890 = 2775$$
$$\text{dryness of steam,} \quad x = 0·9872 \quad \text{Ans. (i)}$$

From steam tables,
20 bar 350°C, $h = 3138$ $v = 0·1386$
Heat energy supplied per kg in superheaters
        = increase of specific enthalpy
        = 3138 — 2775 = 363 kJ/kg   Ans. (ii)

Specific volume of wet steam entering superheaters
        = 0·9872 × 0·099 57 = 0·098 28 m³/kg
% Increase in volume

$$= \frac{0·1386 - 0·09828}{0·09828} \times 100 = 41·02\% \quad \text{Ans. (iii)}$$

29. From $pV = mRT$, initial mass of gas

$$= m = \frac{pV}{RT} = \frac{0 \cdot 58 \times 10^2 \times 0 \cdot 7}{0 \cdot 27 \times (18 + 273)}$$

$$= 0 \cdot 5166 \text{ kg} \quad \text{Ans. (i)}$$

From $pV = mRT$, partial press. of admitted gas

$$= p = \frac{mRT}{V} = \frac{0 \cdot 35 \times 0 \cdot 29 \times (18 + 273)}{0 \cdot 7}$$

$$= 42 \cdot 2 \text{ kN/m}^2 = 0 \cdot 422 \text{ bar}$$

Total pressure = sum of partial pressures
$$= 0 \cdot 58 + 0 \cdot 422 = 1 \cdot 002 \text{ bar Ans. (ii)}$$

30. Referring to Fig. 39 (Chapter 8):

$$T_1 = 32 + 273 = 305 \text{ K} \qquad v_1 = v_4 = 8 \cdot 6$$
$$T_3 = 1650 + 273 = 1923 \text{ K} \qquad v_2 = v_3 = 1$$

$$\frac{T_2}{T_1} = \left\{ \frac{V_1}{V_2} \right\}^{n-1}$$

$$T_2 = 305 \times 8 \cdot 6^{0 \cdot 4} = 721 \cdot 3 \text{ K}$$

Temp. at end of compression $= 721 \cdot 3 - 273 = 448 \cdot 3°\text{C Ans.(i)}$

$$\frac{T_4}{T_3} = \left\{ \frac{V_3}{V_4} \right\}^{n-1}$$

$$T_4 = \frac{1923}{8 \cdot 6^{0 \cdot 4}} = 813 \cdot 2 \text{ K}$$

Temp. at end of expansion $= 813 \cdot 2 - 273 = 540 \cdot 4°\text{C Ans. (ii)}$

$$\text{Theoretical effic.} = \frac{\text{heat supplied} - \text{heat rejected}}{\text{heat supplied}}$$

$$= 1 - \frac{m \times c_v \times (T_4 - T_1)}{m \times c_v \times (T_3 - T_2)}$$

$$= 1 - \frac{T_4 - T_1}{T_3 - T_2} = 1 - \frac{813 \cdot 2 - 305}{1923 - 721 \cdot 3}$$

$$= 1 - \frac{508 \cdot 2}{1201 \cdot 7} = 1 - 0 \cdot 423$$

$$= 0 \cdot 577 \text{ or } 57 \cdot 7\% \text{ Ans. (iii)}$$

31.

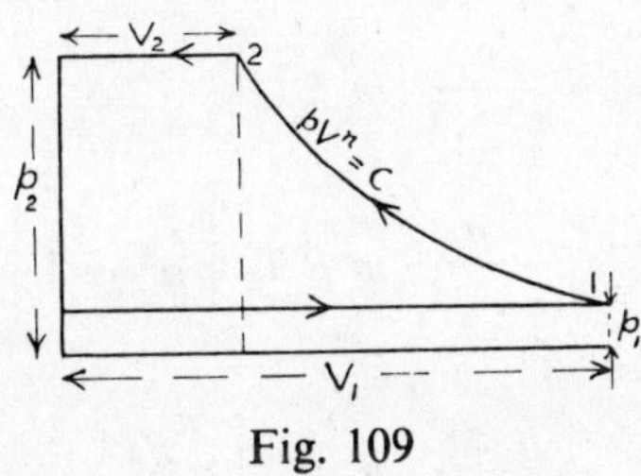

Fig. 109

mean piston speed [m/s] = distance [m] moved by piston per second

$$= 2 \times \text{stroke} \times \text{rev/s}$$

$$\therefore \text{stroke} = \frac{2 \cdot 8}{2 \times 3 \cdot 5}$$

$$= 0 \cdot 4 \text{ m} = 400 \text{ mm} \quad \text{Ans. (i)}$$

$$\text{work per cycle [kJ]} = \frac{\text{work per second [kJ/s = kW]}}{\text{cycles per second}}$$

$$= \frac{13 \cdot 58}{3 \cdot 5} = 3 \cdot 88 \text{ kJ}$$

Referring to Fig. 109, neglecting clearance:

$$\text{Area under compression curve} = \frac{p_1 V_1 - p_2 V_2}{n - 1}$$

this is the work done *BY* the air.

$$\text{work done } ON \text{ the air} = \frac{p_2 V_2 - p_1 V_1}{n - 1}$$

work done on the air per cycle

$$= \text{net area of diagram}$$

$$= \text{compression area} + \text{delivery area} - \text{suction area}$$

$$\frac{p_2 V_2 - p_1 V_1}{n - 1} + p_2 V_2 - p_1 V_1$$

$$\frac{n}{n - 1} (p_2 V_2 - p_1 V_1)$$

$$= \frac{n}{n - 1} \, m R (T_2 - T_1)$$

$$= \frac{n}{n - 1} \, m R T_1 \left[\frac{T_2}{T_1} - 1 \right]$$

$$= \frac{n}{n - 1} \, p_1 V_1 \left[\left\{ \frac{p_2}{p_1} \right\}^{\frac{n-1}{n}} - 1 \right]$$

$$\therefore \; 3 \cdot 88 = \frac{1 \cdot 32}{0 \cdot 32} \times \; \times 110^2 \times V_1 \times \left[\left\{ \frac{10}{1} \right\}^{\frac{0 \cdot 32}{1 \cdot 32}} - 1 \right]$$

$$3 \cdot 88 \times 0 \cdot 32 = 1 \cdot 32 \times 10^2 \times V_1 \times 0 \cdot 748$$

$$V_1 = \frac{3 \cdot 88 \times 0 \cdot 32}{1 \cdot 32 \times 10^2 \times 0 \cdot 748} = 0 \cdot 012 \, 57 \text{ m}^3$$

$$\text{Diameter} = \sqrt{\frac{\text{volume}}{0 \cdot 7854 \times \text{stroke}}} = \sqrt{\frac{0 \cdot 012 \, 57}{0 \cdot 7854 \times 0 \cdot 4}}$$

$$= 0 \cdot 2 \text{ m} = 200 \text{ mm Ans. (ii)}$$

$$\text{Mean eff. press.} = \frac{\text{net area of diagram}}{\text{length of diagram}}$$

$$= \frac{3 \cdot 88}{0 \cdot 012 \, 57} = 308 \cdot 6 \text{ kN/m}^2$$

$$= 3 \cdot 086 \text{ bar Ans. (iii)}$$

32. Referring to Fig. 41 (Chapter 8):

$$\frac{T_2}{T_1} = \left\{ \frac{V_1}{V_2} \right\}^{n-1}$$

$$T_2 = (37 + 273) \times 13^{0.4} = 865 \text{ K}$$

Temp. at end of compression $= 865 - 273 = 592°C$ Ans. (i)

Assume 35 kg of air is compressed and 1 kg of fuel is burned, mass of gases formed $= 35 + 1 = 36$ kg

$$\frac{\text{Heat energy given}}{\text{up by 1 kg fuel}} = \frac{\text{Heat energy received}}{\text{by 36 kg gases}}$$

$$\text{c.v.} = m \times c_\text{P} \times (T_3 - T_2)$$

$$T_3 - T_2 = \frac{42 \times 10^3}{36 \times 1.02} = 1144 \text{ K}$$

$$T_3 = 1144 + 865 = 2009 \text{ K}$$
Temp. at end of combustion $= 2009 - 273 = 1736°C$ Ans. (ii)

$$\frac{V_3}{V_2} = \frac{T_3}{T_2}$$

$$V_3 = 1 \times \frac{2009}{865} = 2.322$$

Ratio of expansion $= \dfrac{V_4}{V_3} = \dfrac{13}{2.322} = 5.598$ Ans. (iii)

33. Referring to Fig. 92 (Chapter 13),
$$V_{a1} = V_1 \sin \alpha_1 = 135 \times \sin 20° = 46.17 \text{ m/s}$$
$$V_{w1} = V_1 \cos \alpha_1 = 135 \times \cos 20° = 126.9 \text{ m/s}$$
$$x = V_{w1} - u = 126.9 - 87 = 39.9 \text{ m/s}$$

$$\tan \beta_1 = \frac{46.17}{39.9} = 1.157$$

$\therefore$ Entrance angle $\beta_1 = 49° \ 10'$ Ans. (i)

$$\beta_2 = \alpha_1 \qquad\qquad V_{r2} = V_1 \qquad\qquad V_{w2} = x$$

Effective change of velocity, $V_w = V_{w1} + V_{w2}$
$$= 126 \cdot 9 + 39 \cdot 9 = 166 \cdot 8 \text{ m/s}$$

$$
\begin{aligned}
\text{Force on blades} &= \text{change of momentum per second} \\
\text{Force[N]} &= \text{mass flow [kg/s]} \times V_w\text{[m/s]} \\
\text{For 1 kg/s, force} &= 1 \times 166 \cdot 8 = 166 \cdot 8\text{N} \\
\text{Power [W} = \text{J/s} = \text{Nm/s]} &= \text{force [N]} \times \text{blade velocity [m/s]} \\
&= 166 \cdot 8 \times 87 \\
&= 1 \cdot 451 \times 10^4 \text{ W} \\
&= 14 \cdot 51 \text{ kW} \quad \text{Ans. (ii)}
\end{aligned}
$$

34. From steam tables, taking h at 25 bar 300°C as the mean value between 20 and 30 bar at the same temperature:

25 bar 300°C, $h = \frac{1}{2}(3025 + 2995) = 3010$
Water at 130°C, $h = 546$
Heat energy given to steam in boiler
$$= \quad 3010 - 546 \quad = 2464 \text{ kJ/kg}$$

$$\text{Fuel consumption} = \frac{33 \times 10^3}{24} = 1375 \text{ kg/h}$$

$$\text{Boiler eff.} = \frac{\text{heat energy transferred to steam [kJ/h]}}{\text{heat energy supplied by fuel [kJ/h]}}$$

$$= \frac{20\ 000 \times 2464}{1375 \times 43 \times 10^3}$$

$$= 0 \cdot 8335 \text{ or } 83 \cdot 35\% \quad \text{Ans. (i)}$$

Actual evaporation per kg of fuel

$$= \frac{20\ 000}{1375} = 14 \cdot 54 \text{ kg steam/kg fuel}$$

From steam tables, h_{fg} at 100°C $= 2256 \cdot 7$
Equivalent evaporation from and at 100°C

$$= 14 \cdot 54 \times \frac{2464}{2256 \cdot 7}$$

$$= 15 \cdot 88 \text{ kg steam/kg fuel} \quad \text{Ans. (ii)}$$

$$\text{Overall eff.} = \frac{\text{heat energy converted into work [kJ/k]}}{\text{heat energy supplied by fuel [kJ/k]}}$$

$$= \frac{2300 \times 3600}{1375 \times 43 \times 10^3}$$

$$= 0\cdot1401 \text{ or } 14\cdot01\% \quad \text{Ans. (iii)}$$

35. Kinetic energy $= \frac{1}{2} m v^2$

Change in kinetic energy per kg of exhaust gases through exchanger due to change of velocity

$$= \tfrac{1}{2}(V_1{}^2 - V_2{}^2) = \qquad \tfrac{1}{2}(140^2 - 10^2) \qquad = 9750 \text{ J/kg}$$

which is converted into heat energy $= 9\cdot75$ kJ/kg

Mass of exhaust gases per kg of fuel

$$= 84 \text{ kg air} + 1 \text{ kg fuel} = 85 \text{ kg gases}$$

Heat energy transferred $=$ mass $\times c_\mathbf{p} \times$ temp. change

$\therefore$ Transfer of heat in exchanger, from 85 kg of gases, to 84 kg of air:

$$85 \times 1\cdot1 \times (300{-}240) + 85 \times 9\cdot75 = 84 \times 1\cdot005 \times \text{temp. change}$$

$$\text{Air temp. rise} = \frac{85 (1\cdot1 \times 60 + 9\cdot75)}{84 \times 1\cdot005} = 76\cdot28°$$

$$\text{Air temp. outlet} = \quad 200 + 76\cdot28 \quad = 276\cdot28°\text{C} \quad \text{Ans.}$$

36. Heat to be taken from water to make ice [kJ/s]

$$= \frac{250}{3600} (4\cdot2 \times 15 + 335 + 2\cdot04 \times 10)$$

$$= 29\cdot05 \text{ kJ/s}$$

Let m[kg/s] $=$ mass flow of refrigerant

Heat absorbed by refrigerant in evaporator [kJ/s]

$$= m \times (0\cdot95 - 0\cdot2) \times 290\cdot2$$

$$= m \times 217\cdot7 \text{ kJ/s}$$

Assuming perfect heat transfer,

$$m \times 217\cdot7 = 29\cdot05$$

$$m = 0\cdot1335 \text{ kg/s} \quad \text{Ans. (i)}$$

Volume flow of refrigerant leaving evaporator and entering compressor $[\text{m}^3/\text{s}] = 0\cdot1335 \times 0\cdot021\,68 \times 0\cdot95$
$$= 2\cdot749 \times 10^{-3}\ \text{m}^3/\text{s}$$

Let d = diameter, stroke = $2d$, assuming 100% volumetric efficiency, volume taken into compressor per second $[\text{m}^3/\text{s}] = 0\cdot7854 \times d^2 \times 2d \times 4\cdot15 = 2\cdot749 \times 10^{-3}$

$$d = \sqrt[3]{\frac{2\cdot749 \times 10^{-3}}{0\cdot7854 \times 2 \times 4\cdot15}}$$

$$= 0\cdot074\,99 \text{ m say 75 mm}$$
$$\left.\text{Stroke} = 2 \times 75 = 150 \text{ mm}\right\} \text{ Ans. (ii)}$$

37. Referring to Fig. 54 (Chapter 9)
$$V_1 = \text{stroke volume} + \text{clearance volume}$$
$$= 1440 + 40 = 1480 \text{ cm}^3$$
$$V_3 = \text{clearance volume} = 40 \text{ cm}^3$$
$$p_1 V_1^{\,n} = p_2 V_2^{\,n}$$
$$1 \times 1480^{1\cdot3} = 5 \times V_2^{\,1\cdot3}$$

$$V_2 = \frac{1480}{\sqrt[1\cdot3]{5}} = 429\cdot1 \text{ cm}^3$$

Volume swept by piston from beginning of compression stroke to point where delivery valves open
$$= V_1 - V_2$$
$$= 1480 - 429\cdot1 = 1050\cdot9 \text{ cm}^3$$

As a fraction of the stroke when delivery valves open

$$= \frac{1050\cdot9}{1440} = 0\cdot7298 \text{ Ans. (i)}$$

$$p_3 V_3^{\,n} = p_4 V_4^{\,n}$$
$$5 \times 40^{1\cdot3} = 1 \times V_4^{\,1\cdot3}$$
$$V_4 = 40 \times \sqrt[1\cdot3]{5} = 138 \text{cm}^3$$

Volume swept by piston from beginning of suction stroke to point when suction valves open
$$= V_4 - V_3$$
$$= 138 - 40 = 98 \text{ cm}^3$$

As a fraction of the stroke when suction valves open

$$= \frac{98}{1440} = 0 \cdot 068\ 05 \quad \text{Ans. (ii)}$$

$$\text{Area under compression curve} = \frac{p_2\ V_2 - p_1\ V_1}{n-1}$$

By using $p_2\ V_2 - p_1\ V_1$ instead of $p_1\ V_1 - p_2\ V_2$ the area represents work done ON the air instead of work done BY the air.

Note that the mean indicated pressure will be obtained from net area $\div$ length, therefore, since the actual work is not required, it is not necessary to convert pressures into kN/m^2 and volumes into m^3.

$$\text{Area under compression curve} = \frac{p_2\ V_2 - p_1\ V_1}{n-1}$$

$$= \frac{5 \times 429 \cdot 1 - 1 \times 1480}{1 \cdot 3 - 1} = 2218 \cdot 3$$

$$\text{Area under delivery line} = p_2 \times (V_2 - V_3)$$
$$= 5 \times (429 \cdot 1 - 40) = 1945 \cdot 5$$
$$\text{Area under expansion curve} = \frac{p_3\ V_3 - p_4\ V_4}{n-1}$$

$$= \frac{5 \times 40 - 1 \times 138}{1 \cdot 3 - 1} = 206 \cdot 7$$

$$\text{Area under suction line} = p_1 \times (V_1 - V_4)$$
$$= 1 \times (1480 - 138) = 1342$$

Net area of diagram
$$= 2218 \cdot 3 + 1945 \cdot 5 - 206 \cdot 7 - 1342$$
$$= 4163 \cdot 8 - 1548 \cdot 7 = 2615 \cdot 1$$
Mean indicated press. = mean height of diagram

$$= \frac{\text{net area}}{\text{length}} = \frac{2615 \cdot 1}{1440} = 1 \cdot 816 \text{ bar} \quad \text{Ans. (iii)}$$

Alternatively, the net area of the diagram, representing work per cycle could be obtained from:

$$\frac{n}{n-1} \, p_1 \, (V_1 - V_4) \left[\left\{ \frac{p_2}{p_1} \right\}^{\frac{n-1}{n}} \right] - 1$$

38. From steam tables,

$$30 \text{ bar } 350°C, \quad h = 3117 \quad s = 6\cdot744$$
$$0\cdot045 \text{ bar}, \quad h_f = 130 \quad s_f = 0\cdot451$$
$$h_{fg} = 2428 \quad s_{fg} = 7\cdot980$$

For the Rankine cycle, isentropic expansion from 30 bar 350°C to 0·045 bar:

Entropy after expansion = Entropy before

$$0\cdot451 + x \times 7\cdot98 = 6\cdot744$$
$$x = 0\cdot7886$$
$$h_2 = h \text{ at } 0\cdot045 \text{ bar}, 0\cdot7886 \text{ dry}$$
$$= 130 + 0\cdot7886 \times 2428 = 2044$$

$$\text{Rankine efficiency} = \frac{h_1 - h_2}{h_1 - h_{f2}} = \frac{3117 - 2044}{3117 - 130}$$

$$= \frac{1073}{2987} = 0\cdot3592 \text{ or } 35\cdot92\% \quad \text{Ans. (i)}$$

$$\text{Actual efficiency} = \frac{\text{heat energy converted into work}}{\text{heat energy supplied by steam}}$$

Heat energy into work [kJ/s = kW] = 5×10^3 kJ/s
Heat energy supplied [kJ/s]

$$= \text{steam consumption [kg/s]} \times (h_1 - h_{f2}) \text{ [kJ/kg]}$$

$$= \frac{22\cdot5 \times 10^3}{3600} \times 2987$$

$$\therefore \text{ Engine effic.} = \frac{5 \times 10^3 \times 3600}{22\cdot5 \times 10^3 \times 2987}$$

$$= 0\cdot2679 \text{ or } 26\cdot79\% \quad \text{Ans. (ii)}$$

$$\text{Efficiency ratio} = \frac{0\cdot2679}{0\cdot3592}$$

$$= 0\cdot7457 \text{ or } 74\cdot57\% \quad \text{Ans. (iii)}$$

39. From steam tables,
$$10 \text{ bar}, \quad h_f = 763 \quad h_{fg} = 2015$$
$$1 \cdot 2 \text{ bar}, \quad \text{sat. temp.} = 104 \cdot 8°\text{C} \quad h_g = 2683$$
$\therefore$ at $1 \cdot 2$ bar $109 \cdot 8°$C, steam is superheated $5°$
Drynesss fraction by separating calorimeter:

$$x_1 = \frac{3 \cdot 03}{3 \cdot 03 + 0 \cdot 113} = 0 \cdot 964$$

Dryness fraction by throttling calorimeter:
Enthalpy before throttling $=$ Enthalpy after
$$763 + x_2 \times 2015 = 2683 + 2 \cdot 02 \times 5$$
$$x_2 \times 2015 = 1930 \cdot 1$$
$$x_2 = 0 \cdot 9579$$

Dryness fraction of sample
$$= 0 \cdot 964 \times 0 \cdot 9579 = 0 \cdot 9234 \quad \text{Ans.}$$

40. Referring to Fig. 42 (Chapter 8),
$$T_1 = \quad 42 + 273 = \quad 315 \text{ K}$$
$$T_4 = 1500 + 273 = 1773 \text{ K}$$
$$V_1 = V_5 = \text{stroke volume} + \text{clearance volume}$$
$$= 0 \cdot 1068 + 0 \cdot 0089 = 0 \cdot 1157 \text{ m}^3$$
$$V_2 = V_3 = \text{clearance volume} = 0 \cdot 0089 \text{ m}^3$$
$$p_1 V_1{}^y = p_2 V_2{}^y$$
$$1 \times 0 \cdot 1157^{1 \cdot 4} = p_2 \times 0 \cdot 0089^{1 \cdot 4}$$

$$p_2 = \left\{ \frac{0 \cdot 1157}{0 \cdot 0089} \right\}^{1 \cdot 4} = 36 \cdot 26 \text{ bar}$$

$$\frac{p_1 V_1}{T_1} = \frac{p_2 V_2}{T_2}$$

$$T_2 = \frac{315 \times 36 \cdot 26 \times 0 \cdot 0089}{1 \times 0 \cdot 1157} = 878 \cdot 8 \text{ K}$$

$$\frac{T_3}{T_2} = \frac{p_3}{p_2}$$

$$T_3 = \frac{878 \cdot 8 \times 45}{36 \cdot 26} = 1090 \text{ K}$$

Heat received at constant volume
$$= m \times c_{\mathrm{V}} \times (T_3 - T_2)$$
(Per kg of air:) $\quad = 1 \times 0.715 \times (1090 - 878.8)$
$$= 151 \text{ kJ/kg}$$
Spec. heat at constant press. $c_{\mathrm{P}} = c_{\mathrm{V}} \times \gamma$
$$= 0.715 \times 1.4 = 1.001$$
Heat received at constant pressure
$$= m \times c_{\mathrm{P}} \times (T_4 - T_3)$$
$$= 1 \times 1.001 \times (1773 - 1090)$$
$$= 683.7 \text{ kJ/kg}$$

$$\text{Ratio:} \ \frac{\text{heat received at const. vol.}}{\text{heat received at const. press.}}$$

$$= \frac{151}{683.7} = 0.2209:1 \quad \text{Ans.}$$

41. From steam tables,
$$\text{Water } 175.4°C, \quad h_f = 743$$
$$\text{Water } 100°C, \quad h_f = 419.1 \quad h_{fg} = 2256.7$$
Enthalpy released by 1 kg of water when pressure falls due to escape valve being opened
$$= 743 - 419.1 = 323.9 \text{ kJ}$$
This is absorbed by some of the water which evaporates it into steam at atmospheric pressure.
Mass of steam evaporated

$$= \frac{323.9}{2256.7} = 0.1435 \text{ kg}$$

Water remaining $= 1 - 0.1435 = 0.8565 \text{ kg} \quad \text{Ans. (i)}$

In second experiment, for equal mass of water to be flash-evaporated, an equal release of enthalpy must take place when the water again falls in temperature to 100°C, hence 323.9 kJ is to be released by 0.8565 kg of water:
Let h_{f2} = enthalpy [kJ/kg] of sat. water at new temperature,
$$0.8565(h_{f2} - 419.1) = 323.9$$

$$h_{f2} = \frac{323.9}{0.8565} + 419.1 = 797.3 \text{ kJ/kg}$$

From steam tables, 797.3 is sufficiently close to the listed value of 798 to be taken as such, this corresponds to a pressure of 12 bar and temperature 188°C, therefore,
temp. to which water should be heated $= 188°C \quad \text{Ans. (ii)}$

42.

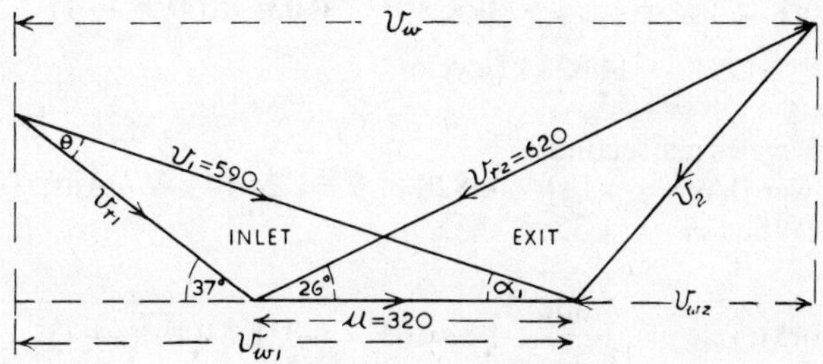

Fig. 110

By sine rule referring to inlet triangle:

$$\frac{320}{\sin \theta} = \frac{590}{\sin(180° - 37°)}$$

$$\sin \theta = \frac{320 \times 0.6018}{590} = 0.3263$$

$$\theta = 19° 3' \qquad \alpha_1 = 37° - 19° 3' = 17° 57'$$
$$v_{w1} = 590 \times \cos 17° 57' = 561.4 \text{ m/s}$$
$$v_{w2} = 620 \times \cos 26° - 320 = 237.3 \text{ m/s}$$

Effective change of velocity $v_w = v_{w1} + v_{w2}$
$$= 561.4 + 237.3 = 798.7 \text{ m/s}$$

Force on blades [N] = mass flow [kg/s] × change of velocity [m/s]
$$= 0.075 \times 798.7 = 59.91 \text{ N} \quad \text{Ans. (i)}$$

Power [W = J/s = N m/s] = force [N] × blade velocity [m/s]
$$= 59.91 \times 320$$
$$= 1.917 \times 10^4 \text{ W}$$
$$= 19.17 \text{ kW} \quad \text{Ans. (ii)}$$

43. $V_1 = 0.7854 \times 0.3^2 \times 0.45 = 0.031\,81 \text{ m}^3$

$$\text{Work per cycle} = \frac{n}{n-1} p_1 V_1 \left[\left\{ \frac{p_2}{p_1} \right\}^{\frac{n-1}{n}} - 1 \right]$$

When $n = 1.15$:

$$\text{Work/cycle} = \frac{1 \cdot 15}{0 \cdot 15} \times 1 \times 10^2 \times 0 \cdot 031\ 81(4^{\frac{0 \cdot 15}{1 \cdot 15}} - 1)$$

$$= 4 \cdot 828 \text{ kJ/cycle}$$

At 5 cycles per second:

Power [kW $=$ kJ/s] $= 4 \cdot 828 \times 5 = 24 \cdot 14$ kW   Ans. (i)

When $n = 1 \cdot 35$:

$$\text{Work/cycle} = \frac{1 \cdot 35}{0 \cdot 35} \times 1 \times 10^2 \times 0 \cdot 031\ 81(4^{\frac{0 \cdot 35}{1 \cdot 35}} - 1)$$

$$= 5 \cdot 299 \text{ kJ/cycle}$$

Power $= 5 \cdot 299 \times 5 = 26 \cdot 5$ kW   Ans. (ii)

$$\% \text{ increase} = \frac{26 \cdot 5 - 24 \cdot 14}{24 \cdot 14} \times 100 = 9 \cdot 777\% \quad \text{Ans. (iii)}$$

An alternative solution to the above is to find the value of V_2 in each case, from $p_1 V_1{}^n = p_2 V_2{}^n$ then calculate the work per cycle from:

$$\frac{n}{n-1}(p_2 V_2 - p_1 V_1)$$

44. From steam tables,

18 bar, $t_s = 207 \cdot 1°\text{C}$ $h_{fg} = 1912$

Enthalpy drop per kg of steam

$$= (1 - 0 \cdot 985) \times 1912 = 28 \cdot 68 \text{ kJ/kg}$$

Rate of heat loss [J/s] per unit length

$$= \frac{1200 \times 28 \cdot 68 \times 10^3}{3600 \times 25} = 382 \cdot 4 \text{ J/s}$$

and this is to be equal to $\dfrac{2\pi\lambda(T_1 - T_2)\,\text{J/s}}{\ln\ r_2/r_1}$

$$\frac{2\pi \times 0 \cdot 13 \times (207 \cdot 1 - 35)}{\ln r_2/r_1} = 382 \cdot 4$$

$$\ln \left\{\frac{r_2}{r_1}\right\} = \frac{2\pi \times 0 \cdot 13 \times 172 \cdot 1}{382 \cdot 4} = 0 \cdot 3676$$

From natural (Naperian) log tables,

0·3676 is the natural log of 1·444

$$\therefore \frac{r_2}{r_1} = 1\cdot444 \qquad r_2 = 1\cdot444 \times 70 = 101\cdot1 \text{ mm}$$

$$\text{Thickness} = r_2 - r_1 = 101\cdot1 - 70 = 31\cdot1 \text{ mm Ans.}$$

45. From steam tables,

    0·5 bar, $v_g = 3\cdot239$ 0·9 bar, $v_g = 1\cdot869$

Spec. vol. at 0·5 bar 0·9 dry $= 0\cdot9 \times 3\cdot239 = 2\cdot915 \text{ m}^3/\text{kg}$

  „   „   „ 0·9 bar 0·95 dry $= 0\cdot95 \times 1\cdot869 = 1\cdot775 \text{ m}^3/\text{kg}$

Let $V\,[\text{m}^3] = $ volume of vessel

$$\text{Initial mass of steam} = \frac{V}{2\cdot915} \text{ kg}$$

$$\text{Final mass of steam} = \frac{V}{1\cdot775} \text{ kg}$$

Difference is 0·26 kg, therefore,

$$\frac{V}{1\cdot775} - \frac{V}{2\cdot915} = 0\cdot26$$

$$2\cdot915V - 1\cdot775V = 0\cdot26 \times 1\cdot775 \times 2\cdot915$$
$$1\cdot14V = 0\cdot26 \times 1\cdot775 \times 2\cdot915$$
$$V = 1\cdot18 \text{ m}^3 \text{ Ans.}$$

46. From steam tables,

               20 bar 400°C, $h = 3248$

  1·4 bar, $t_s = 109\cdot3$ $h_f = 458$ $h_{fg} = 2232$

  0·04 bar, $h_f = 121$ $h_{fg} = 2433$

WITHOUT FEED HEATING:

h at 0·04 bar 0·85 dry

$$= 121 + 0\cdot85 \times 2433 = 2188$$

Enthalpy drop through turbine/kg

$$= 3248 - 2188 = 1060$$

Heat supplied to steam in boiler

$$= 3248 - 121 = 3127$$

$$\text{Thermal eff.} = \frac{\text{heat energy converted into work}}{\text{heat energy supplied}}$$

$$= \frac{1060}{3127} = 0.3389 \text{ or } 33.89\%$$

WITH FEED HEATING:

0·134 kg of steam at 1·4 bar is tapped off per kg of supply steam, leaving $(1 - 0.134) = 0.866$ kg to complete its path through the engine.

Considering heater and hotwell as one system,

$$\text{Enthalpy at entry} = \text{Enthalpy at exit}$$
$$0.134\,(458 + x \times 2232) + 0.866 \times 121 = 1 \times 458$$
$$61.38 + 299x + 104.8 = 458$$
$$299x = 291.82$$
$$\text{dryness fraction at 1·4 bar, } x = 0.976$$

h at 1·4 bar 0·976 dry

$$= 458 + 0.976 \times 2232 = 2636$$

Enthalpy drop through turbine/kg

$$= 1 \times (3248 - 2636) + 0.866\,(2636 - 2188)$$
$$= 612 + 388 = 1000$$

Heat supplied to steam in boiler

$$= 3248 - 458 = 2790$$

$$\text{Thermal eff.} = \frac{1000}{2790} = 0.3584 \text{ or } 35.84\%$$

Comparison: With feed heating, $\eta = 35.84\%$ $\Big\}$ Ans.
Without feed heating, $\eta = 33.89\%$

47. $T_1 = 19 + 273 = 292 \text{ K}$
$T_2 = 508 + 273 = 781 \text{ K}$

$$\frac{p_1 V_1}{T_1} = \frac{p_2 V_2}{T_2}$$

$$V_2 = \frac{1.01 \times 0.125 \times 781}{36 \times 292} = 0.009\,38 \text{ m}^3$$

$$p_1 V_1^n = p_2 V_2^n$$
$$1.01 \times 0.125^n = 36 \times 0.009\,38^n$$

$$n \times \log \left\{ \frac{0 \cdot 125}{0 \cdot 00\,938} \right\} = \log \left\{ \frac{36}{1 \cdot 01} \right\}$$

$$n \times 1 \cdot 1247 = 1 \cdot 5520$$
$$n = 1 \cdot 38 \quad \text{Ans. (i)}$$

Alternatively, n could be obtained from

$$\frac{T_1}{T_2} = \left\{ \frac{p_1}{p_2} \right\}^{\frac{n-1}{n}}$$

$$R = c_P - c_V = 1 \cdot 005 - 0 \cdot 718 = 0 \cdot 287 \text{ kJ/kg K}$$
$$p_1 V_1 = m R T_1$$

$$m = \frac{1 \cdot 01 \times 10^2 \times 0 \cdot 125}{0 \cdot 287 \times 292}$$

$$= 0 \cdot 1506 \text{ kg} \quad \text{Ans. (ii)}$$

$$\begin{array}{c}\text{Work done}\\ \text{by the air}\end{array} = \frac{p_1 V_1 - p_2 V_2}{n - 1}$$

$$= \frac{m R (T_1 - T_2)}{n - 1}$$

$$= \frac{0 \cdot 1506 \times 0 \cdot 287 \times (292 - 781)}{1 \cdot 38 - 1}$$

$$= -55 \cdot 62 \text{ kJ} \quad \text{Ans. (iii)}$$

The minus sign indicates that the work is done ON the air. Increase in internal energy

$$E_2 - E_1 = m \times c_V \times (T_2 - T_1)$$
$$= 0 \cdot 1506 \times 0 \cdot 718 \times (781 - 292)$$
$$= 52 \cdot 86 \text{ kJ} \quad \text{Ans. (iv)}$$

$$\text{Heat supplied} = \begin{array}{c}\text{increase in}\\ \text{internal energy}\end{array} + \begin{array}{c}\text{external}\\ \text{work done}\end{array}$$

$$= 52 \cdot 86 + (-55 \cdot 62) = -2 \cdot 76 \text{ kJ}$$
$$\text{Ans. (v)}$$

The minus sign indicates that the transfer of heat is from the air to its surrounds.

48. ip of 4 cylinders $= p_m ALn \times 4$
where $n = $ rev/s $\div 2$ for a four-stroke engine
$$\text{ip} = 7\cdot45 \times 10^2 \times 0\cdot7854 \times 0\cdot32^2 \times 0\cdot48 \times 2 \times 4$$
$$= 230 \text{ kW}$$
$$\text{bp [kW]} = T[\text{kN m}] \times \omega[\text{rad/s}]$$
$$= 6 \times 0\cdot96 \times 2\pi \times 4 = 144\cdot8 \text{ kW}$$

$$\text{Ind. thermal effic.} = \frac{\text{Heat into work in cylinders [kJ/h]}}{\text{Heat energy supplied [kJ/h]}}$$

$$= \frac{230 \times 3600}{49\cdot5 \times 44\cdot5 \times 10^3}$$

$$= 0\cdot3758 \text{ or } 37\cdot58\% \quad \text{Ans. (i)}$$

$$\text{Brake thermal effic.} = \frac{\text{Heat into work at brake [kJ/h]}}{\text{Heat energy supplied [kJ/h]}}$$

$$= \frac{144\cdot8 \times 3600}{49\cdot5 \times 44\cdot5 \times 10^3}$$

$$= 0\cdot2367 \text{ or } 23\cdot67\% \quad \text{Ans. (ii)}$$

Heat energy carried away by cooling water [kJ/h]
$$= \text{mass} \times \text{spec. ht.} \times \text{temp. rise}$$
$$= 77 \times 60 \times 4\cdot2 \times (47 - 14)$$

As a percentage of the heat supplied

$$= \frac{77 \times 60 \times 4\cdot2 \times 33}{49\cdot5 \times 44\cdot5 \times 10^3} \times 100$$

$$= 29\cdot07\% \quad \text{Ans. (iii)}$$

The remainder of the heat losses may be attributed to the heat carried away in the exhaust gases
$$= 100 - (37\cdot58 + 29\cdot07) = 33\cdot35\%$$

Friction and pumping losses
$$= 37\cdot58 - 23\cdot67 = 13\cdot91\%$$

Heat balance diagram:

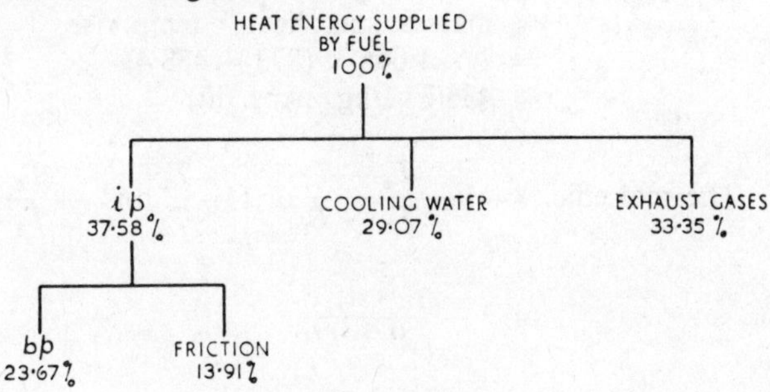

Fig. 111

49. Referring to Fig. 94 (Chapter 13),
$$T_1 = 16 + 273 = 289 \text{ K}$$
$$T_3 = 600 + 273 = 873 \text{ K}$$

$$\frac{T_2}{T_1} = \left\{\frac{p_2}{p_1}\right\}^{\frac{\gamma-1}{\gamma}} \text{ where } \frac{\gamma-1}{\gamma} = \frac{0\cdot4}{1\cdot4} = \frac{2}{7}$$

$$T_2 = 4\cdot3^{\frac{2}{7}} \times 289 = 438\cdot4 \text{ K}$$

Temperature at compressor outlet
$$= 438\cdot4 - 273 = 165\cdot4°\text{C}\quad\text{Ans. (i)}$$

$$\frac{T_4}{T_3} = \frac{T_1}{T_2} \text{ because pressure ratios are equal}$$

$$T_4 = \frac{873 \times 289}{438\cdot4} = 575\cdot4 \text{ K}$$

Alternatively T_4 could be obtained from

$$\frac{T_4}{T_3} = \left\{\frac{p_4}{p_3}\right\}^{\frac{\gamma-1}{\gamma}}$$

Temperature at turbine outlet
$$= 575\cdot4 - 273 = 302\cdot4°\text{C}\quad\text{Ans. (ii)}$$

Heat supplied per kg of working fluid

$$= \text{mass} \times \text{spec. heat} \times \text{temp. rise}$$
$$= 1 \times 1{\cdot}005 \times (873 - 438{\cdot}4)$$
$$= 436{\cdot}7 \text{ kJ/kg} \quad \text{Ans. (iii)}$$

Thermal effic. $= 1 - \dfrac{T_4 - T_1}{T_3 - T_2}$ or $1 - \dfrac{T_1}{T_2}$ or $1 - \dfrac{T_4}{T_3}$

$$\text{or } 1 - \frac{1}{r_p^{(y-1)/y}}$$

$$1 - \frac{T_1}{T_2} = 1 - \frac{289}{438{\cdot}4} = 1 - 0{\cdot}6592$$

$$= 0{\cdot}3408 \text{ or } 34{\cdot}08\% \quad \text{Ans. (iv)}$$

50. ·
$$1 \text{ mm Hg} = 133{\cdot}3 \text{ N/m}^2$$
$$\text{Condenser press.} = (762 - 714) \times 133{\cdot}3$$
$$= 6400 \text{ N/m}^2 = 0{\cdot}064 \text{ bar}$$

From steam tables, when temp. is 32·9°C,

$$t_s = 32{\cdot}9, \qquad p = 0{\cdot}05 \text{ bar}, \qquad v_g = 28{\cdot}2 \text{ m}^3/\text{kg}$$
$$\text{Partial press. of air} = \text{total press.} - \text{steam press.}$$
$$= 0{\cdot}064 - 0{\cdot}05 = 0{\cdot}014 \text{ bar}$$

From $pV = mRT$

$$\text{Mass of air [kg/min]} = \frac{p \times V \text{ [m}^3/\text{min]}}{RT}$$

$$= \frac{0{\cdot}014 \times 10^2 \times 3{\cdot}82}{0{\cdot}287 \times (32{\cdot}9 + 273)}$$

$$= 0{\cdot}060\ 91 \text{ kg/min} \quad \text{Ans. (i)}$$

Mass of vapour associated with this air

$$= \frac{3{\cdot}82}{28{\cdot}2} = 0{\cdot}1355 \text{ kg/min} \quad \text{Ans. (ii)}$$

INDEX

REED'S MARINE ENGINEERING SERIES

The series covers the full range of subjects for all grades of the Department of Trade Certificates of Competency in Marine Engineering. The books will be extremely useful to marine engineer cadets and other engineering students studying on Technician Education Council (and SCOTEC) Courses. Material is presented from first principles with many diagrammatic sketches and worked solutions to examples.

Vol. 1. MATHEMATICS
Vol. 2. APPLIED MECHANICS
Vol. 3. HEAT & HEAT ENGINES
Vol. 4. NAVAL ARCHITECTURE
Vol. 5. SHIP CONSTRUCTION
Vol. 6. BASIC ELECTROTECHNOLOGY
Vol. 7. ADVANCED ELECTROTECHNOLOGY
Vol. 8. GENERAL ENGINEERING KNOWLEDGE
Vol. 9. STEAM ENGINEERING KNOWLEDGE
Vol. 10. INSTRUMENTATION & CONTROL SYSTEMS
Vol. 11. ENGINEERING DRAWING
Vol. 12. MOTOR ENGINEERING KNOWLEDGE

These books are obtainable from all Nautical Opticians, Chartsellers and Booksellers and are published by:

THOMAS REED PUBLICATIONS LIMITED
36-37 Cock Lane, London EC1A 9BY